Green Chemistry Strategies for Drug Discovery

**RSC Drug Discovery Series**



*Titles in the Series:*

1: Metabolism, Pharmacokinetics and Toxicity of Functional Groups
2: Emerging Drugs and Targets for Alzheimer's Disease; Volume 1
3: Emerging Drugs and Targets for Alzheimer's Disease; Volume 2
4: Accounts in Drug Discovery
5: New Frontiers in Chemical Biology
6: Animal Models for Neurodegenerative Disease
7: Neurodegeneration
8: G Protein-Coupled Receptors
9: Pharmaceutical Process Development
10: Extracellular and Intracellular Signaling
11: New Synthetic Technologies in Medicinal Chemistry
12: New Horizons in Predictive Toxicology
13: Drug Design Strategies: Quantitative Approaches
14: Neglected Diseases and Drug Discovery
15: Biomedical Imaging
16: Pharmaceutical Salts and Cocrystals
17: Polyamine Drug Discovery
18: Proteinases as Drug Targets
19: Kinase Drug Discovery
20: Drug Design Strategies: Computational Techniques and Applications
21: Designing Multi-Target Drugs
22: Nanostructured Biomaterials for Overcoming Biological Barriers
23: Physico-Chemical and Computational Approaches to Drug Discovery
24: Biomarkers for Traumatic Brain Injury
25: Drug Discovery from Natural Products
26: Anti-Inflammatory Drug Discovery
27: New Therapeutic Strategies for Type 2 Diabetes: Small Molecules
28: Drug Discovery for Psychiatric Disorders
29: Organic Chemistry of Drug Degradation
30: Computational Approaches to Nuclear Receptors

31: Traditional Chinese Medicine
32: Successful Strategies for the Discovery of Antiviral Drugs
33: Comprehensive Biomarker Discovery and Validation for Clinical Application
34: Emerging Drugs and Targets for Parkinson's Disease
35: Pain Therapeutics; Current and Future Treatment Paradigms
36: Biotherapeutics: Recent Developments using Chemical and Molecular Biology
37: Inhibitors of Molecular Chaperones as Therapeutic Agents
38: Orphan Drugs and Rare Diseases
39: Ion Channel Drug Discovery
40: Macrocycles in Drug Discovery
41: Human-based Systems for Translational Research
42: Venoms to Drugs: Venom as a Source for the Development of Human Therapeutics
43: Carbohydrates in Drug Design and Discovery
44: Drug Discovery for Schizophrenia
45: Cardiovascular and Metabolic Disease: Scientific Discoveries and New Therapies
46: Green Chemistry Strategies for Drug Discovery

*How to obtain future titles on publication:*
A standing order plan is available for this series. A standing order will bring delivery of each new volume immediately on publication.

*For further information please contact:*
Book Sales Department, Royal Society of Chemistry, Thomas Graham House, Science Park, Milton Road, Cambridge, CB4 0WF, UK
Telephone: +44 (0)1223 420066, Fax: +44 (0)1223 420247,
Email: booksales@rsc.org
Visit our website at www.rsc.org/books

# *Green Chemistry Strategies for Drug Discovery*

Edited by

**Emily A. Peterson**
*Amgen, Stow, Massachusetts, USA*
*Email: epeterso@amgen.com*

**Julie B. Manley**
*Guiding Green LLC, Sanford, Michigan, USA*
*Email: Juliemanley@guidinggreen.com*

ROYAL SOCIETY OF CHEMISTRY

THE QUEEN'S AWARDS FOR ENTERPRISE: INTERNATIONAL TRADE 2013

RSC Drug Discovery Series No. 46

Print ISBN: 978-1-84973-961-0
PDF eISBN: 978-1-78262-265-9
ISSN: 2041-3203

A catalogue record for this book is available from the British Library



Published by The Royal Society of Chemistry,
Thomas Graham House, Science Park, Milton Road,
Cambridge CB4 0WF, UK

Registered Charity Number 207890

For further information see our web site at www.rsc.org

Printed and bound by CPI Group (UK) Ltd, Croydon, CR0 4YY

# Preface

*Green Chemistry Strategies for Drug Discovery* is authored largely by veterans of the pharmaceutical industry, primarily in discovery, to address the incorporation of green chemistry principles into the fast-paced environment of drug discovery. The purpose of the book is to provide actionable strategies, case studies, and tools as a practical guide for both academic and industrial laboratories wanting to know how to start introducing greener techniques and, importantly, where to channel efforts for greatest return on investment. Recognizing that the incorporation of green chemistry into drug discovery is perhaps hindered most by resistance to cultural change, many of the examples provided are aimed at achieving incremental improvements that lead to the largest positive outcomes. As such, we anticipate and hope that, in the future, the actions recommended in this book will become routine and there will be a readiness to incorporate innovative technologies and tackle new challenges aimed at making our science more sustainable.

This book acts as a resource that could be utilized in its entirety or as a reference by topic area. In addition to expected subjects such as reaction optimization, high-throughput screening, analysis, purification, and solvent selection, the book addresses burgeoning fields such as continuous processing in drug discovery and green chemistry in biological drug discovery. Chapters also provide unique perspectives on green chemistry as it relates to patent protection and the electronic lab notebook, as well as the business case for the incorporation of green chemistry specifically in the drug discovery phase of pharmaceutical research. The majority of chapters are written by authors with experience in the pharmaceutical industry who can speak from an informed position, not only about what should be done, but what in fact has been accomplished. All authors represent a wealth of experience in the chosen subject area, and we greatly appreciate their willingness to share their expertise in this book.

RSC Drug Discovery Series No. 46
Green Chemistry Strategies for Drug Discovery
Edited by Emily A. Peterson and Julie B. Manley

Published by the Royal Society of Chemistry, www.rsc.org

We would like to especially thank Dr Berkeley "Buzz" Cue, Jr for his invaluable insight, perspective, and time to share his expertise with us. As a silent partner throughout this process, he also knew when encouragement would be more effective than critique. He has directly and indirectly influenced a significant transition in the sustainability of the pharmaceutical industry, and we are honored to have his guidance and friendship.

Finally, we would like to thank our families for supporting us through the duration of this project, which took more time and focus away from the most important things in life. During the development of this book, we experienced the death of several close family members, but also the birth of a new generation. We know those who have passed would have been proud of this accomplishment, and are rewarded by knowing the book enables a more sustainable future for generations to come.

Emily A. Peterson and Julie B. Manley

# *Contents*




RSC Drug Discovery Series No. 46
Green Chemistry Strategies for Drug Discovery
Edited by Emily A. Peterson and Julie B. Manley



Published by the Royal Society of Chemistry, www.rsc.org

CHAPTER 1

# *Introduction: The Five Ws of Pharmaceutical Green Chemistry*

JULIE B. MANLEY

Guiding Green LLC, 457 E. Mier Road, Sanford, MI 48657, USA
Email: juliemanley@GuidingGreen.com

## 1.1 Introduction

Louis Pasteur said, "Chance favors the prepared mind." This chapter is designed to prepare the reader with the foundation upon which to build green chemistry into the business of drug discovery. Understanding green chemistry and its importance is a starting point, and being able to communicate it to the target audience is a necessity. By reviewing pharmaceutical green chemistry in the context of the essential journalism questions nicknamed the Five Ws (What? Why? Who? Where? When?), this chapter will provide an intentionally succinct perspective to act as the infrastructure for the invaluable chapters to follow. The Five Ws will prepare the reader to integrate green chemistry into drug discovery, and make successful integration more seamless and effective.

## 1.2 What is Green Chemistry?

First and foremost, green chemistry is chemistry, the scientific discipline of arranging molecules to create new materials and products; yet its focus is on the intentional integration of source and hazard reduction into the design of

RSC Drug Discovery Series No. 46
Green Chemistry Strategies for Drug Discovery
Edited by Emily A. Peterson and Julie B. Manley

Published by the Royal Society of Chemistry, www.rsc.org

matter. By focusing on the design of materials at the molecular level, innovations are more efficient, cost-effective, safer, and environmentally preferable. Historically, environmental benefits were a side effect of optimizing efficiency and minimizing cost. Green chemistry turns that notion on its head and says that by intentionally designing a more sustainable process, the business needs will be met and even exceeded.

Green chemistry is commonly defined as the design of chemical processes and products to minimize the use and/or generation of hazardous materials.[1] It is further clarified by a set of principles intended to provide a cohesive framework for the design of chemicals with reduced intrinsic hazard. The 12 principles begin with the recognition that it is more efficient to prevent waste from being generated in the first place than to treat it later (Principle 1). It is also more cost-effective to do so; the materials being purchased would be consumed rather than incurring a second cost on the same material for disposal. The principles address all aspects of the chemical lifecycle from the selection of safer materials (Principle 5 and Principle 12), renewably sourced where feasible (Principle 7), and their efficient use in the process (Principle 2). In the design of the chemistry, the principles emphasize the use of less hazardous chemical syntheses (Principle 3), reducing the need for derivatives (Principle 8), using catalysis where possible (Principle 9), incorporating real-time process monitoring (Principle 11), and minimizing energy use by considering ambient conditions when feasible (Principle 6). The principles also address end user considerations including designing the product to be effective while minimizing toxicity (Principle 4). Finally, in consideration of the end of the product life, the principles address the need to design for safe degradation in the environment (Principle 10). While some of these principles may not seem relevant to the drug discovery setting, decisions made in discovery can ultimately have a significant impact on the marketed product. The following chapters are intended to provide the reader with a more thorough understanding of their practical implementation in drug discovery.

For the current purposes, it is important to recognize the implicit challenge with implementing the 12 Principles in their entirety in any one process, and to appreciate this challenge not as a hindrance, but as an opportunity to continue to innovate.[2] Even technologies recognized with the US Presidential Green Chemistry Challenge Award rarely, if ever, meet all 12 principles at any one time. Similarly, a process recognized as an effective implementation of green chemistry could also be further improved as evidenced by Merck's sitagliptin, the active ingredient in Januvia™, being recognized in both 2006 and 2010 (with Codexis) with the Presidential Green Chemistry Challenge Award.[3] The principles are a framework upon which to design, and to make informed decisions when a trade off between principles is inevitably needed.

One could argue that green chemistry is less a scientific field than it is specification for performance characteristics. Green chemistry describes how to incorporate design for the environment into current scientific methods. In 2005, metathesis was recognized with the Nobel Prize in Chemistry as a "great step forward for green chemistry".[4] The technology received the highest honor globally in chemistry, not an environmental award, not a green chemistry award. Green chemistry is about doing chemistry more efficiently, safer, and more cost-effectively than it is now. Medicinal chemists, process chemists, analytical chemists, biochemists, and so on are not green chemists; they are scientists in their respective disciplines doing green chemistry. Job descriptions are not written to hire a green chemist *per se*; they seek qualified candidates able to perform the essential job functions. Arguably, knowledge of green chemistry, in addition to the targeted education and experience, assures the person is capable of utilizing his or her expertise to design and synthesize medicines efficiently, while minimizing cost and environmental impact, thereby meeting the short- and long-term goals of the company. Green chemistry is not a scientific field unto itself; it is the intentional integration of source and hazard reduction into chemistry. Paul Anastas, one of the fathers of green chemistry, has himself even been quoted, "I always say that we will know when green chemistry was successful when the term green chemistry goes away because that is simply the way that we always do chemistry."[5]

## 1.3 Why Should the Pharmaceutical Industry Incorporate Green Chemistry?

Sustainability, defined as meeting the needs of today without compromising the ability of future generations to meet their needs,[6] was once a more commonly used vocabulary word for long-term financial stability than environmental stewardship. For the past 30 years, stability is not a term many would use to describe the pharmaceutical industry. Mergers and acquisitions have reduced what was once 110 companies to about 30 companies today, and that number is continually changing even as this book is being published.[7] Figure 1.1 illustrates the history of AstraZeneca and Pfizer as examples to demonstrate the effect of mergers and acquisitions. At the time of writing, these two companies were engaged in communications for a possible merger.

R&D spending has been on the rise with approximately $51.1 billion spent in 2013, as compared to half that amount in 2000, and $1.2B in 1980, yet only two of ten marketed drugs return revenues that match or exceed the R&D costs.[8] Restructuring has become the norm to manage these challenges. Whether it is outsourcing R&D or production, or spinning off companies like the Abbott surprising spin off of the R&D segment resulting in the creation of AbbVie,[9,10] companies are downsizing and decreasing R&D spending throughout the industry. Lilly projected R&D spending to decrease 15–20% and Merck reduced headcount by 20% in 2013–2014 and minimized risk by acquiring experimental drugs.[11]

Companies need to do more with less, and green chemistry provides more for less.[12] It is well established that the pharmaceutical industry generates a substantial amount of waste per kilogram of active pharmaceutical ingredient produced. Estimates indicate an average of over 100 kg material is used per kg

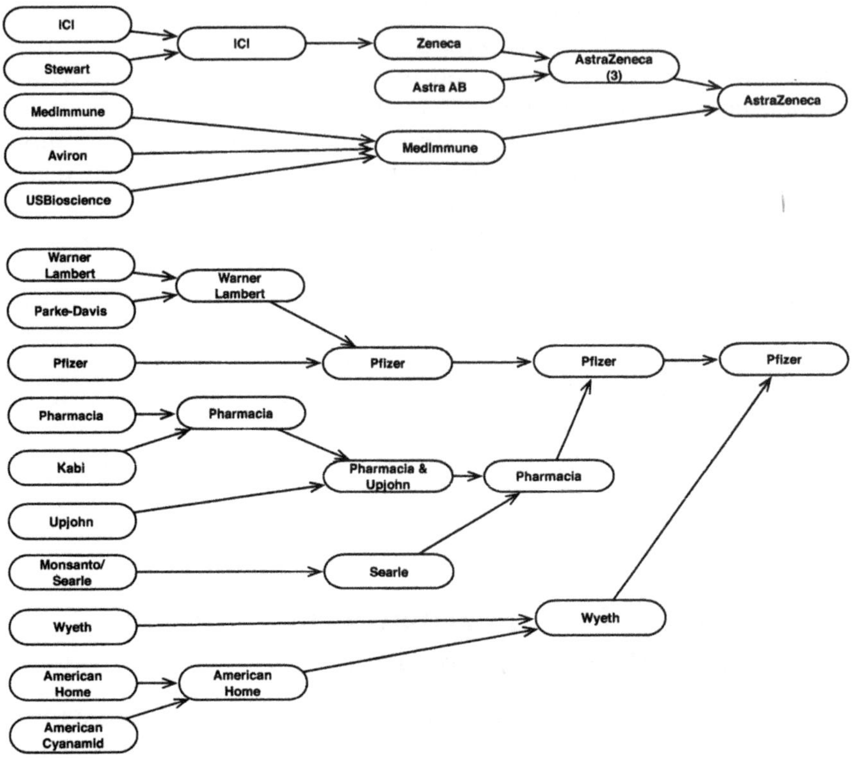


**Figure 1.1** Merger and acquisition history for AstraZeneca and Pfizer. Excerpt reproduced with permission from ref. 7.

product produced (and even in the thousands for pre-clinical processes).[13] With green chemistry, this has been shown to decrease to ~20 kg and even as low as single digits for some commercial processes. By utilizing the 12 principles, materials are used more efficiently, generating less waste and fewer hazards, lowering the standard cost for an active pharmaceutical ingredient. The use of green chemistry principles in drug discovery results in a faster production cycle time, which creates a competitive advantage.

Chemistry and innovation are the core of the pharmaceutical business. Bringing these together to discover and develop safe and effective medicines to help improve lives of patients is the objective. Achieving this goal cost-effectively with minimal environmental impact is the requirement. Green chemistry is the mechanism to meet these needs; it is an innovative, non-regulatory, economically driven approach toward sustainability:[14]

> *"The core of what we do here is to define transformative medicine that will help the patient. The goal is doing chemistry that gives equal or better results and in a way that benefits the environment."*
>
> – Bruce Roth, Vice President, Drug Discovery, Genentech[15]

By viewing the entire life cycle of material and energy processes as an opportunity for design innovation, green chemistry enables the design of drug candidates to not just minimize unintended consequences, but more importantly to empower sustainability. Efficient and selective utilization of resources during the discovery, development, and manufacture of medicines enables the opportunity to meet the needs of today without limiting future generations to achieve theirs.

Chapter 13 will provide a more detailed analysis on the business case for green chemistry in drug discovery, highlighting the advantages for corporate profitability.

## 1.4 Who is Doing Green Chemistry?

In the early 2000s, green chemistry was talked about in the pharmaceutical industry, and some companies had demonstrated successes but many companies were struggling to understand how to incorporate it into the business.[16] Under the leadership of Dr Paul Anastas, then Director of the American Chemical Society's Green Chemistry Institute®, and Dr Buzz Cue, retired Vice President of Pharmaceutical Sciences at Pfizer, the ACS GCI Pharmaceutical Roundtable was launched as a non-competitive partnership between the Institute and the pharmaceutical industry to catalyze the integration of green chemistry and engineering in the industry. It started with just three companies, Lilly, Merck, and Pfizer, in 2005; other companies were interested but not yet able to justify participation within their organization. Nine years later in 2014, the Roundtable has 16 member companies, namely, Amgen, AstraZeneca, Boehringer Ingelheim, Bristol-Myers Squibb, Codexis, Cubist Pharmaceuticals, DSM Pharmaceutical Products, Dr Reddy's, GlaxoSmithKline, Johnson & Johnson, Lilly, Merck, Novartis, Pfizer, F. Hoffmann-La Roche Ltd, and Sanofi, and their respective subsidiaries including but not limited to Genentech, and MedImmune. The Roundtable is certainly not the only organization facilitating green chemistry, nor is it intended to imply that all pharmaceutical companies implementing green chemistry are part of the Roundtable. However, the membership of the Roundtable paints a picture that is representative of a majority of the global pharmaceutical industry having at least a basic level of commitment and participation in green chemistry.

Joining an organization may demonstrate commitment but it does not always correlate to active engagement. However, in the Roundtable, companies are not just listening; they are doing. Companies actively participate in benchmarking exercises, research collaborations, tool development, publications, and, more importantly, take the outcomes from the efforts and integrate them into their respective organizations as appropriate. Process mass intensity (PMI), a metric defined by the Roundtable to address the amount of material used in a process per kilogram of active pharmaceutical ingredient (API), is now used broadly throughout the industry. AstraZeneca sets PMI targets for their active pharmaceutical ingredients by the time of commercial launch, and has reported as much as a 90% PMI reduction during the development phase.[17] Lilly similarly implements a methodology to judge suitability for commercialization and encourages the reduction of hazardous material usage, increasing material efficiency, and evaluating chemistry and chemical alternatives. A Lilly Environmental Development Review in 2013 identified a solvent reuse opportunity that was worth up to $5 million annually in recovered solvent and would decrease greenhouse gas emissions by more than 83%.[18]

Recognizing the amount of outsourcing employed in the industry, the Roundtable companies are collectively considering how to engage the supply chain. It would be impossible to meet their mission of catalyzing green chemistry in the global pharmaceutical industry if they only considered what went on within their own walls. In fact, several years after the Roundtable launched, the membership scope was expanded to include contract manufacturers and research organizations as well as generic manufacturers. Many of the tools developed by the Roundtable, including the solvent selection guide and PMI calculator, are available publicly and some companies provide them directly to their suppliers. What if, in the future, greenness of a supplier is incorporated into the preferred supplier profile? Understanding that green chemistry provides an economic incentive along with the environmental and safety benefits makes this a reasonable consideration. A supplier using green chemistry should be producing the desired product more efficiently, with lower costs, and reduced environmental and safety footprint. With that perspective, it really becomes a question of when green chemistry is incorporated into the supplier profile, not if.

Going deeper into the supply chain, chemical manufacturers are also involved in green chemistry implementation in the pharmaceutical industry. As an example, the Grignard reaction is commonly used to forge carbon–carbon bonds, although the reaction has serious safety and environmental concerns. Recent efforts initiated by the pharmaceutical industry evaluated the reagents in less hazardous solvents.[19] Sigma-Aldrich, among others, subsequently applied the learnings and listened to market drivers by providing Grignard reagents in greener solvents such as 2-methyltetrahydrofuran (2-MeTHF).

The study evaluating the aforementioned reagents was conducted as a collaboration between the Roundtable and Professor Wei Zhang from the University of Massachusetts Boston. Academicians have a critical role in the ability for industry to implement green chemistry. With academia addressing basic research on reactions, materials, and conditions with industrial relevance, industry can utilize the findings to introduce more sustainable alternatives with lower risk. Collaboration between industry and academia provides proof of concept that helps the company implement the technology with a higher level of confidence. Similarly, by incorporating green chemistry into the curriculum and academic research environment, students learn the principles of green chemistry and become the informed job applicant mentioned previously.

In short, to answer the question of who is doing green chemistry, a broad spectrum of the global pharmaceutical industry including the supply chain is incorporating green chemistry into their business, not just because it is the "right" thing to do, but because it is also right for business.

## 1.5 Where is Green Chemistry Being Applied?

Consistent with the earlier message that green chemistry is more similar to performance criteria than a field unto itself, green chemistry can be implemented in any segment of the pharmaceutical industry. Historically efforts have focused on the small molecule pharmaceutical development, recognizing the large amount of material, predominantly solvent, used to manufacture one kilogram of API. As biologics have become more prevalent, studies have shown that large molecules are not as consistent with the 12 principles as had once been assumed.[20] Although utilizing less solvent, the significant water usage, related energy requirements, and use of disposables leave a lot of room for improvement. These challenges are being addressed as a subgroup within the ACS GCI Pharmaceutical Roundtable dedicated to the needs of green chemistry in the biopharma business. Similarly, Bristol-Myers Squibb's transformation into a next-generation biopharma leader commits to a strong alignment with green and sustainable business practices.[21] A later chapter will address the opportunities for green chemistry in biopharma more directly.

Without question, process chemistry has been the predominant application of green chemistry in the industry. This is not surprising as it is the "low hanging fruit", as some may say, with the greatest potential for impact. Positive changes in efficiency and hazard reduction in a process with foreseeable large volume production are more easily justified. Examples of process chemistry improvements are abundant and documented cost and material savings are significant, including Lilly's convergent route to LY2624803•$H_3PO_4$, an investigational new drug candidate in phase II clinical trials, where the original process obtained through acquisition was not amenable to large-scale manufacture.[22] The implementation of a flow process reduced PMI from over 1000 before chromatography to a net PMI of 59, representing a 94% PMI reduction.

Green chemistry in drug discovery is a fairly new horizon. Led by medicinal chemists, the Roundtable started its own medicinal chemistry subgroup in 2012. It is an active group working to overcome the challenges with implementing green chemistry in a discovery organization. The remainder of this book is dedicated to providing tools, case studies, and more detail on how to apply the 12 principles into drug discovery.

Finally, when addressing the question of where green chemistry is being implemented within the business, it is worth taking a moment to consider the people leading the efforts. Although it would be advantageous to have senior management lead the integration of green chemistry into the corporation, it is not an essential requirement. Many efforts initiate as a grassroots program led by the very scientists designing the molecules. Leadership expert Robin S. Sharma describes this leadership from within:

> *Basically, to lead without a title is to derive your power within the organization not from your position, but from your competence, effectiveness, relationships, excellence, innovation, and ethics.*[23]

Management can empower their staff to incorporate the 12 principles, but the scientists in the laboratory designing and testing the drug candidates are the ones who do it. The person closest to the process needs to recognize the opportunities to consider more efficient reactions, to select less hazardous materials, and, importantly, to communicate those choices to fellow colleagues. People learn most effectively by doing, not by telling. Grassroots efforts integrate green chemistry by doing.

A bottom up approach can be effective in gaining a groundswell of support for green chemistry, but eventually, to be sustainable, it should be tied with leadership support by senior management.

Management expert Peter F. Drucker is quoted as saying "Management is doing things right; leadership is doing the right things." The decision to integrate the 12 principles into the business at its earliest opportunity is doing the right thing.

## 1.6 When Should Green Chemistry be Implemented?

The interface between discovery and development is closer and more transparent than ever before. With the acquisition of experimental drug candidates, the line practically disappears. As discussed previously, R&D budgets are getting smaller, and there is a need to do more with less. The chemistry used in the first synthesis of a compound can have a disproportionately large influence over the chemistry of an ongoing project and the subsequent chemistry of compounds in development.[24] To realize the greatest benefits, green chemistry principles should be integrated into the drug development process at its earliest stages, in drug discovery, and every phase thereafter. Green chemistry is a continuous endeavor to innovate. As many companies do, Pfizer recognizes this continuum is critical to advancing scientific innovation, and commits to integrating environmental and health and safety considerations throughout the research, discovery, development, and manufacture of their products.[25]

"Right first time" is commonly associated with emphasis on quality, but the same mantra could be said for designing sustainability in from the onset. With discovery organizations designing the first synthetic route, the direction put forth at this earliest stage can have a significant long-term impact on the environmental profile of the product, as well as cost, scalability, and efficiency. In the pharmaceutical industry, second or third generation processes are filed for any number of reasons. A recent study evaluated the regulatory approval times for process changes and found a disparity among approval times from regulatory agencies resulting in the potential need to carry inventory for both first and second generation processes for months to several years until all regulatory agencies granted approval.[26] Thus, the environmental and economic benefits of a green chemistry process change may not be realized for a significant period of time following the original product being introduced to the market.

Johnson & Johnson (J&J), among other companies, recognize that waiting to incorporate green chemistry into a second or third generation process not only delays the ecological impact, but it is also time-consuming and significantly more expensive. J&J's focus is to design a sustainable chemical process the first time. Abiraterone acetate, the active ingredient in ZYTIGA™, a treatment for prostate cancer, was recognized with an internal company award for reducing the number of solvents from eight to two, utilizing 64% less raw material and 78% less water.[27]

To summarize, the time to implement green chemistry is now. There is no perfect application of the 12 principles; however, intentionally applying them to the design of materials at their earliest phase of development, in discovery, will optimize the impact of the decisions. Second generation processes may be further improvements, but the regulatory approvals further delay implementation and realization of the benefits.

## 1.7 How is Green Chemistry Integrated into Drug Discovery?

Although not formally one of the Five Ws, "how" has arguably greater importance than the questions already addressed. This book is intended to be a resource to facilitate the integration of green chemistry in medicinal chemistry organizations. All but one chapter (which is by a recognized government authority) are written by authors with experience in the pharmaceutical industry who can speak from an informed position and share their knowledge in addressing green chemistry implementation in drug discovery. The strategies presented could be used in their entirety or in individual segments as may be needed. From a more thorough analysis of the business case for green chemistry to a realistic picture of the barriers to implementation, the following chapters provide a perspective from peers who have addressed the same challenges. Several chapters are dedicated to providing tools to enable implementation specifically in drug discovery, including techniques for reaction setup and workup, choices of solvents, reagents, and their reuse or recycling, use of analysis and purification tools to reduce time and cost, use of high-throughput screening for more rapid and more informed decisions, and an understanding of the applicability of continuous processing in drug discovery. A chapter on application of the principles in biologics drug development was deemed essential to cover the expanding field and to illustrate how the principles are applied in a biologics setting. Finally, additional topics were included to provide a broader perspective for the medicinal chemist to better appreciate the interrelationship between the work in discovery and its impacts. These chapters provide a foundation on toxicology and environmental regulations, a perspective on the complexity of electronic lab notebooks, and, importantly, an understanding of green chemistry innovations relating to patent protection.

In summary, the review of the Five Ws in pharmaceutical green chemistry has set the stage for the chapters that follow providing the tools needed for successful implementation. The ability to answer the Five Ws enables the reader to succinctly communicate the importance of green chemistry, making successful integration more probable and less a game of chance:

*What is green chemistry?*
Green chemistry is chemistry, the scientific discipline of arranging molecules to create new materials and products; it places emphasis though on the intentional integration of source and hazard reduction into the design of matter.

*Why should the pharmaceutical industry incorporate green chemistry?*
Chemistry, innovation, and a mission to improve lives are the core of the pharmaceutical business. Green chemistry brings these together with a positive direct impact on the bottom line and the environment, meeting the needs of the business today without compromising the ability of future generations to meet their needs.

*Who is doing green chemistry?*
A broad spectrum of the global pharmaceutical industry, including the supply chain, is incorporating green chemistry into their business, not just because it is the "right" thing to do, but because it is also right for business. Green chemistry is further enabled by innovations and collaborations with academia, government, and other organizations.

*Where is green chemistry being applied?*
Green chemistry is applied by anyone influencing the design of the chemical product. Although historically emphasized in process chemistry in the pharmaceutical industry, it is now increasingly being applied in discovery and large molecule biologics.

*When should green chemistry be integrated into the process of pharmaceutical discovery and development?*
Now. There is no perfect application of the 12 principles; however, intentionally applying them during every step of drug discovery and development will optimize the impact and enable further innovation.

## References

1. P. T. Anastas and J. C. Warner, *Green Chemistry: Theory and Practice*, Oxford University Press, New York, 1998.

2. J. B. Manley, in *Scalable Green Chemistry*, ed. S. G. Koenig, Pan Stanford Publishing Pte Ltd, Singapore, 2013, pp. 1–24.
3. http://www2.epa.gov/green-chemistry/presidential-green-chemistry-challenge-award-recipients-industry-sector#pharmaceuticals (Accessed December 1, 2014).
4. Press Release: The Nobel Prize in Chemistry 2005. Nobelprize.org, November 14, 2010. http://nobelprize.org/nobel_prizes/chemistry/laureates/2005/press.html (Accessed November 1, 2014).
5. E. Gershon, Q&A: Climate change is already here, says 'father of green chemistry', Yale News, February 27, 2014. http://news.yale.edu/2014/02/27/qa-climate-change-already-here-says-father-green-chemistry (Accessed November 24, 2014).
6. Our Common Future, Report of the World Commission on Environment and Development, World Commission on Environment and Development, 1987. Published as Annex to General Assembly document A/42/427, Development and International Co-operation: Environment August 2, 1987.
7. D. Davidovic, *The History of Bio-Pharma Industry M&As, Lessons Learned and Trends to Watch*, on PM360 online, 23 May 2014. http://www.pm360online.com/the-history-of-bio-pharma-industry-mas-lessons-learned-and-trends-to-watch/ (Accessed November 1, 2014).
8. Pharmaceutical Research and Manufacturers of America, 2014 Biopharmaceutical Research Industry Profile, Washington, DC, April 2014.
9. http://articles.chicagotribune.com/2012-12-30/business/ct-biz-1230-bf-abbott-spin-20121230_1_longtime-abbott-executive-abbvie-abbott-laboratories (Accessed November 1, 2014).
10. http://dealbook.nytimes.com/2011/10/19/abbott-to-split-into-two/?_r=0 (Accessed November 1, 2014).
11. http://online.wsj.com/articles/SB10001424052702303918804579108991879368028 (Accessed February 23, 2015).
12. A. Larsen and M. Meier, in *Green Techniques for Organic Synthesis and Medicinal Chemistry*, ed. W. Zhang and B. W. Cue, Jr., John Wiley & Sons, Ltd., 2012.
13. R. K. Henderson, J. Kindervater and J. B. Manley, Lessons Learned through Measuring Green Chemistry Performance: The Pharmaceutical Experience, American Chemical Society Green Chemistry Institute Pharmaceutical Roundtable, 2008, http://portal.acs.org/portal/PublicWebSite/greenchemistry/industriainnovation/roundtable/CTP_005585 (Accessed December 17, 2014).
14. J. B. Manley, P. T. Anastas and B. W. Cue, *J. Cleaner Prod.*, 2007, **16**, 743.
15. *Better Chemistry*, http://www.gene.com/stories/better-chemistry (Accessed December 1, 2014).
16. http://www2.epa.gov/green-chemistry/green-chemistry-program-nomination-table (Accessed November 1, 2014).
17. http://www.labtalk.astrazeneca.com/hot-topics/green-chemistry-at-astrazeneca/ (Accessed November 1, 2014).

18. http://www.lilly.com/Responsibility/environmental-sustainability/Pages/product-stewardship.aspx (Accessed November 1, 2014).
19. A. Kadam, M. Nguyen, M. Kopach, P. Richardson, F. Gallou, Z. Wane and W. Zhang, *Green Chem.*, 2013, **15**, 1880.
20. S. V. Ho, J. M. McLaughlin, B. W. Cue and P. J. Dunn, *Green Chem.*, 2010, **12**, 755.
21. http://www.bms.com/sustainability/go_green/Pages/default.aspx (Accessed November 1, 2014).
22. http://www2.epa.gov/green-chemistry/green-chemistry-program-nomination-table (Accessed September 8, 2014).
23. http://gulfnews.com/life-style/people/motivational-guru-robin-sharma-on-leadership-1.983647 (Accessed November 1, 2014).
24. M. C. Bryan, B. Dillon, L. G. Hamann, G. J. Hughes, M. E. Kopach, E. A. Peterson, M. Pourashraf, I. Raheem, P. Richardson, D. Richter and H. F. Sneddon, *J. Med. Chem.*, 2013, **56**(15), 6007.
25. http://www.pfizer.com/responsibility/protecting_environment/green_journey (Accessed September 20, 2014).
26. P. J. Dunn, *Green Chem.*, 2013, **15**, 3099.
27. http://www.janssen-verslag-samenleving-2012.be/en/Ecological-sustainability/Green-chemistry-and-green-technology (Accessed September 20, 2014).

CHAPTER 2

# Barriers to Adopting Green Chemistry in Drug Discovery

HELEN F. SNEDDON

Green Chemistry Performance Unit, GlaxoSmithKline, Medicines Research Centre, Gunnels Wood Road, Stevenage, Hertfordshire, UK, SG1 2NY
Email: helen.f.sneddon@gsk.com

## 2.1 Introduction

The barriers to the adoption of green chemistry in drug discovery in many ways echo barriers that have been identified to the adoption of green chemistry in industry more widely.[1–3] In medicinal chemistry, the barriers can be broadly classified as economic, technical, and cultural, with cultural barriers posing by far the greatest challenge.

Economic and technical barriers pertaining to solvents, reagents, time, and technologies are briefly covered, with some suggestions as to how some of these may be addressed. The cultural barriers mainly result from a lack of awareness of green chemistry both within the industry and in the wider chemistry community, including academia, and the perception that incorporation of green chemistry could slow down the pace of research. Specific concerns arising from each of the 12 principles of green chemistry[4] are considered, as is the possibility of confusion over communication, especially over how reagents, solvents, or processes are labelled as "green" or which metrics should be used. Some recommendations for overcoming these barriers and embedding a green chemistry culture are discussed.

RSC Drug Discovery Series No. 46
Green Chemistry Strategies for Drug Discovery
Edited by Emily A. Peterson and Julie B. Manley

Published by the Royal Society of Chemistry, www.rsc.org

## 2.2 Economic Barriers

There is considerable debate about the cost of developing a single drug, and the methodology that should be used when performing such a calculation. One recent estimate, dividing total R&D spend by drugs approved gave a range for major pharmaceutical companies between \$4 and \$11 billion per drug,[5] another estimate put the figure between \$315 million and \$2.8 billion, with an average of \$1.1 billion.[6] What is not disputed is that drug development is an expensive process, with the great majority of the expense deriving from clinical trials rather than chemistry. The main drive to reduce these costs focuses on improving the industry's rate of attrition. However, money saved through efficient drug discovery operations is also rightly seen as money that can be spent on more drug discovery and development to bring more drugs to market faster, and benefit more patients.

The annual worldwide production of active pharmaceutical ingredient (API) has variously been estimated at greater than 65 million kg per year,[7] and greater than 100 million kg per year,[8] and Cue has predicted[7] (using typical mass intensities from the ACS GCI Pharmaceutical Roundtable (ACS GCIPR) companies and waste disposal costs) that more than 10 billion kg per year of waste is produced with an annual disposal cost of nearly \$20 billion. Factoring in annual new drug applications (NDAs), candidate survival rates for each stage of R&D, and the ranges of weights of APIs produced in each R&D stage, Cue estimates that between 200,000 and 2 million kg of waste is produced annually from the drug discovery stage of the whole pharmaceutical industry, with a further 150,000 to 1.5 million kg produced at a preclinical stage.[9]

There is understandably considerable unease at any initiative that might threaten to incur increased costs in day to day operations in drug discovery, or later when a potential drug reaches development or even manufacture. It is not easy to assuage unease, because as the diversity of issues considered in the 12 principles shows, green chemistry is a wide ranging concept. However, there is clearly potential for savings to be made by reducing the amount and hazardous nature of the waste produced at the drug discovery stage.

Some solvents and reagents recommended on the basis of their environmental sustainability credentials might be more expensive (such as 2-MeTHF), some might be less costly (such as ethanol). Others might initially appear more expensive but be more economical once costs of disposal (especially chlorinated solvent alternatives), or later costs spent reworking a route (especially alternatives to solvents which are authorized by REACH (Registration, Evaluation, Authorization and Restriction of Chemicals) legislation),[10] are taken into account. Other greener options may never be financially viable until such time as increased demand makes their current price drop through competition or legislation, or rises in the price of crude oil make the existing option unattractive. New equipment such as reverse phase medium pressure liquid chromatography (MPLC) or supercritical fluid chromatography (SFC)[11,12] may require an initial financial outlay but yield superior separation results as well as sustainability benefits.

The costs of developing new approaches or technologies such as synthetic biochemistry or flow chemistry must be balanced against numerous potential advantages, such as the possibilities of accessing templates that might not be accessed another way. Such long-term strategic decisions are beyond the remit of this chapter, but "drop-in" replacements, where one solvent or reagent might be substituted by another potentially greener option are easier for drug discovery to assess and it is certainly not the case that a premium always has to be paid for greener options.

### 2.2.1 Solvents

The cost individual chemistry groups will pay for a given solvent may fluctuate depending on any bulk order discounts or preferential pricing agreements with different suppliers, as well as on the size of bottles/drums in which it is typically ordered, so exact comparisons are often not possible.

Even harder to quantify are the associated costs of using each solvent. 2-MeTHF for example has low toxicity[13] and is believed to result in greater overall process economy, as a result of its limited water-miscibility, and ability to form an azeotrope with water on distillation.[14] On scale, once time considerations and costs of handling aqueous waste are factored in, these financial benefits may outweigh its current increased purchasing cost compared to THF. On a discovery scale, the lower melting point and higher boiling point of 2-MeTHF may occasionally be an advantage in allowing reactions across a wider temperature range, and allowing advantageous higher temperature aqueous separations.[15] In addition, the improved ease of washing and work-up when used as a reaction solvent may result in significant time savings. Such considerations may justify the up-front purchase cost.

In drug discovery, chromatography is by far the largest use of solvent, and hydrocarbons are the most commonly used solvent in drug discovery chromatography systems. Hexane has a lower occupational exposure threshold of 50 ppm (USA, TLV (threshold limit value)) and 20 ppm (EU) and *n*-hexane is metabolized firstly to 2-hexanol, and then to 2,5-hexanediol by cytochrome P450s. 2,5-Hexanediol may be further oxidized to neurotoxic 2,5-hexanedione.[16] Cyclohexane and heptane are significantly less toxic, with occupational exposure values of 300 ppm (USA, TLV) for cyclohexane and 400 ppm (USA, TLV) and 500 ppm (EU) for heptane. The cost difference, and operational differences between *n*-hexane and safer hydrocarbon solvents of similar polarity, is negligible, and the barriers in this case may be as simple as overcoming a resistance to change.

Other cost arguments may be more subtle. Chlorinated solvents such as dichloromethane may be relatively inexpensive to buy, but their use can incur additional costs. For example, when solvents with low boiling points, such as dichloromethane (40 °C), are stored in warmer climates, the requirements for chilled storage to avoid evaporation can be costly in terms of both energy and money. Chlorinated solvent waste typically requires scrubbing on incineration, and the replacement of carbon filters required to do this on a manufacturing scale can easily contribute a significant percentage of an operational running cost. Dichloromethane (DCM) is frequently used as a major component of chromatography mobile phases; for such uses the well-validated alternative heptane : EtOAc : EtOH[17] is actually less expensive than DCM : MeOH, even before disposal costs are included.

Development chemists have already started to move away from chlorinated solvents. For example, dichloromethane was the joint second most common solvent in GSK pilot plants between 1990 and 2000, but had fallen to eighth most popular by 2005.[18] Drug discovery chemists must not only balance the immediate advantages and disadvantages of dichloromethane and alternative solvents, but also the probability of success, and the trade-off between their own extra time to find and validate an alternative, *versus* potential future work by development chemists to explore such alternatives in the event of a candidate being transferred down the pipeline.

### 2.2.2 Reagents

The scale on which reagents are used is orders of magnitude smaller than that of solvents, and as such the cost of individual reagents may be subject to much less scrutiny, especially when the sheer diversity of reactions carried out in drug discovery is considered. However, for all the undeniable diversity of reactions explored, some transformations remain common, and a recent analysis of chemistry research across three of the largest pharmaceutical companies (GSK, AstraZeneca, and Pfizer) indicated that *N*-acylation to prepare amides was the most common transformation, accounting for 16% of all reactions performed.[19]

Given the volume of amide bond formations occurring within a drug discovery environment it is inevitable that chemists revert to a few favored reagents, commonly those with broadest scope. Aryltriazole coupling reagents generally have low atom economy, and can rarely be safely scaled up as a result of their shock sensitivity. It may therefore interest medicinal chemists to learn that in the case of amide bond formation some of the options currently judged to be "greener",[20] such as CDI and T3P®, not only have reasonably broad scope, and can be readily scaled, but are also considerably less expensive.

### 2.2.3 Recommendations for Overcoming Economic Barriers

The possibility for the money saved by green chemistry and operational activities to be reinvested into drug discovery should be emphasized to chemists, to increase engagement with campaigns to encourage the closing of fume hoods when not in use, and switching off of all unnecessary equipment at the end of every working day, or at weekends.

If medicinal chemists are down to their last 10 mg of a compound they will be understandably reluctant to try a chemistry with which they are less

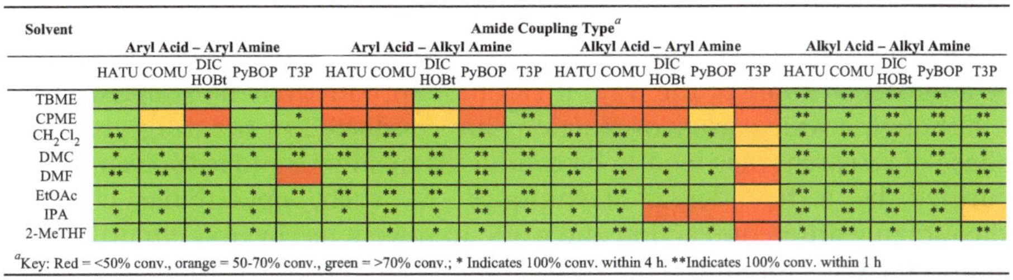


**Figure 2.1** Amidation solvent and reagent screen representation.[21] Reprinted from ref. 21 with permission from The Royal Society of Chemistry.

familiar. The scope of greener reagents should therefore be explored as far as possible when chemists have a more plentiful supply of material and the scope of greener reagents should be clearly communicated, to build confidence for future use. One example of this is the amide formation grid,[21] which shows that the inexpensive reagent T3P® works well with all classes of acid and amine except for coupling alkyl acids to aryl amines, in the reasonably priced "greener" solvent EtOAc (Figure 2.1).

Where some reagents or solvents may initially be more costly, chemists should be encouraged to consider if they are still more expensive once factors such as waste handling costs are taken into account. If prices are still prohibitive, reminders should be set to check in the future whether prices have decreased to a feasible level.

The aphorism "time is money" clearly applies in drug discovery as elsewhere, and time considerations feed into several technical barriers to the adoption of green chemistry.

## 2.3 Technical Barriers

Drug discovery chemists typically perform a wide variety of reactions on very small scale, typically making between 10 and 50 mg of final compound, or maybe a few grams once there is particular interest in a given compound. The environmental impact of each individual reaction is relatively small, and one barrier is that some well validated green chemistry approaches (such as solvent recycling to improve process mass intensity) are not appropriate to be applied in this context in the way it might be once the final API is known.

Optimization of every reaction at a drug discovery stage would not only be impossible from a time and productivity perspective, but also counterproductive in terms of improving overall green credentials of this stage of operation, as attrition is so high that most reactions once tried will never be repeated, and optimization would represent wasted time, energy, and materials. However, for commonly performed reactions, such as Suzuki couplings, it is beneficial to be able to recommend a standard set of conditions with greener solvents and lower catalyst loadings. Such an improved initial start point can be extremely effective in shaping the overall final route.

The perception that green chemistry at a drug discovery scale might involve individual tweaking of multiple small-scale operations can itself be a barrier to some of the ways it can more productively be adopted.

Although a wide variety of reactions are conducted in the drug discovery phase of research, they tend to be conducted in a few standard solvents, and purified by a few default techniques. As such, operational changes encouraging different reaction and work up solvent use by first intent,[22–25] or more efficient[26] or different[17,27–30] solvent use in purification, can have impact across a wider range of reactions.

Similarly the vast majority of medicinal chemistry reactions fall into <20 transformation classes, and broad generalized guidelines on preferred reagents[20] for use by first intent not only minimize the generation of hazardous waste on the drug discovery scale, but also increase the probability that the route to the compound that eventually is scaled-up already utilizes more benign reagents.

Once a template has been identified as being likely to lead to a candidate, and it is clear that a lot of similar chemistry will be carried out to give analogues, it can be appropriate to start to consider the atom efficiency or process mass intensity of the route. Such early efforts can have a noticeable impact even in drug discovery, when chemists have the need to scale up one or more analogues for more advanced testing. The advantage of doing so at this time, in parallel with the drug discovery work, and in consultation with development colleagues, is that it is inexpensive and low risk to explore alternatives on a small scale, and should minimize delays once there is commitment to a given molecule.

Many recycling approaches may be easier to implement on large scales where all material has arisen from the same process, making it easier to know the precise composition of the material to be recycled. However, this is not an insurmountable barrier. Medicinal chemistry laboratories have been able to successfully implement catalyst recycling schemes where multiple sources of precious metal catalyst containing waste are combined, and smelted to allow recovery. Likewise solvent recovery is possible – both on an informal case by case basis, reusing solvent for an individual column for example, and on a more formalized basis collecting common solvent streams from multiple laboratories. These solvent streams can then be either distilled onsite or passed on to a separate waste handler who can then either sell the purified solvents back to the laboratory (tolling), or sell on to another customer requiring lower purity solvents, depending on the quality of the solvent and nature of the agreement. A more in-depth discussion of recycling in the laboratory is provided Daniel Richter in Chapter 8 of this book.

Other technical barriers are associated with the considerable time pressure that drug discovery chemists are under to find whether a given template is worth pursuing, or to optimize a given lead to generate a pre-candidate. In some cases, greener alternatives may not be well precedented, or it might take more time to find protocols or reagents.

### 2.3.1 Recommendations for Overcoming Technical Barriers

Some technical barriers can be overcome by:

- Greater sharing of knowledge and equipment between development and discovery groups can allow rapid screening of different reaction conditions.
- Educating chemists as to where information can be found as to the scope of different reagents and solvents, such as in-house solvent and reagent guides, or in-house databases of both green and process-scale chemistry.
- Recommending standard sets of (greener) initial reaction conditions by default before optimization is required.
- Recommending standard (greener) work up and chromatography solvents by default before optimization is required.
- Reminding chemists of filtering options available when using literature search tools such as SciFinder® or Reaxys® to help speed the finding of suitable precedents.
- Making greener reagents more readily available by having a cache in a central location within a drug discovery department.

While in the short term little can be done about reactions where greener alternatives are scarce or limited in scope, chemists can highlight the nature of such reactions to encourage further research in these industrially relevant areas.[31]

## 2.4 Cultural Barriers

### 2.4.1 Awareness of the 12 Principles within Drug Discovery

Limited awareness as to the nature of green chemistry poses one of the biggest barriers to its integration into day-to-day activities.

Green, or sustainable, chemistry as a field has been growing in popularity since arguably its inception following the publication of Rachel Carson's *Silent Spring* in 1962[32] and its more widespread recognition in the 1990s.[4] The 12 principles of green chemistry are designed to apply to all aspects of chemistry, have been analyzed and discussed from a variety of perspectives,[33–35] and there are now dozens of courses on green chemistry worldwide.[36] However, it is still not integrated into all undergraduate and graduate chemistry programs, and many new drug discovery chemists may not have had the opportunity to study green chemistry in the university setting. Furthermore, existing medicinal chemists may not only not have had the opportunity to study it but may also have been under the apprehension that this was the responsibility of their colleagues in development when their compounds progressed, or the responsibility of their colleagues in environment, health and safety (EHS) as far as it impacted the operational aspects of drug discovery.

The assumption that the sustainability implications of pharmaceutical chemistry should be the responsibility of those in development dealing with large volume, clinical trial routes is understandable. Development routes (and the manufacturing processes arising from them) are performed on scales orders of magnitude larger than those in medicinal chemistry and have proportionately larger sustainability impacts. In addition, the rate of attrition of compounds in medicinal chemistry is far higher than in development. (Success rates of drugs entering the clinic were recently estimated at ~10%, with Phase III success rates estimated to be ~60%.[37]) However, as this book will argue, such an approach misses a relatively simple opportunity to use the chemistry knowledge of those drug discovery chemists who know best the synthetic challenges of their series, at a time when a simple change might be least expensive or easiest to explore. Early choices may also have their impacts magnified as, once a route is known to be successful, time constraints or human nature may be such that there is increased bias to stick with something that is known to work, possibly avoiding optimization (and maybe even the need for bridging toxicology studies) if it is already fit for purpose. Additionally, as has already been acknowledged from an economic perspective, the total impact of drug discovery across the whole industry is significant.

Until such time as green chemistry is part of the mainstream curriculum in all undergraduate and graduate courses, and embedded in all medicinal chemistry departments, it is likely that one of the main cultural barriers to adopting green chemistry in drug discovery may well be only patchy familiarity with what green chemistry means, and how it may be implemented in this context.

For those drug discovery chemists who are familiar with the 12 principles of green chemistry, as laid out by Anastas and Warner,[4] (Figure 2.2)[38] there are aspects of drug discovery as it is commonly practiced today which may seem incompatible with these principles. However, further analysis of some of these commonly perceived conflicts of interest may show that it is indeed possible to incorporate green chemistry ideals into a drug discovery setting.

#### 2.4.1.1 *Prevent Waste: Design Chemical Syntheses to Prevent Waste, Leaving No Waste to Treat or Clean Up*

There may be little incentive on a small scale to prevent waste. Drug discovery chemistry groups may not be responsible for, and therefore not aware of, the cost of treating their waste. Ultimately it would be ideal if solvents were transparently priced such that waste disposal costs were included in the purchase cost. However, such a change is likely to be difficult if not impossible to implement with multiple suppliers, waste handlers, and purchasing systems, and the monetary cost may not truly reflect other costs, such as carbon footprint.

**Figure 2.2** The 12 principles of Green Chemistry.[38a]
Reprinted with permission from ref. 38*b*. Copyright 2013 American Chemical Society.

There are, however, simple ways in which green chemistry initiatives in medicinal chemistry can influence chemists to be more aware of the waste they generate. These can be operational approaches such as encouraging reuse of columns, recycling of fraction tubes, or metal catalysts. These operational approaches can have an additional aim, explicit or implied, that chemists should be more aware of other aspects of waste. Additional information sharing or educational approaches can be used for encouraging colleagues from later stages of the business to share case studies where waste has been a problem for different routes, and, ideally, how these problems have been solved. For example, issues with troublesome metal salts requiring energy intensive specialty handling on scale can be highlighted,[39] encouraging chemists earlier in the organization to explore alternatives to these routes where possible.

From a more fundamental perspective, reducing the daily API dose required will ultimately reduce the amount of API that needs to be manufactured, and the associated waste produced. While some technological approaches can assist in achieving this goal, for example by substantially improved oral bioavailability for low solubility/high permeability compounds,[40] drug discovery chemists have a key role to play in finding candidates with high potency in the optimal physicochemical space. Such physicochemical considerations should not only minimize waste by minimizing the dose required, but should also maximize the likelihood of success of a given candidate,[41] thereby also minimizing waste from ultimately unsuccessful routes. This represents one way in which many of the key goals of medicinal chemists are already aligned with the goals of green chemistry.

#### *2.4.1.2 Design Safer Chemicals and Products: Design Chemical Products to be Fully Effective, yet have Little or No Toxicity*

Drug discovery chemists are already trying to optimize multiple chemical and biological parameters. Consideration of environmental toxicity may be an additional request on an already lengthy list, and one where limited information may be available as to how to achieve it.

However, notably, such a goal is not necessarily entirely unaligned with the existing goals of the medicinal chemist, who is already likely to be aiming to hit a single target as effectively as possible. As discussed, clearly the lower the dose that can be achieved, the smaller the overall footprint of any ultimate manufacturing process is going to be, and it is possible (though not necessarily a given) that lower dose drugs may pose less of a burden in waste water when the issues of pharmaceuticals in the environment are considered. Toxicity in the environment tends to track with lipophilicity, as persistent compounds have been found to be more likely to bioaccumulate in lipids in organisms.[42] In this instance, interestingly, the recent drive for less lipophilic drug candidates[43] can be seen as being entirely aligned with this principle.

#### *2.4.1.3 Design Less Hazardous Chemical Syntheses: Design Syntheses to Use and Generate Substances with Little or No Toxicity to Humans and the Environment*

Medicinal chemists are already trying to generate substances with a minimum of toxicity concerns. However, when it comes to the reagents and solvents being used in syntheses, greener alternatives do not always exist, and when they do their scope may not be adequately delineated. Chemists may not always have the time to explore the greener options, and in some instances those options might be more expensive.

Information as to which solvents and reagents have fewer hazards is available. Solvent guides have been published by several companies, including GSK,[22] Pfizer,[23] and Sanofi[24] with an American Chemical Society Green Chemistry Institute Pharmaceutical Roundtable (ACS GCIPR) guide[25] also available and in the process of being updated. Pfizer[23] and GSK[20] have published reagent guides, and an ACS GCIPR guide is in preparation. Where guides are not available, material safety data sheets (MSDSs) can be used to a certain extent to gauge the relative sustainability merits of different reagents.

The scope of preferred solvent and reagent alternatives, however, remains a major barrier to their adoption in drug discovery. Efforts are underway to counter this, including tables of when common reactions such as amide coupling[21] and reductive amination[44] are most likely to work (Chapter 3) and a range of papers on the scope of alternative chromatography options[17,27–30] (Chapter 4), but further research and dissemination of information about successful alternatives is necessary.

As previously acknowledged, green chemistry is not yet widely adopted by many large academic synthesis groups. Not only are these the groups that train many medicinal chemists and shape their default behaviors but they also generate literature precedents that are relied upon when incorporating newer transformations into drug discovery routes. Greater incorporation of green chemistry considerations into mainstream chemistry should therefore facilitate easier design of less hazardous chemical syntheses.

#### *2.4.1.4 Use Renewable Feedstocks: Use Raw Materials and Feedstocks that are Renewable rather than Depleting. Renewable Feedstocks are often Made from Agricultural Products or are the Wastes of other Processes; Depleting Feedstocks are Made from Fossil Fuels (Petroleum, Natural Gas, or Coal) or are Mined*

On small scale, where multiple different starting materials are being explored this is rarely a viable option for any reaction component other than solvents. Renewable solvents are limited in number, and may be more expensive, until such point as increased usage increases the scale on which they are being made.

Solvents are by far the largest component used in any synthesis, with an industry benchmarked average of solvents comprising 56% of the process mass intensity (PMI) of a final active pharmaceutical ingredient (API) route (Figure 2.3).[45] Notably, it is typical for reactions to be run at an increasing concentration as they are scaled up. With process refinement and optimization, the total use of solvent tends to decrease across the development phases towards product launch. Not only are reactions typically more dilute earlier in the R&D pipeline, but they may require solvent intensive purification protocols. As such, the volumes of solvent used in drug discovery for the production of a given amount of final compound are proportionally many times greater than would be expected considering an appropriate fraction of a development process.

Accordingly if renewable materials are to be considered anywhere in drug discovery, it should be in the preparation of common solvents. Ethanol can be derived from fermentation and 2-MeTHF can be synthesized by catalytic hydrogenation of furfural,[46] and the potential isolation of furfural or levulinic acid from the acid-catalyzed digestion of $C_5$ sugars in biomass (such as corn cobs or bagasse) has already been developed.[47] Citrus waste can give rise to (*R*)-(+)-limonene, which can be further converted into *p*-cymene,[48] and several other less widely used biomass derived solvents are becoming available. Studies are being carried out on the performance of reagents in some renewable solvents, for example the recent ACS GCIPR funded research into Grignard reactions in different solvents.[49] As data on the properties and scope of these newer solvents is generated and disseminated, and as the price of oil derived solvents increases, it is likely that more renewable

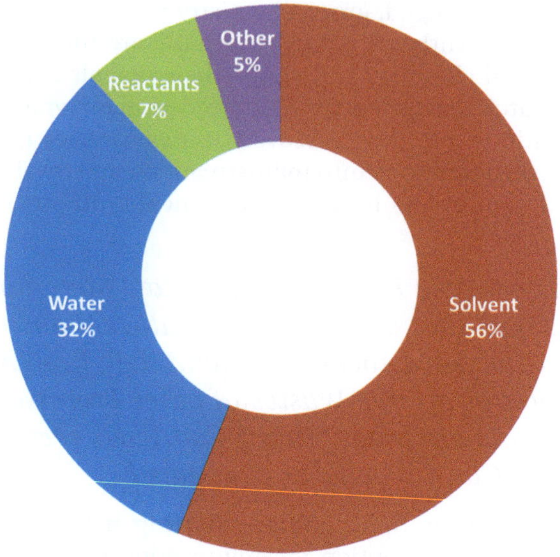


**Figure 2.3** Composition by mass of types of material used to manufacture an API (American Chemical Society Green Chemistry Institute Pharmaceutical Roundtable Benchmarking 2006 & 2008).[45]
Reprinted with permission from ref. 45. Copyright 2011 American Chemical Society.

solvents will become commonplace in drug discovery. It is also likely that, as renewable solvents grow in popularity, competition may ultimately drive down the price.

The topic of bio-derived solvents can risk polarizing chemists. It is a sensitive area as end-users of the solvents (and indeed of the products made using the solvents) want to be certain that the solvents truly are made from waste agricultural products, and are not competing with food crops, and driving up food prices. This is a very real concern, and it is to be hoped that greater transparency on sourcing and life-cycle analysis of bio-derived solvents is adopted as an industry standard, just as with oil-derived solvents, where it is to be hoped that these are obtained with a minimum of environmental damage consistent with that process.

#### *2.4.1.5 Use Catalysts, not Stoichiometric Reagents: Minimize Waste by Using Catalytic Reactions. Catalysts are Used in Small Amounts and can Carry out a Single Reaction Many Times. They are Preferable to Stoichiometric Reagents, Which are Used in Excess and Work Only Once*

Catalytic approaches are already frequently used in drug discovery, and medicinal chemists are generally keen to exploit new chemistry coming out of academia, especially if that chemistry offers access to new structures and areas of chemical space, or allows easier access to an existing template of interest. As many such recent advances involve catalysis, drug discovery chemists are often already on board with this principle.

Metal based catalysis and organocatalysis can raise different concerns. With metal based catalysis, the abundance of the metal in question (often crudely reflected in the price) may be key, as the extraction of rarer elements may not only generate more waste,[50] but also lead to concerns as to how viable a route may be a decade or two decades hence, as rare elements become even scarcer.[51] Some catalysts may have ligands which themselves require complex multi-step synthesis and are not always easily recoverable and reusable. Organocatalysts are often used in higher catalyst loadings, and may also require complex multi-step synthesis.

It may be inappropriate to optimize an individual catalytic step on small scale before a candidate has been selected, and reuse of catalysts is very rare in a drug discovery environment. The recycling of catalysts, however, is possible, where precious metal enriched waste is segregated from other waste streams, and processed separately to allow recovery of the precious metals, and this can be done cost effectively on medicinal chemistry scales. Academics should be encouraged to consider the recoverability of their catalysts right from the outset. In particular with homogeneous catalysts, consideration should be given to the ease with which they can be separated from reaction mixtures by extractive work up to avoid the need for chromatography.

#### 2.4.1.6 *Avoid Chemical Derivatives: Avoid Using Blocking or Protecting Groups or any Temporary Modifications if Possible. Derivatives use Additional Reagents and Generate Waste*

If medicinal chemists are seeking to make analogues to explore the structure–activity relationship (SAR) around a particular template, then this principle may not strictly apply. It is likely to be more efficient, both in terms of time and resources, to prepare a bulk supply of a template blocked with orthogonal protecting groups, which can then be elaborated to scope out the SAR, than it would be to design "greener" routes to each target from scratch. Effort can then be saved for optimizing the route to the most promising compound.

However, not all chemistry conducted in drug discovery is designed with the synthesis of multiple analogues in mind. For individual syntheses, or even remakes of compounds that may once have been accessed *via* multiply-protected analogue-friendly approaches, this is still a valuable principle to be considered in the interests of efficiency and expediency. If protecting groups are unavoidable, issues of atom economy and considerations of the solvents, reagents, and conditions required should be evaluated.

#### 2.4.1.7 *Maximize Atom Economy: Design Syntheses so that the Final Product Contains the Maximum Proportion of the Starting Materials. There Should be Few, if Any, Wasted Atoms*

In the drug discovery arena the imperative is to make the desired target, as quickly as possible, to gain data that will eventually lead the team to the candidate molecule. Detailed considerations of the atom economy of each step may risk detracting from this key goal. While many of the more preferred reagents in guides or in the literature may be more atom efficient than conventional reagents, as previously discussed, there may be a lack of information as to the scope of these reagents.

Some more atom efficient reagent alternatives are already known to be relatively wide ranging in scope. Hydrogenations, surely the ultimate in atom efficiency, are already likely to be a preferred method of reduction for the drug discovery chemist. For amide formation, carbonyl diimidazole (CDI) and propylphosphonic anhydride (T3P®) are relatively robust and are more atom efficient than uronium based coupling reagents such as 1-[bis(dimethylamino)methylene]-1*H*-1,2,3-triazolo[4,5-*b*]pyridinium 3-oxide hexafluorophosphate (HATU) and *N*,*N*,*N*′,*N*′-tetramethyl-*O*-(benzotriazol-1-yl)uronium hexafluorophosphate (HBTU). T3P® and CDI also do not carry the same risk of explosion as aryltriazole reagents.

The challenge remains to discover, expand, and publicize the scope of other atom efficient approaches, to further increase the tool box of greener reagents.

#### 2.4.1.8 *Use Safer Solvents and Reaction Conditions: Avoid Using Solvents, Separation Agents, or other Auxiliary Chemicals. If these Chemicals are Necessary, Use Innocuous Chemicals*

On a small scale, it is difficult and frequently impossible to explore solventless reactions. Solvent may be needed to handle or transfer small quantities of substrates or reagents effectively. It may also be needed to dissipate heat and enable reactions to take place safely. Separation is likewise often unavoidable on a small scale, where compounds have to be purified before testing, and auxiliary or co-reagents are essential in many transformations.

Although these key components cannot always be avoided in drug discovery, chemists can be encouraged to run reactions at an appropriate concentration for the scale at which they are operating, which in most cases is more concentrated than a simple scale up from the first trial reaction would suggest. While certain reactions (*e.g.* ring closing metathesis to minimize cross-metathesis, or reactions where the solvent is critical to dissipating heat in the case of an exothermic reaction) may always have to be run at a dilute concentration, it is typical for the number of volumes of solvent per volume of reagent to decrease on scaling up. Ten volumes of solvent per volume of starting material is still an approximate rule of thumb for many large-scale reactions. For drug discovery chemists more used to thinking in reaction molarity (and common concentrations of 0.001–0.1 M), it is worth considering that for a compound of molecular weight 250 (assuming density = 1), 10 volumes is 0.4 M). Likewise, volumes of solvent required for purification decrease as scale increases, dramatically in cases where crystallizations become viable over chromatography for example. Some purifications can be avoided by telescoping of reactions, made more efficient by using crystallization, or optimized approaches to chromatography[26] (Chapter 4).

The request to use innocuous solvents, separation agents, or auxiliaries where they are unavoidable is still relevant. Solvent guides are already embraced in many drug discovery departments, and pre-packed (and therefore safer) columns have already largely replaced free-flowing silica as a separation agent.

#### *2.4.1.9 Increase Energy Efficiency: Run Chemical Reactions at Ambient Temperature and Pressure Whenever Possible*

The drive to explore the widest possible volume of chemical space in the search for the best compound to interact with a particular target will necessarily relegate the temperature and pressure of the chemistry required to a secondary concern. Certain templates may be ruled out on tractability grounds if the chemistry is too difficult, but if a series looks particularly promising, or is the only viable series, it is not uncommon to assume that a better route may be achieved with further optimization later, and whatever conditions necessary will be utilized.

Microwave heating is popular in many drug discovery laboratories for the speed with which it enables reactions to be conducted on a small scale. Although microwave processing on larger scales is certainly possible, especially in flow mode,[52] it is not yet prevalent in the pharmaceutical industry and it is usual for reaction conditions to have to be converted back into conventional reflux heating when being scaled up.

In cases where extremes of temperature and pressure are not required, and speed is not a critical factor, chemists can benefit from running reactions at less extreme conditions, for example, 5 °C less than reflux, which has been shown to result in up to a six-fold decrease in energy consumption.[53] Energy saving when considering chemical reactions must always be considered as part of the bigger picture. The energy required for safe air handling in a laboratory environment is orders of magnitude greater than the energy requirements of individual reactions on a medicinal chemistry scale, and this is one area where behavioral campaigns, encouraging chemists to close fume hood sashes completely when not in use, can dovetail with other green chemistry concerns.

#### 2.4.1.10 *Design Chemicals and Products to Degrade after Use: Design Chemical Products to Break Down to Innocuous Substances after Use so that They Do Not Accumulate in the Environment*

This is a secondary consideration when chemists are already seeking to maximize several orthogonal properties to find something that has the right potency, pharmacokinetics, and pharmacodynamics. It is acknowledged that active pharmaceutical ingredients and active metabolites can enter the water treatment system if they are insufficiently metabolized in patients, with the percentage excreted varying from 5% to 100% (in the case of metformin[54]) of the initial dose. Kummerer has looked at degradability studies of 2200 drugs and found that more than 25% should not persist in the environment because they are readily biodegradable.[55] However, many drugs have been found, through the Shelf Life Extension Plan,[56] to be stable well beyond the assigned expiration date, and this stability to heat, acidic and basic conditions, humidity, oxidation, and light is likely to also result in stability to the major processes required for biodegradation. There have been calls for medicinal chemists to design new chemical entities (NCEs) to rapidly degrade after they enter the environment,[57] but with current information it is not currently possible to predict accurately which structures are then susceptible to breakdown by photolysis, hydrolysis, and biotransformation by environmental bacteria.

However, just as has already been noted when considering toxicity, the drive for less lipophilic drugs may result in drugs that are more likely to be metabolized in the environment. If pharmaceuticals in the environment (PIE) become increasingly regulated and if progress is made into the pharmacophores susceptible to biodegradation by common bacteria in the environment, this will then sit firmly within the remit of the drug discovery chemist – with an optimal structure neither too ready nor too resistant to such bacterial metabolism.

#### 2.4.1.11 *Analyze in Real Time to Prevent Pollution: Include In-process Real-time Monitoring and Control During Syntheses to Minimize or Eliminate the Formation of By-products*

Advice on pollution prevention by in-line analysis or otherwise is often not really appropriate for multiple small-scale reactions as conducted in drug discovery. However, it is worth noting that flow chemistry is growing in popularity in the pharmaceutical industry[58,59] (Chapter 6) and the possibility for in-process monitoring and adjustment of reaction conditions such as temperature or residence time in these processes may in some cases allow for easier optimization of reaction conditions.

On a practical day-to-day basis, this principle can be readily applied in reminding chemists to monitor chromatography in real time to stop the run (and wastage of solvent) after the product has been collected.

#### *2.4.1.12 Minimize the Potential for Accidents: Design Chemicals and Their Forms (Solid, Liquid, or Gas) to Minimize the Potential for Chemical Accidents Including Explosions, Fires, and Releases to the Environment*

A strong safety culture within the industry means drug discovery departments are already likely to be aligned to this principle, though some chemists may initially be surprised to see it listed under the context of Green Chemistry. There is, however, still a need for drug discovery chemists to have a better understanding of the role of process safety, so the practical implications of process considerations such as heat of reaction and potential for "runaway" reactions can be factored in and addressed as soon as possible.

Notably, drug discovery is not a homogenous, or static process. What may appear an insurmountable inconsistency between the aims of medicinal chemistry and the aims of green chemistry, may not appear as such in a different group or company, when working on a different chemical series, or within the constraints of different regulations. Chemistry techniques change, and technology approaches such as synthetic biology or flow chemistry may yet revolutionize day to day research on some projects.

Some cultural barriers may arise from the loose definition of what constitutes "Green Chemistry". Even once chemists are familiar with the 12 principles, they are an approach for considering chemistry rather than a series of goals or a checklist. Indeed, it is not unusual that optimizing a route for one of the principles involves a trade-off against one or more of the others. Beyond the immediate pharmaceutical industry environment, limited awareness of green chemistry may also pose barriers.

### 2.4.2 Awareness of the 12 Principles in the Wider Chemistry Community

The literature has many examples of publications where an alternative methodology is proclaimed as being "greener", but closer inspection reveals that it may only be better in a single aspect, and there may be trade-offs in other areas. In other cases there are insufficient data to determine whether the methodology really represents an improvement. Wider awareness in the chemistry community, and more rigorous reviewing of green chemistry claims in manuscripts submitted to all journals, not just those specializing in green chemistry, is essential for this to improve.

A related issue can arise when new solvent and reagent alternatives are developed, but insufficient toxicology or globally harmonized system (GHS) risk phrase data is available. This lack of data is not surprising. Some testing may only take place after a material is manufactured on large scale, and some safety data may take many years to be discovered. One might expect as a general trend that with time and experience of using individual materials more is learned and more potential risks are found. This is indeed generally found to be the case, and examples include *tert*-butyl methyl ether (TBME), which was until recently recommended as an alternative to dichloromethane,[22] but has since been found to be a potential endocrine disrupter,[60] and has accordingly been listed by ChemSec on the substitute it now (SIN) List as something REACH (Registration, Evaluation, Authorisation and Restriction of Chemicals) should consider as a substance of very high concern (SVHC) solely due to its endocrine disrupting properties.[61] Another example is limonene, which was once promoted as a result of its bio-derived origins,[48] but which has been found to cause sensitization by skin contact.[62] However, notably, there are counter examples where some solvents and reagents have been labelled as having certain risks and hazards that have later been revised on closer examination of the data. One such example is 4-(4,6-dimethoxy-1,3,5-triazin-2-yl)-4-methylmorpholinium chloride (DMTMM), which until recently was listed as having an R40 (Limited evidence of a carcinogenic effect) risk phrase, but has since passed a range of mutagenicity tests[63] and had the R40 phrase (and corresponding GHS phrase) removed from its data sheets.

This apparent shifting of goalposts can undermine the credibility of green chemistry teams if they recommend an alternative that is later found to have serious shortcomings. The ambiguities of the area may lead some chemists to become disillusioned or disengaged with the approach or consider it "greenwashing", political, or subjective. Clear communication is therefore required, as is application of a precautionary principle, where an absence of data is not necessarily assumed to be good news.

### 2.4.3 Communication, within Drug Discovery and Beyond

Depending on the quality, communication might be considered as either a barrier to adopting green chemistry or a solution to some of the issues previously described. To counteract the challenges posed by the limited awareness of green chemistry, and cynicism arising from contradictory papers or claims, clear and open communication is essential.

A continuing education culture within chemistry departments can assist in communicating what green chemistry is, and what it is not. Chemists can be educated as to the benefits that green chemistry can bring, and the ambiguities can be acknowledged and to a certain extent embraced. There is a strong argument to be made that some ambiguities or issues for debate can even assist with the overall communication process, encouraging chemists to engage with the process, rather than simply being passive recipients of what may appear politically driven communication, where it may be dismissed as something that we must be seen to be doing. Assorted green chemistry myths can be debunked (such as bio-based alternatives always being greener, or water always being the best solvent), and internal green chemistry examples can be highlighted that may cover a multitude of different approaches and acknowledging the strengths and weaknesses of all.

As well as embracing ambiguity, however, rigor needs to be shown in assessing individual solvents, reagents or routes, and the methodology behind such assessments needs to be clearly communicated. Transparency is essential – decisions can only be made based on the chemists' best interpretation of the data at the time, and tracking the references and rationale behind that decision is important for anyone following that work. It is entirely possible that future data, *e.g.* discovery of toxicology data on some of the newer, as yet untested solvents, might change the assessment as to which chemistry is and is not green, but this can only be tracked if the logic behind the decisions being made is clearly recorded, and if cases where assumptions have had to be made are clearly differentiated from cases where hard data are available.

Green chemistry is a rapidly growing discipline. While the appearance of new data potentially undermining existing messages may be challenging at times, it should be remembered that the ability to change hypotheses, and therefore in these cases recommendations as to which solvents or reagents to use, is at the very heart of science and should be embraced.

Communication should also encompass academia, as a two-way conversation. In earlier discussion of the cultural barriers to adopting green chemistry in drug discovery, the limited exposure that many chemists get to green chemistry before joining industry was highlighted. This is especially important since practices that chemists learn as undergraduates, PhD students, and PostDocs can stay with them for the rest of their working lives, and, in contributing to the chemistry culture in the laboratory, can also shape the actions of their colleagues. Industry needs to inform academia of the information and skills in green chemistry that they would like their next generation of recruits to be taught. Academia likewise needs to engage with industry, perhaps asking for their assistance in offering industry-focused courses or discovering more about real industry problems to which their research might be applied. Progress is being made in this dialogue. A public–private consortium between the European Union (EU) and European Federation of Pharmaceutical Industries and Associations (EFPIA) companies, called the Innovative Medicines Initiative: Chemistry for the 21st Century (IMI CHEM21), is working on a two-way educational package, designed to better educate chemists still at universities as to the green chemistry needs of the pharmaceutical industry, and to bring chemists in industry up to speed with some of the newer developments in green chemistry coming out of academia. Similarly, the ACS GCIPR has published a widely quoted list of topics where further research is needed,[31] and gives out annual grants supporting work into problem areas identified by their member companies.

In the time pressured environment of drug discovery, with limited quantities of intermediate available, it is inevitable that chemists will seek the closest literature precedence to their desired reaction that they can find. Unfortunately, there are more "less-green" examples in the literature than "green" examples. Many non-preferred solvents such as dichloromethane or dimethylformamide have long histories of being widely used and may well have been the first solvent in which a reaction was tried in the past. The existence of literature precedence in a given solvent does not necessarily mean that that is the optimal solvent for that reaction, but faced with limited time and material it is understandable that people may stick closely to what is known, thereby compounding the wealth of precedent in those solvents. Dialogue with academics should emphasize the need to share which solvents have been screened, and other journals should be encouraged to follow the lead that has been taken by OPRD (*Organic Process Research and Development*),[64] *Green Chemistry*,[65] and *ChemSusChem*[66] in only accepting papers with reactions in certain non-preferred solvents if they have been shown not to work in more preferable alternatives.

Equally, when newer greener approaches are shown in the literature, the desire for a quick publication showcasing the best aspects of a researcher's work can lead to methodologies being demonstrated on simple substrates that may lack the complexity of a typical drug discovery target, and potential incompatibilities or limitations may not be explored or reported.[67] As medicinal chemists are typically looking for as close a precedent as possible to have greatest confidence in a methodology's likelihood of success, they are unlikely to adopt new approaches without such scope being published.

There is no easy solution to these wider chemistry culture problems. Using literature searching tools that permit filtering of hits by solvent, as Reaxys® does, allows for easier "green minded" searching of what is already published, though further optimization of some filters may be useful. Encouraging internal sharing of both successes and failures, and ready searching of electronic lab notebooks, provide further opportunities that might yet benefit other projects, as does sharing basic research into the scope and limitation of common reagents in common solvents.[21,44]

### 2.4.4 Definitions and Metrics

Disagreements regarding definitions as to what constitutes a "green" reagent or solvent, or which metrics should be used, have previously been cited as a barrier in introducing green chemistry to industry in general.[1] At present this has yet to pose a real problem in drug discovery; however, care must be taken as described above with open and transparent communication, especially regarding ambiguities or where data is known to not yet be fully available.

Once a clear algorithm for assessing solvent, reagents, or routes as "green" has been communicated, solvents and reagents can be assigned as "green" or not, and tools can be used to encourage chemists to use the greener alternatives. Such tools might be as simple as solvent[22–25] and reagent[20,23] guides, or more specific tools for purification purposes such as equi-elutropic series comparing the polarity of different mixtures of solvents, or even PCA (principle component analysis) maps showing which solvents might have similar properties to each other. In some instances DoE (design

**Table 2.1** Comparison of the *E*-factors of different aspects of the chemical industry.[69] Adapted from ref. 68 with permission from The Royal Society of Chemistry.

| Industry | Output (tonnes) | *E*-factor |
|---|---|---|
| Oil | $10^5$–$10^8$ | <0.1 |
| Bulk chemicals | $10^4$–$10^6$ | <5 |
| Fine chemicals | $10^2$–$10^4$ | 5–50 |
| Pharmaceuticals | 10–$10^3$ | 25–100 |

of experiments) might even be appropriate in a drug discovery setting, allowing identification of greener conditions. For example, before embarking on a large array, a simple three component ten reaction DoE optimization screen might allow a chemist to get more successful reactions from the planned array, and these early (partially) optimized conditions may then echo down through the rest of the lead optimization procedure and beyond.

After sharing of greener alternatives and tools to encourage their use, metrics on the usage of greener solvents and reagents can then be collated, and used to incentivize changes in behavior. Such metrics can be incorporated in institutional or departmental goals and use of certain classes of reagents and solvents can be rewarded, or restricted as appropriate.

The *E*-factor[68] (environmental (*E*)-factor = [total waste (kg) per product (kg)]) and process mass intensity (PMI)[45] (PMI = total mass in a process or process step (kg) per mass of product (kg)) are metrics that can be used to compare the environmental impact of alternative routes to a given compound, or used to drive overall efficiency targets, and benchmark against other routes within the industry, or even against other industries (Table 2.1).[69] Such metrics may not be as appropriate in drug discovery departments as in later stages of the organization where routes are closer to being optimized. However, communication as to the metrics which are considered at later stages of development can still assist when early stage chemists envisage which aspects of their chemistry might be able to be optimized on scale, and which might not.

### 2.4.5 Recommendations for Embedding a Green Chemistry Culture within Drug Discovery

Ultimately the success or failure of any attempt to embed green chemistry principles within a drug discovery environment will depend on the culture within a given department. The following questions should be considered when determining whether a green chemistry effort will be a success:

- Are the tangible benefits of green chemistry (*e.g.* easier work ups using 2-MeTHF, less streaky chromatography using alternatives to dichloromethane) clearly communicated?
- Is support given at all levels of the organization for sustainability?

- Are drug discovery chemists encouraged to take ownership of their routes, and the potential impacts they might have if scaled up without major changes (perhaps because of time pressures)?
- Are drug discovery chemists encouraged to explore alternative approaches not narrowly within the confines of the "day-job"?
- Do leaders show that sustainability is important to them, and incentivize their teams to consider sustainability, perhaps through inclusion in objectives, and acceptance of occasional extra time needed to work out green conditions? Are awards and recognition given to chemists that implement green chemistry?
- Is green chemistry visible? Are there opportunities to share potential greener options which work – and, crucially, steer others away from those which do not? Such opportunities might be through databases, seminars, or internal publications.
- Are green chemistry metrics visible and transparent? Do they strike the right balance between simple and simplistic? Many of the metrics applicable for later stage development are inappropriate in a drug discovery context, but simple measures of solvent used, reagents used, and metrics on routes which do pass into development can still be useful in driving changes in behavior, if their limitations are acknowledged.
- Are contract research organizations (CROs) held to the same standards as internal chemists, ensuring a consistency of approach?

One of the best ways to create a culture within drug discovery where green chemistry can thrive is to ensure that it is owned and promoted by drug discovery chemists who know best the challenges and opportunities of the area. Clearly, colleagues from Environment, Health and Safety and Development departments have a wealth of knowledge they can provide, but without a feel for day to day operations from a medicinal chemist's eye view, it may be hard for such ideas to gain traction.

A careful balance must be struck between pure chemistry and operational approaches. Too much emphasis on practical approaches such as recycling or streamlined ordering might cause some drug discovery chemists to disengage or even doubt the scientific rigor of the chemistry which is being recommended. Likewise, too much emphasis on individual improved reactions might lead some chemists to conclude that green chemistry is too "ivory tower" and focused on the minutiae of a single reaction while ignoring higher impact, if less glamorous, "quick wins" on a day-to-day basis.

## 2.5 Opportunities

To conclude, while there are many barriers to the adoption of green chemistry in drug discovery, there are also opportunities and benefits associated with embedding such an approach early in R&D. Increased knowledge of greener alternatives explored at this relatively inexpensive and low risk stage can feed into other projects later in development. Choosing more wisely the first time a reaction is carried out might avoid time and money spent later reworking a synthesis to avoid the generation of hazardous waste in development. Confidence in alternatives can be built in advance of the use of some reagents or solvents potentially being restricted by legislation. And, as is inevitably the case, where gaps are found in the scope of the current alternatives, academic research into new alternative procedures can be encouraged and funded.

New approaches to synthetic strategy, designed with green chemistry principles in mind, such as non-precious metal catalysis, C–H activation, biocatalysis, and synthetic biology, may yet lead not only to more efficient syntheses of existing templates but also to ready access to new areas of chemical space.

## References

1. K. J. M. Matus, W. C. Clark, P. T. Anastas and J. B. Zimmerman, *Environ. Sci. Technol.*, 2012, **46**, 10892.
2. M. B. Satterfield, C. E. Kolb, R. Peoples, G. L. Adams, D. S. Schuster, H. C. Ramsey, E. Stechel, F. Wood-Black, R. J. Garant and M. A. Abraham, *Environ. Sci. Technol.*, 2009, **43**, 4221.
3. *Green Chemistry in the Pharmaceutical Industry*, ed. P. J. Dunn, A. S. Wells and M. T. Williams, Wiley-VCH, Weinheim, 2009, p. 349.
4. P. T. Anastas and J. C. Warner, *Green Chemistry: Theory and Practice*, Oxford University Press, New York, 1998, p. 30.
5. http://www.forbes.com/sites/matthewherper/2012/02/10/the-truly-staggering-cost-of-inventing-new-drugs/ InnoThink Center For Research In Biomedical Innovation; Thomson Reuters Fundamentals via FactSet Research Systems (accessed week of May 30th, 2014).
6. http://www.reuters.com/article/2012/12/04/us-pharmaceuticals-rd-idUSBRE8B300W20121204.
7. B. W. Cue Jr., in *Green Chemistry in the Pharmaceutical Industry*, ed. P. J. Dunn, A. S. Wells and M. T. Williams, Wiley-VCH, Weinheim, 2009, ch. 20, p. 557 (accessed week of May 30th, 2014).
8. S. Trybus and N. Bush, Systematic quantification of micro reactor potential in pharma and specialty chemicals, Presented at the 3rd International Symposium on Green Processing in the Pharmaceutical and Fine Chemical Industries, Boston, October 1st, 2010.
9. B. W. Cue Jr., in *Green Chemistry in the Pharmaceutical Industry*, ed. P. J. Dunn, A. S. Wells and M. T. Williams, Wiley-VCH, Weinheim, 2009, ch. 20, p. 562.
10. http://www.hse.gov.uk/reach/ (accessed March 15th, 2015).
11. C. J. Welch, W. R. Leonard Jr., J. O. DaSilva, M. Biba, J. D. Albaneze-Walker, W. Henderson, B. Laing and D. J. Mathre, *LCGC North Am.*, 2005, **23**(1), 16.
12. A. Rajendran, *J. Chromatogr. A*, 2012, **1250**, 227.

13. V. Antonucci, J. Coleman, J. B. Ferry, N. Johnson, M. Mathe, J. P. Scott and J. Xu, *Org. Process Res. Dev.*, 2011, **15**, 939.
14. Azeotrope at 89.4% 2-MeTHF, 10.6% water (http://www.pennakem.com/pdfs/methf5.pdf) (accessed March 15th, 2015).
15. Solubility of water in 2-MeTHF: 0 °C, 4.0%; 19.3 °C, 4.1%; 39.6 °C, 4.3%; 70.6 °C, 5.0%. Solubility of 2-Me-THF in water: 0 °C, 21%; 19.3 °C, 14.4%; 39.6 °C, 9.2%; 70.6 °C, 6.0% (http://www.pennakem.com/pdfs/methf5.pdf) (accessed March 15th, 2015).
16. S. R. Clough and L. Mulholland, Hexane, *Encyclopedia of Toxicology*, Elsevier, 2nd edn, 2005, vol. 2, p. 522.
17. J. P. Taygerly, L. M. Miller, A. Yee and E. A. Peterson, *Green Chem.*, 2012, **14**, 3020.
18. D. J. C. Constable, C. Jimenez-Gonzalez and R. K. Henderson, *Org. Process Res. Dev.*, 2007, **11**(1), 133.
19. S. D. Roughley and A. M. Jordan, *J. Med. Chem.*, 2011, **54**, 3451.
20. J. P. Adams, C. M. Alder, I. Andrews, A. M. Bullion, M. Campbell-Crawford, M. G. Darcy, J. D. Hayler, R. K. Henderson, C. A. Oare, I. Pendrak, A. M. Redman, L. E. Shuster, H. F. Sneddon and M. D. Walker, *Green Chem.*, 2013, **15**, 1542.
21. D. S. MacMillan, J. Murray, H. F. Sneddon, C. Jamieson and A. J. B. Watson, *Green Chem.*, 2013, **15**, 596.
22. R. K. Henderson, C. Jiménez-González, D. J. C. Constable, S. R. Alston, G. G. A. Inglis, G. Fisher, J. Sherwood, S. P. Binks and A. D. Curzons, *Green Chem.*, 2011, **13**, 854.
23. K. Alfonsi, J. Colberg, P. J. Dunn, T. Fevig, S. Jennings, T. A. Johnson, H. P. Kleine, C. Knight, M. A. Nagy and D. A. Perry, *Green Chem.*, 2008, **10**, 31.
24. D. Prat, O. Pardigon, H.-W. Flemming, S. Letestu, V. Ducandas, P. Isnard, E. Guntrum, T. Senac, S. Ruisseau, P. Cruciani and P. Hosek, *Org. Process Res. Dev.*, 2013, **17**, 1517.
25. http://surveys.acs.org/se.ashx?s=04BD76CC0E5496A7 (accessed week of May 30th, 2014).
26. E. A. Peterson, B. Dillon, I. Raheem, P. Richardson, D. Richter, R. Schmidt and H. F. Sneddon, *Green Chem.*, 2014, **16**, 4060.
27. D. S. MacMillan, J. Murray, H. F. Sneddon, C. Jamieson and A. J. B. Watson, *Green Chem.*, 2012, **14**, 3016.
28. D. G. Drueckhammer, S. Q. Gao, X. Liang and J. Liao, *ACS Sustainable Chem. Eng.*, 2012, 87.
29. A. N. Hobbs and R. J. Young, *Drug Discovery Today*, 2013, **18**(3–4), 148.
30. F. M. Chardon, N. Blaquiere, G. M. Castanedo and S. G. Koenig, *Green Chem.*, 2014, **16**, 4102.
31. D. J. C. Constable, P. J. Dunn, J. D. Hayler, G. R. Humphrey, J. L. Leazer Jr., R. J. Linderman, K. Lorenz, J. Manley, B. A. Pearlman, A. Wells, A. Zaks and T. Y. Zhang, *Green Chem.*, 2007, **9**, 411.
32. R. Carson, [1st. Pub. Houghton Mifflin, 1962], *Silent Spring*, Mariner Books, 2002, ISBN 0-618-24906-0.

33. J. H. Clark and P. Smith, Basic Principles of green chemistry, *Innovations of Pharmaceutical Technology*, 2005, p. 94.
34. N. Winterton, *Green Chem.*, 2001, **3**, G73.
35. S. Y. Tang, R. A. Bourne, R. L. Smith and M. Poliakoff, *Green Chem.*, 2008, **10**, 268.
36. http://www.acs.org/content/acs/en/greenchemistry/students-educators/academicprograms.html (accessed March 15th, 2015).
37. M. Hay, D. W. Thomas, J. L. Craighead, C. Economides and J. Rosenthal, *Nat. Biotechnol.*, 2014, **32**(1), 40.
38. (a) ACS Green Chemistry Institute®. Green Chemistry Pocket Guides. http://bit.ly/ACSGCIPocketGuides (accessed January 15, 2013); (b) M. C. Bryan, *et al.*, *J. Med. Chem.*, 2013, **56**(15), 6007.
39. M. K. Jha, V. Kumar and R. J. Singh, *Resour., Conserv. Recycl.*, 2001, **33**, 1.
40. D. T. Friesen, R. Shankar, M. Crew, D. T. Smithey, W. J. Curatolo and J. A. S. Nightingale, *Mol. Pharm.*, 2008, **5**(6), 1003.
41. R. J. Young, D. V. S. Green, C. N. Luscombe and A. P. Hill, *Drug Discovery Today*, 2011, **16**, 822.
42. A. J. Hendriks, A. van der Linde, G. Cornelissen and D. Sijm, *Environ. Toxicol. Chem.*, 2001, **20**(7), 1399.
43. M. Hann, *Med. Chem. Commun.*, 2011, **2**, 349.
44. F. I. McGonagle, D. S. MacMillan, J. Murray, H. F. Sneddon, C. Jamieson and A. J. B. Watson, *Green Chem.*, 2013, **15**, 1159.
45. C. Jimenez-Gonzalez, C. S. Ponder, Q. B. Broxterman and J. B. Manley, *Org. Process Res. Dev.*, 2011, **15**, 912.
46. K. Yan, G. Wu, T. Lafleur and C. Jarvis, *Renewable Sustainable Energy Rev.*, 2014, **38**, 663.
47. D. F. Aycock, *Org. Process Res. Dev.*, 2007, **11**, 156.
48. J. H. Clark, D. J. Macquarrie and J. Sherwood, *Green Chem.*, 2012, **14**, 90.
49. A. Kadam, M. Nguyen, M. Kopach, P. Richardson, F. Gallou, Z.-K. Wan and W. Zhang, *Green. Chem.*, 2013, **15**, 1880.
50. http://www.stillwaterpalladium.com/concentrating.html, last accessed March 24, 2014.
51. Armin Reller of the University of Augsburg and Tom Graedel of Yale University, http://intercontinentalcry.org/infographic-how-long-will-it-last/ (accessed week of May 30th, 2014).
52. R. Morschhäuser, M. Krull, C. Kayser, C. Boberski, R. Bierbaum, P. A. Püschner, T. N. Glasnov and C. O. Kappe, *Green Process. Synth.*, 2012, **1**(3), 281.
53. C. Hargreaves, Reducing the energy burden of active pharmaceutical manufacture, The 11th Annual Green Chemistry and Engineering Conference, Poster June 26–29, 2007.
54. G. G. Graham, J. Punt, M. Arora, R. O. Day, M. P. Doogue, J. Duong, T. J. Furlong, J. R. Greenfield, L. C. Greenup, C. M. Kirkpatrick, J. E. Ray, P. Timmins and K. M. Williams, *Clin. Pharmacokinet.*, 2011, **50**(2), 81.
55. *Green and Sustainable Pharmacy*, ed. K. Kummerer and M. Hempel, Springer-Verlag, Berlin, 2010, p. 135.

56. https://www.shelflife.hq.dla.mil/fda_slep.asp (accessed March 15th, 2015).
57. J. M. Fortunak, *Future Med. Chem.*, 2009, **1**, 571.
58. L. Malet-Sanz and F. Susanne, *J. Med. Chem.*, 2012, **55**, 4062.
59. S. V. Ley, *Chem. Rec.*, 2012, **12**, 378.
60. G. J. Moser, D. C. Wolf, M. Sar, K. W. Gaido, D. Janszen and T. L. Goldsworthy, *Toxicol. Sci.*, 1998, **41**, 77.
61. http://www.chemsec.org/what-we-do/influencing-public-policy/endocrine-disrupters/edcs-on-the-sin-list (accessed March 15th, 2015).
62. N. S. Chandrashekar and S. R. Hiremath, *Biol Pharm Bull.*, 2008, **31**(4), 656.
63. http://www.sigmaaldrich.com/MSDS/MSDS/DisplayMSDSPage.do?country=US&language=en&productNumber=74104&brand=ALDRICH&PageToGoToURL=http%3A%2F%2Fwww.sigmaaldrich.com%2Fcatalog%2Fproduct%2Faldrich%2F74104%3Flang%3Den (accessed week of May 30th, 2014).
64. T. Laird, *Org. Process Res. Dev.*, 2012, **16**(1), 1.
65. M. Poliakoff, W. Leitner and S. Ruthven, *Green Chem.*, 2010, **12**, 15.
66. G. M. Kemeling, *ChemSusChem*, 2012, **5**(12), 2291.
67. A. Nadin, C. Hattotuwagama and I. Churcher, *Angew. Chem., Int. Ed.*, 2012, **51**(5), 1114.
68. R. A. Sheldon, *Green Chem.*, 2007, **9**, 1273.
69. R. A. Sheldon, *Chem. Ind.*, 1992, 903.

CHAPTER 3

# *Toward a Green Laboratory: One Reaction at a Time*

LEANNA E. SHUSTER*[a] AND ANIKÓ M. REDMAN[b]

[a] GlaxoSmithKline, 1250 South Collegeville Road, Collegeville, PA 19426, USA; [b] GlaxoSmithKline, 5 Moore Drive, Research Triangle Park, NC 27709, USA
*Email: leanna.e.shuster@gsk.com

## 3.1 Introduction

Within the typical medicinal chemistry laboratory, timelines are tight and chemists must quickly produce compounds to enable key decisions. As covered earlier in Chapter 2, this rushed environment can pose a barrier to "green" thinking. These difficulties, however, can be overcome, and implementing green chemistry during the early stages of drug discovery can have a considerable impact. While the scale of most reactions in medicinal chemistry is quite small, thousands of analogs per project may be generated and the gradual chemical usage can add up to significant quantities, especially when viewed across the scope of the global pharmaceutical industry. If these syntheses utilize reagents with environment, health and safety (EHS) or waste handling concerns, the added risks along with resultant financial implications are worth examining in more depth.

In addition to this cumulative factor, there is value in creating a culture where greener options are considered by first intent. The synthetic route devised by an early stage hit triage group is often carried forward into lead optimization. Some changes to optimize yields of the final route may occur once a lead is identified. If that compound moves into development, many

RSC Drug Discovery Series No. 46
Green Chemistry Strategies for Drug Discovery
Edited by Emily A. Peterson and Julie B. Manley

Published by the Royal Society of Chemistry, www.rsc.org

companies do not have the resources to consider further changes to the candidate selection route to enable preclinical and early stage clinical studies, given the rate of attrition at this stage. If a medicinal chemistry route gives material of sufficient purity, it is often deemed "fit for purpose" and will be expanded to kilogram scale. If early small-scale experiments investigated the use of greener alternatives, the larger scale impact of utilizing reagents and solvents which carry EHS risks or waste handling difficulties could potentially be eliminated.

The ability to consult in-house resources is extremely valuable for medicinal chemists. Whether accessing laboratory notebooks, consulting a corporate database, or receiving information *via* word-of-mouth, there is a wealth of knowledge to be found within a pharmaceutical company's chemistry infrastructure. However, the actual resources available to individual chemists tend to be limited to their role within the company, as discovery and development efforts are often compartmentalized from each other. As a result, medicinal chemists are often unaware of process chemistry improvements, even as they relate directly to their own projects. Additional communication can lead to reduction of EHS risks of drug discovery and improving the efficiency of future projects.

Literature sources are also essential in sourcing synthetic chemistry data. Many of the chemicals considered as greener alternatives have only recently been popularized. As a result, there are limited academic or pharmaceutical publications on the topic and it can be difficult to find a greener precedent that fits the desired transformation on a specific substrate. Academic groups can help by validating new methodologies on more complex substrate types and discussing total syntheses in a green chemistry context. Many examples of green chemistry principles can already be found in pharmaceutical development publications, in part because their operational scale necessitates extra attention to EHS and regulatory concerns. As medicinal chemistry departments begin to adopt similar practices, it will be important to communicate results to the wider chemistry community in external publications. When looking for new greener approaches in the literature, a useful resource is the periodic review of "Green Chemistry Articles of Interest to the Pharmaceutical Industry".[1] These are published by members of the ACS Green Chemistry Institute Pharmaceutical Roundtable (ACS GCIPR) in the journal *Organic Process and Research Development* and were used to source several of the examples herein (Schemes 3.10, 3.14, and 3.15 below).

## 3.2 Synthetic Route Design

The main objective of a medicinal chemistry route is to quickly produce target structures in reasonably high yields, as well as creating opportunities for efficient analog generation *via* advanced intermediates. When considering the consequences of a planned route, general assumptions can be made by examining such factors as anticipated yields, the numbers of steps, the roles of protecting groups, stoichiometry, and molecular weights of reagents. In addition, tools such as solvent and reagent guides, discussed in forthcoming sections, may be consulted. Once the route is fully developed, more rigorous analyses are possible. Of the many types of metrics calculations available, some take into account only the amounts of reagents and reactants, while others incorporate data on all materials, including solvents and water.[2] Some metrics try to address the limitations of mass-based methods to obtain a more holistic overview by including safety and environmental risks, toxicity, cost of reaction components, and energy consumption.

Given the objective in early drug discovery research, some of the details needed to complete full metrics calculations for a synthetic scheme are typically not fully documented. While the amounts of reagents, reaction solvents, and isolated yields are usually reported, the specifics for the rest of the operations can vary greatly. For extractive workups, filtrations, and other processing steps, amounts and exact concentrations of solvents, salts, filtration aids, and other materials are not always included. While experimental procedures regarding purifications sometimes include the size of a column, volumes of solvents used for chromatography or crystallization are generally not disclosed. Specific amounts of aqueous solutions and drying agents used in the workup process are typically not measured. In addition, water and energy consumption of the operations are not monitored. Purchase and disposal costs of all materials, including laboratory consumables, are also not taken into account at this stage. Metrics using only amounts of reagents and products may be employed to decide between synthetic routes, but this overlooks the much larger impact of solvent and water usage. On the other hand, attempts to carry out more encompassing calculations require assumptions to compensate for missing data, which can call the validity of the results into question.

Although two of the twelve principles of green chemistry are reduction of derivatives and atom economy, early discovery efforts may at times be greener when using a protecting group strategy.[3] This allows analogs to be easily generated from common intermediates, thus using less resources, generating less waste, and leading to faster structure-activity relationship (SAR) development. However, once larger scale preparation of a particular molecule is needed, the elimination of protecting groups is desirable to improve both atom economy and overall yield. Lessons learned from synthetic methods not requiring protecting groups can be most efficiently applied at this junction.[4]

When devising a new synthetic route, medicinal chemists often rely on software such as SciFinder® or Reaxys® to find relevant literature precedents. While these tools are extremely useful, it is not always intuitive for a user to identify a greener procedure for any given transformation. Indeed, filtering a large list of search results in favor of greener options can prove time-consuming and may provide irrelevant results. For example, a search for a greener reaction solvent for a particular transformation may return hits in which the solvent appears either in a different synthetic step of the cited synthesis or is used only in reaction workup. In addition to structure and reaction searches, searches by keyword can also prove challenging. It is not unusual for search terms such as "green chemistry" to yield irrelevant

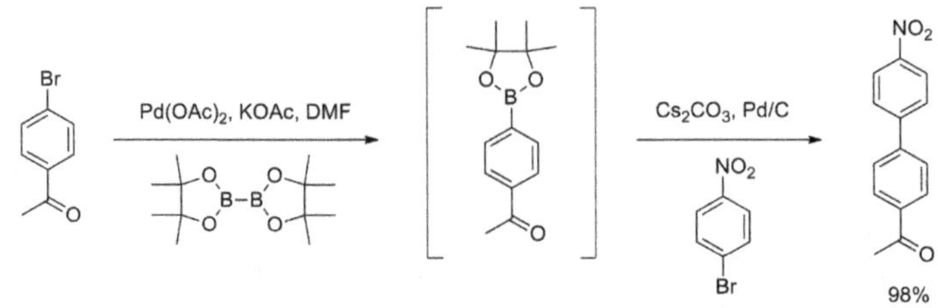


**Scheme 3.1** One-pot Suzuki procedure for reagent generation and coupling.

results, for example discussing green fluorescent protein, referring to green-colored portions of a figure, or even citing Greene's classic volume on protecting groups. Reaxys® offers the ability to filter reaction search results by solvent categories, which include "red", "yellow", and "green" options, in addition to classifications by boiling point range and some chemical classes. This is a useful first step in incorporating green chemistry components, and could possibly be improved by allowing the end user to customize the solvent categories. It is to be hoped that other similar software programs will offer similar functionalities.

When reagents and operations are similar, shorter synthetic routes tend to lead to lower resource consumption and waste generation. Combination of multiple synthetic steps into a single process or avoiding intermediate purification steps can further improve a route's efficiency. If reaction profiles are clean and all reagents and intermediates are chemically compatible, using the same solvent for successive transformations reduces the amount of materials needed and minimizes waste by eliminating workup and purification. For example, the generation of arylboronates and their subsequent use in palladium coupling can be accomplished in this fashion (Scheme 3.1).[5]

## 3.3 Planning Your Reaction

### 3.3.1 Solvent Choices

Solvents contribute the largest percentage to the carbon footprint of most reaction mixtures. The majority of them are derived *via* high energy methods from petroleum, which may also lead to future sustainability problems as global supplies decrease. Multiple companies and organizations have now developed solvent selection guides (Figures 3.1–3.3) to help evaluate potential alternatives to less preferred solvents.[6–8] These guides are based on an extensive compilation of data, including physiochemical properties, toxicity, reactivity, environmental impacts, safety profiles, and regulatory restrictions. The results of these efforts from different companies are largely in agreement, although some differences do remain, most notably the assessments of acetonitrile, 2-MeTHF, chlorobenzene, and *N*-methylpyrrolidone (NMP), for which the final rankings are not in agreement as a result of different scoring criteria.[9]

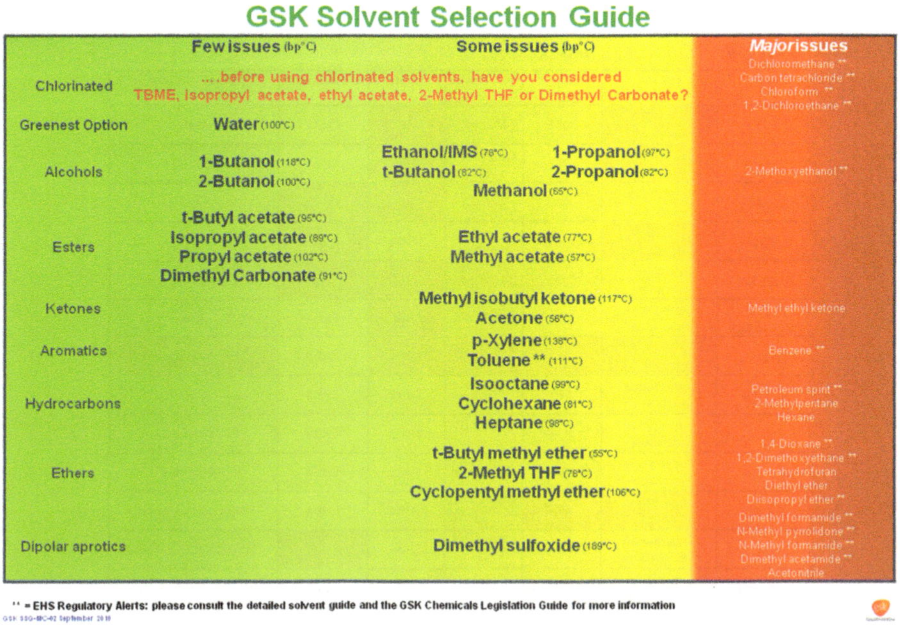


**Figure 3.1** Medicinal chemistry version of GlaxoSmithKline's Solvent Selection Guide.[6]
Reproduced from ref. 6.

| Preferred | Usable | Undesirable |
|---|---|---|
| Water | Cyclohexane | Pentane |
| Acetone | Heptane | Hexane(s) |
| Ethanol | Toluene | Di-isopropyl ether |
| 2-Propanol | Methylcyclohexane | Diethyl ether |
| 1-Propanol | TBME | Dichloromethane |
| Ethyl Acetate | Isooctane | Dichloroethane |
| Isopropyl acetate | Acetonitrile | Chloroform |
| Methanol | 2-MeTHF | NMP |
| MEK | THF | DMF |
| 1-Butanol | Xylenes | Pyridine |
| *t*-Butanol | DMSO | DMAc |
| | Acetic Acid | Dioxane |
| | Ethylene Glycol | Dimethoxyethane |
| | | Benzene |
| | | Carbon tetrachloride |

**Figure 3.2** Pfizer solvent selection guide for medicinal chemistry.[7]
Reproduced from ref. 7.

In recent years, the list of chemicals considered as reaction media continues to grow, with a definite emphasis on greener alternatives. These solvents span a range of "green space", from those which are bio-derived to

(a)

Solvents guide — HYDROCARBONS: OVERVIEW — SANOFI

| Name | Overall evaluation | ICH limit (ppm) | Occ. health | Safety | Environment | Other concern |
|---|---|---|---|---|---|---|
| Pentane | Banned | 5000 | OEBV2 | SHB5 | EHB4 | VOC |
| Hexane | Substitution requested | 290 | OEBV3 G1 Sk | SHB5 | EHB4 | Isomers |
| Heptane | Substitution advisable | 5000 | OEBV2 | SHB5 | EHB5 | Cost |
| Octane | Substitution requested | Not listed | not available | SHB4 | EHB5 | Cost |
| Cyclohexane | Substitution advisable | 3880 | OEBV2 | SHB5 | EHB5 | High freezing point |
| Methyl-cyclohexane | Substitution advisable | 1180 | OEBV2 | SHB5 | EHB4 | None |
| Benzene | Banned | 2 | OEBV5 G1 Sk | SHB5 | EHB3 | CMR (carcinogenic) |
| Toluene | Substitution advisable | 890 | OEBV3 G1 Sk | SHB4 | EHB3 | None |
| Xylene | Substitution advisable | 2170 | OEBV3 G1 Sk | SHB3 | EHB3 | Isomers |
| Cumene | Substitution advisable | 2000 | OEBV3 Sk | SHB3 | EHB4 | Peroxides |
| Tetralin | Substitution requested | 100 | OEBV2 Sk | SHB3 | EHB4 | High boiling point |
| Limonene | Substitution advisable | Not listed | OEBV3 G1 Sk | SHB4 | EHB5 | Degradation over time |
| Turpentine | Substitution advisable | Not listed | not available | SHB4 | not available | Degradation over time |

(b)

Solvents guide — APROTIC POLAR SOLVENTS: OVERVIEW — SANOFI

| Name | Overall evaluation | ICH limit (ppm) | Occ. health | Safety | Environment | Other concern |
|---|---|---|---|---|---|---|
| Acetonitrile | Recommended | 410 | OEBV3 Sk | SHB3 | EHB1 | Availability |
| Propionitrile | Substitution requested | Not listed | OEBV4 | SHB3 | n. d. | Decomposition gives HCN |
| Dimethyl-formamide | Substitution requested | 880 | OEBV4 G2Sk | SHB2 | EHB1 | CMR (reprotoxic) |
| Dimethyl-acetamide | Substitution requested | 1090 | OEBV4 G2Sk | SHB2 | EHB1 | CMR (reprotoxic) |
| N-Methyl-pyrolidone | Substitution requested | 530 | OEBV3 G2Sk | SHB3 | EHB1 | CMR (reprotoxic); high BP |
| Dimethyl-ethyleneurea | Substitution advisable | Not listed | OEBV2 G1 | SHB2 | EHB2 | Hygroscopic, high BP |
| Dimethyl-propyleneurea | Substitution advisable | Not listed | OEBV3 G1 | SHB3 | EHB2 | Hygroscopic, high BP, cost |
| Propylene carbonate | Substitution advisable | Not listed | OEBV2 | SHB1 | EHB1 | High BP |
| Dimethyl-sulfoxide | Substitution advisable | 5000 | OEBV2 Sk | SHB4 | EHB2 | Many incompatibilities, odor |
| Sulfolane | Substitution advisable | 160 | OEBV2 | SHB1 | EHB2 | High BP and MP |
| Nitromethane | Banned | 50 | OEBV3 | SHB5 | EHB3 | High energy of decomposition |

**Figure 3.3** Solvent guides from Sanofi-Aventis, organized by chemical class. For additional solvent classes, please see the full ref. 8.
Reprinted with permission from ref. 8. Copyright 2014 American Chemical Society.

ones from traditional sources but with fewer EHS or waste management risks. To encourage chemists to try novel solvents, GSK has put together a Green Solvent NMR Guide,[10] providing previously unpublished NMR chemical shifts for greener options, thus enabling easy identification of peaks from residual solvents in analytical samples.

In the realm of biomass generated solvents, new choices are emerging, the properties and risks of which will have to be assessed more fully. Citrus waste can give rise to (*R*)-(+)-limonene, which can be further converted into *p*-cymene (see Figure 3.4 for structures). Although both have shown utility as reaction solvents for simple transformations including esterification and amide formation,[11] risk of sensitization and toxicity to the aquatic environment are of concern. Multiple examples of *p*-cymene usage within the medicinal chemistry setting can be found in the literature, including an aromatization example from AstraZeneca scientists (Scheme 3.2).[12] Other biomass derived solvents purported to be green include ethyl lactate,[13] dimethyl isosorbide,[14] γ-valerolactone,[15] furfural,[16] and tetrahydrofurfuryl alcohol (structures shown in Figure 3.4).[17] These each have been shown to have utility as reaction solvents, but their appearance in medicinal chemistry literature is infrequent.

Glycerol can be produced from biodiesel waste and has been used as a solvent for decades across a wide range of synthetic transformations.[18] Glycerol is considered greener because it is not flammable, exhibits very low toxicity, and is readily biodegradable.[18] It has been shown to give excellent yields across a range of Heck and Suzuki reactions.[19] In an aza-Michael reaction (Scheme 3.3) glycerol acts as a promoter of the reaction in

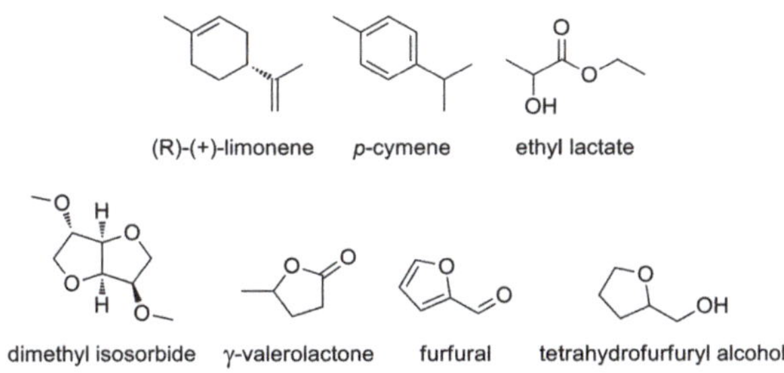


**Figure 3.4** Chemical structures of biomass-derived solvents discussed in the text.

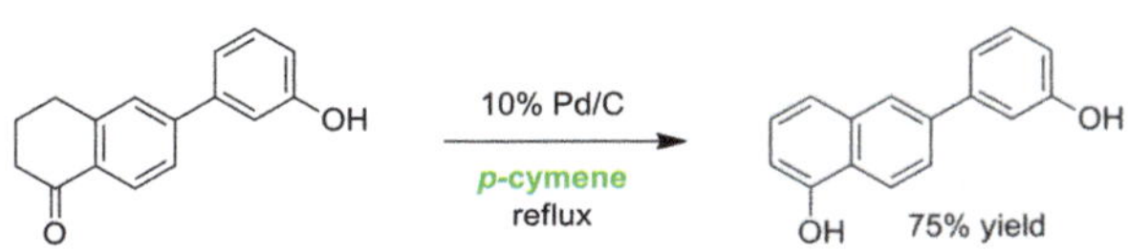


**Scheme 3.2** Aromatization using *p*-cymene.[12]

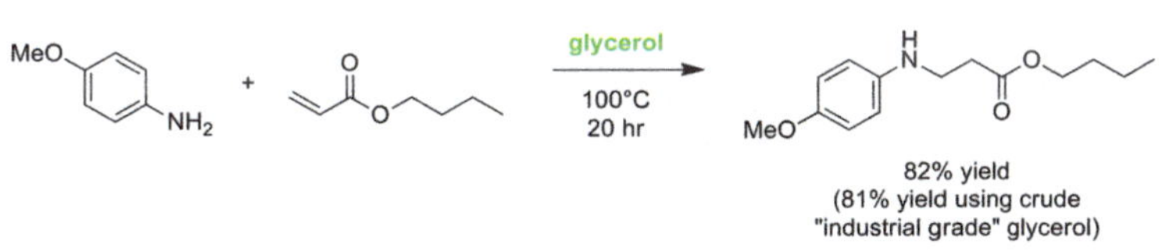


**Scheme 3.3** Aza-Michael reaction using glycerol as promoter and solvent.[20]

the absence of catalyst.[20] It is also useful in extractive workups, as it is immiscible with non-polar hydrocarbon solvents and dissolves a wide variety of polar substrates.[18] Glycerol itself can be further converted into various derivatives, including 1,2-propanediol, 1,3-propanediol, diacetin, and triacetin, which have all been shown to be useful as reaction solvents.[21,22]

Carbonates, both linear and cyclic, are generally thought of as a greener and more sustainable class of solvents. Cyclic examples such as propylene and ethylene carbonate can easily be sourced from carbon dioxide and either ethylene or propylene gas, in a process that is completely atom efficient.[23] Although carbonate solvents are not very common in pharmaceutical discovery research, there are multiple reports where they have been successfully employed. An intramolecular Claisen–Schmidt cyclization was reported by Takeda scientists using dimethyl carbonate as the reaction medium (Scheme 3.4).[24] Diethyl carbonate was utilized by a Schering-Plough discovery group as solvent in multiple amide coupling experiments.[25] Merck chemists showed that ethylene carbonate worked well for a range of tandem Heck-lactamization reactions to form naphthyridinone systems (Scheme 3.5).[26]

Ether solvents are often of particular concern due to their potential to form explosive peroxides. *tert*-Butyl methyl ether (TBME) and cyclopentyl methyl ether (CPME) exhibit significantly slower rates of oxidative decomposition as compared to diethyl ether, THF, or diisopropyl ether.[27] TBME has been reported by medicinal chemistry groups as a solvent for transformations as diverse as iridium-catalyzed borylation[28] (Scheme 3.6)

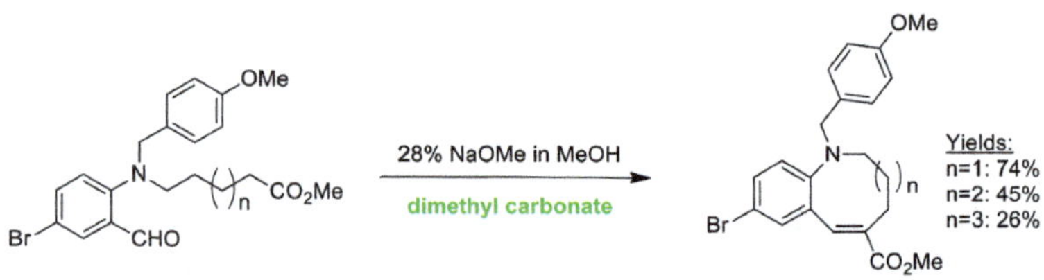


**Scheme 3.4** Claisen–Schmidt cyclization in dimethyl carbonate, from a Takeda publication.[24]

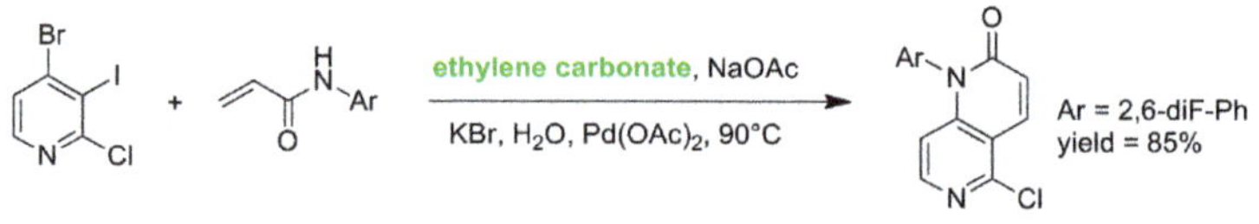


**Scheme 3.5** Merck example of tandem Heck-lactamization in ethylene carbonate.[26]

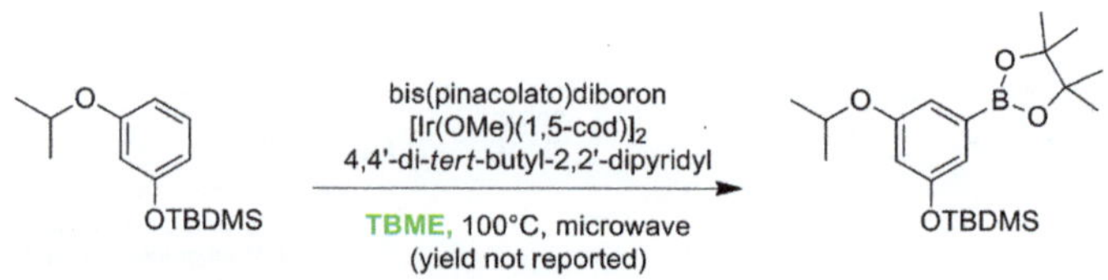


**Scheme 3.6** Borylation in TBME from Pfizer discovery chemists.[28]

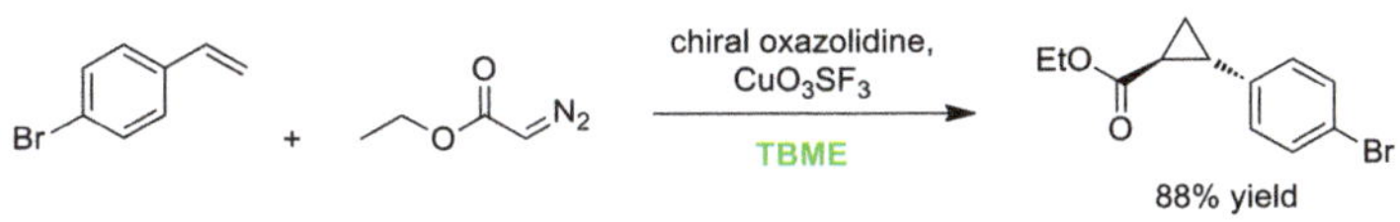


**Scheme 3.7** 1,3-Dipolar addition in TBME from Merck discovery chemists.[29]

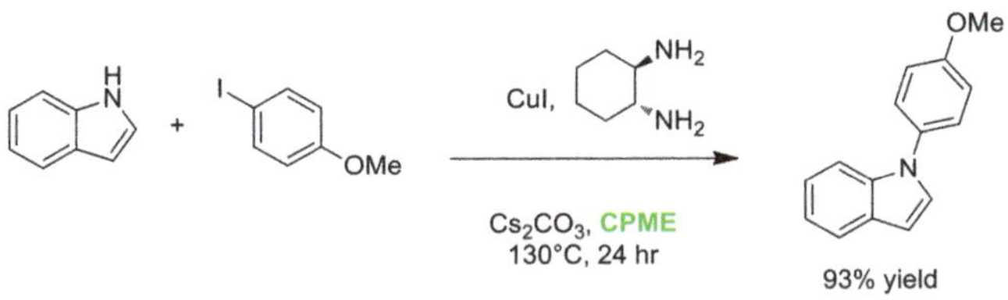


**Scheme 3.8** Development example of Cu-catalyzed process in CPME from Merck.

and 1,3-dipolar addition[29] (Scheme 3.7). Although TBME is favorable from an operational safety perspective, some studies suggest that it is an endocrine disruptor.[30] If this data is fully validated and verified, TBME is unlikely to be recommended from a green chemistry standpoint.

There are reports of CPME usage in pharmaceutical development (Scheme 3.8) which can serve as examples for future early discovery projects.[31] CPME is also shown in academic publications as a solvent for a range of radical reactions.[32] Although CPME is not yet widely employed in medicinal chemistry settings, it is becoming increasingly popular as a solvent for commercially purchased solutions. Hydrochloric acid in CPME is available from major vendors as a potential replacement for hydrochloric acid in dioxane. Takeda discovery scientists have demonstrated the utility of this solution for deprotection of Boc-amines in multiple publications.[33]

The physiochemical properties of 2-methyltetrahydrofuran (2-MeTHF) have made it popular for large-scale processes, and its use is gradually increasing in early discovery. While THF is typically acquired from petroleum byproducts, 2-MeTHF can be manufactured *via* the hydrogenation of furfural obtained from agricultural waste such as corn cobs, bagasse, oat hulls, and sugar cane waste.[34] 2-MeTHF has been found to be an excellent replacement for THF in many conditions, and actually imparts benefits for Grignard reactions.[34] It has also been shown in examples of Suzuki coupling,[35] ester hydrolysis,[36] and amide coupling.[36] The example in Scheme 3.9 shows the use of this more preferred solvent for multiple steps; the authors note the advantageous replacement of DMF for amide coupling, and the benefit of immiscible media to enable phase separation in the final step. 2-MeTHF can also substitute for dichloromethane (DCM)[37] and other chlorinated solvents, the usage of which is a significant green chemistry concern.[34] 2-MeTHF also has the added benefit of having a higher boiling point than either DCM or THF (Table 3.1). Given its limited water miscibility and high polarity, 2-MeTHF can be used as an extractive solvent to recover polar compounds from aqueous media. Both 2-MeTHF and CPME have low toxicity,[38] thus limiting potential

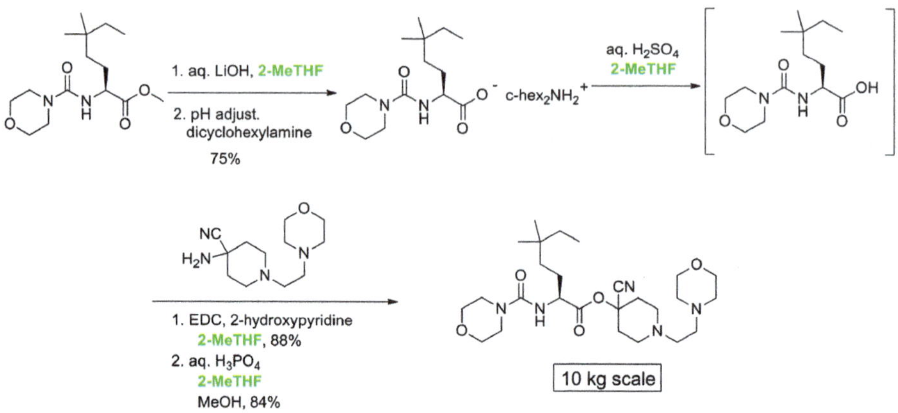


**Scheme 3.9** Telescoped Boehringer-Ingelheim synthesis of a cathepsin S inhibitor using 2-MeTHF as reaction solvent.[36]

**Table 3.1** Comparison of properties of 2-MeTHF, THF, and DCM.

| Property | 2-MeTHF | THF | DCM |
|---|---|---|---|
| Boiling point (°C) | 80 | 66 | 39.8 |
| Density @20 °C (g $mL^{-1}$) | 0.855 | 0.888 | 1.32 |
| Dielectric constant | 7.0 | 7.4 | 10.7 |
| Hansen total solubility parameter | 8.9 | 9.4 | 9.9 |

health concerns. However, it is important to note that unstabilized 2-MeTHF is peroxidizable, as are most ethers, and therefore the same safety precautions apply.

Dipolar aprotic solvents such as *N,N*,-dimethylformamide (DMF), dimethyl sulfoxide (DMSO), *N*-methylpyrrolidinone (NMP), acetonitrile, and *N,N*-dimethylacetamide (DMAc) are used extensively as reaction media in medicinal chemistry laboratories. Their physical properties make them excellent solubilizing media and their high boiling points allow for heating across a wide range of temperatures. Of the most commonly used dipolar aprotic solvents, DMSO is the least toxic.[6] However, its reactivity makes it difficult to consider as solvent for many transformations, and its ability to efficiently carry dissolved materials through the skin is an occupational hazard.[39] Sulfolane may be a greener option because of its low skin penetration, making it less of an occupational hazard, although it is known to be more toxic than DMSO.[40] Due to the lack of greener dipolar aprotic alternatives, the ACS GCIPR has recently funded an academic project in this research area.

Some reactions traditionally carried out in dipolar aprotic media can instead utilize solvents from other chemical classes. In place of DMF or DMAc, the use of *n*-pentanol or isopentanol was demonstrated for a range of substrates in a Pd-catalyzed direct coupling of aryl halides with heteroaromatics (Scheme 3.10).[41]

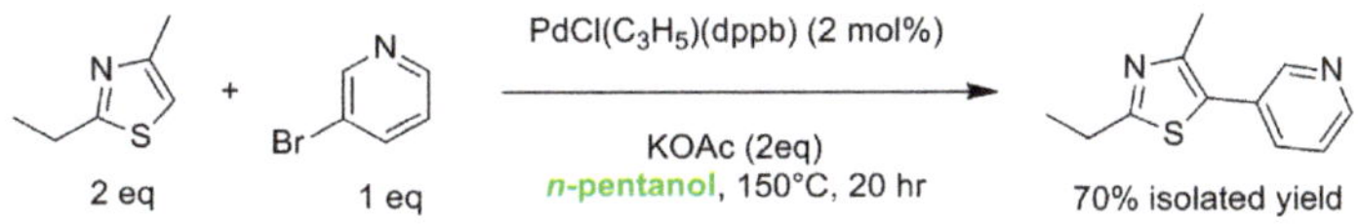


**Scheme 3.10** Direct arylation using alternatives to diprotic apolar solvents.[41]

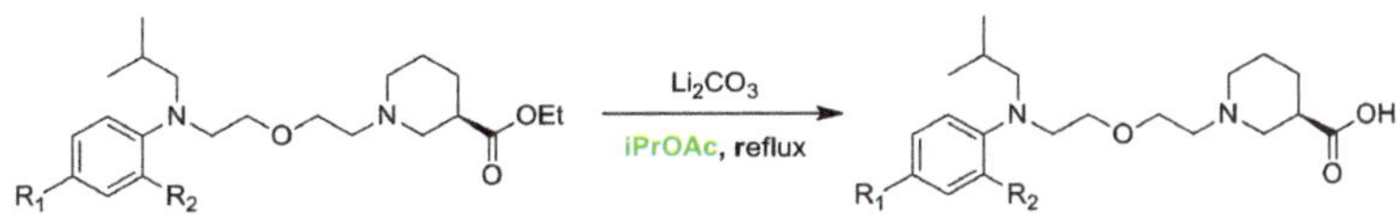


**Scheme 3.11** Novo Nordisk medicinal chemistry example of isopropyl acetate for ester hydrolysis.[44] (N.B. The starting materials were not fully isolated from the previous step, so yields specifically for the hydrolysis reactions were not reported.)

**Scheme 3.12** Novartis example using isopropyl acetate for *N*-acylation.[46]

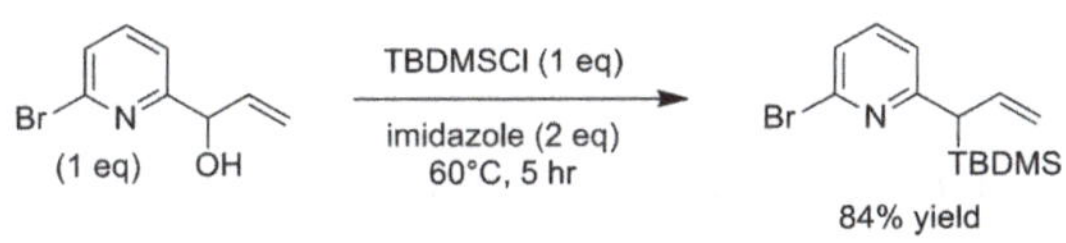


**Scheme 3.13** Solvent free *O*-silylation example.[47]

The susceptibility of ethyl acetate to hydrolysis and transesterification limits its use as a reaction medium, but the lower reactivity and higher boiling point of isopropyl acetate allow for a wider range of reaction conditions. Isopropyl acetate has been used as solvent by multiple medicinal chemistry groups, including examples from Abbott,[42] Boehringer Ingelheim,[43] Novo Nordisk,[44] Lilly,[45] and Novartis.[46] These examples span a wide range of chemical transformations, including nucleophilic displacement,[42] bromination,[43] ester hydrolysis (Scheme 3.11),[44] amide coupling,[45] and *N*-acylation (Scheme 3.12).[46]

When evaluating the choice of reaction media, solvent-free methods should also be considered. While techniques such as mill grinding do exist to affect chemical reactions on a mixture of solids, these are usually impractical in a medicinal chemistry setting. However, in situations where one or more of the reaction components is a liquid, it may be possible to avoid the use of a solvent. An elegant example is the silylation of alcohols, which has been shown to give high yields for even sterically hindered substrates (Scheme 3.13).[47]

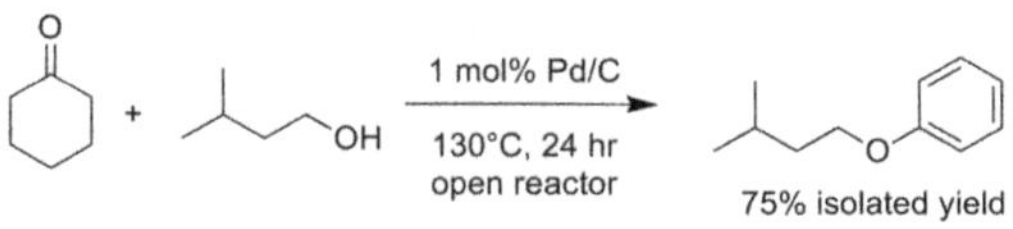


**Scheme 3.14** Dehydrogenative alkylation of cyclohexanone with isopentanol.[48]

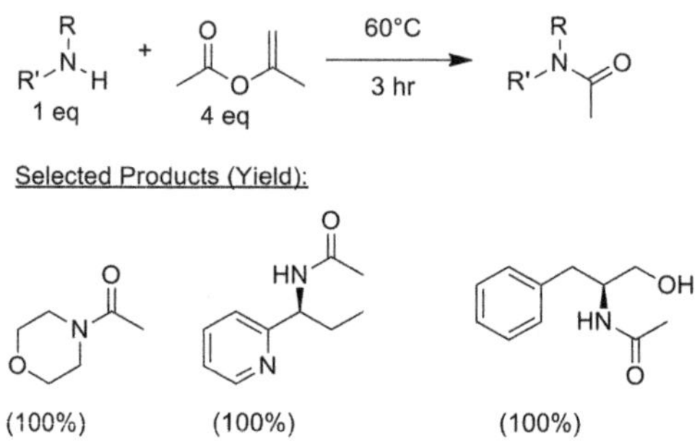


**Scheme 3.15** Solvent- and catalyst-free amine acylation.[49]

The dehydrogenative arylation shown in Scheme 3.14 takes place catalytically without added solvent using air as the oxidant.[48] A solvent and catalyst free method for acetylation of amines has been demonstrated using stoichiometric isopropenyl acetate (Scheme 3.15).[49] Additional academic examples remain to be demonstrated on more complex molecules of pharmaceutical interest.

Given the typical time pressures of medicinal chemistry departments, scientists often find it difficult to justify testing greener solvents as reaction media without reasonable confidence of a successful outcome. Such attempts are made even more difficult because the most common pharmaceutical synthetic transformations have extensive literature precedent in non-preferred solvents. It is therefore important that future researchers conduct solvent screens and share the results of these experiments. Along these lines, Watson *et al.* have published "solvent guides" for reductive amination[50] and amide coupling,[51] showing in color-coded grid format the results of independently screening both solvent and reagent for these popular transformations (see Figure 3.5 for reductive amination).

## 3.3.2 Reagent and Methodology Choices

In addition to evaluation of solvents, multiple efforts are underway to provide guidance regarding reagent choices for the most common chemical transformations. Both GlaxoSmithKline[52] and Pfizer[7] have published suggested green reagent guides, with GSK choosing a similar "traffic-light" approach to that used in their solvent guide, and Pfizer using a Venn diagram to highlight "wide utility" and "scalability" in addition to "greenness" (Figure 3.6). While some of the guides, such as olefin hydrogenation, show

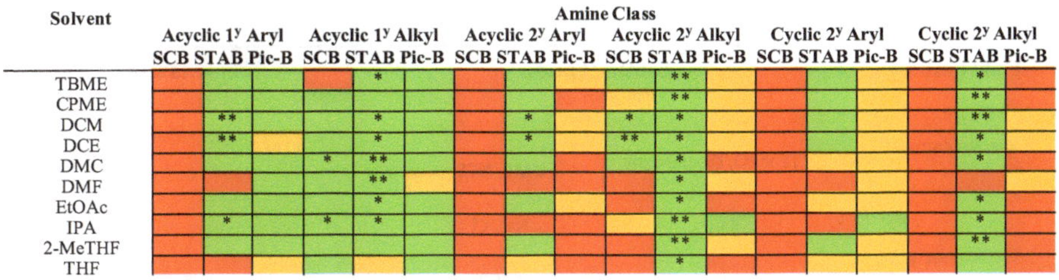


**Figure 3.5** Reductive amination solvent and reagent screen representation.[50] (Key: red ≤ 50% conv., orange = 50–70% conv., green ≥ 70% conv.; * indicates 100% conv. within 4 h; ** indicates 100% conv. within 1 h.) SCB = $NaBH_3CN$, STAB = $NaBH(OAc)_3$, Pic–B = picoline–borane complex. Reproduced from ref. 50.

that the most commonly used reagents fall into the green category, others highlight areas where additional research may be warranted. In the case of deoxychlorination (Figure 3.7), only one option could be identified with "few issues" related to green chemistry. Additional academic research is needed to find greener options with wide scope and utility.

Many chemicals, especially those more recently published, have not been subjected to extensive testing to determine such details as toxicity, biodegradation, environmental effects, life cycle analysis, and a full range of physiochemical properties. As the data become available, and additional novel reagents periodically emerge in the literature, existing guides will need to be revised and reagent recommendations adjusted accordingly.

Where no relevant guides are available, the Material Safety Data Sheet (MSDS) provides detailed information on chemical properties and risks, and can be used to compare reagent or solvent options. However, a more meticulous examination may find that these documents are not standardized and their content can vary greatly based on the issuing company, date, and country of origin. This is currently being addressed through a new system called *The Globally Harmonized System of Classification and Labeling of Chemicals* (GHS)[53] aiming to eliminate these variations.

As highlighted by the ninth of Anastas and Warner's original 12 principles of green chemistry,[3] it is preferable to use reagents in catalytic quantities whenever possible. Indeed, the range of transformations that can be carried out catalytically continues to expand. Amide formation is one of the top synthetic transformations in the medicinal chemistry setting[54] and typically uses stoichiometric amounts of coupling reagents that often possess significant known health and safety hazards. However, there are a range of greener catalytic methods. These include coupling of amines with carboxylic acids,[55,56] amines with alcohols,[57,58] and amines with esters.[59] Some of these methods are known to be applicable to relatively complex substrates, such as the heterocycle and amino acid derivative examples from Jamieson *et al.* (Scheme 3.16).[59]

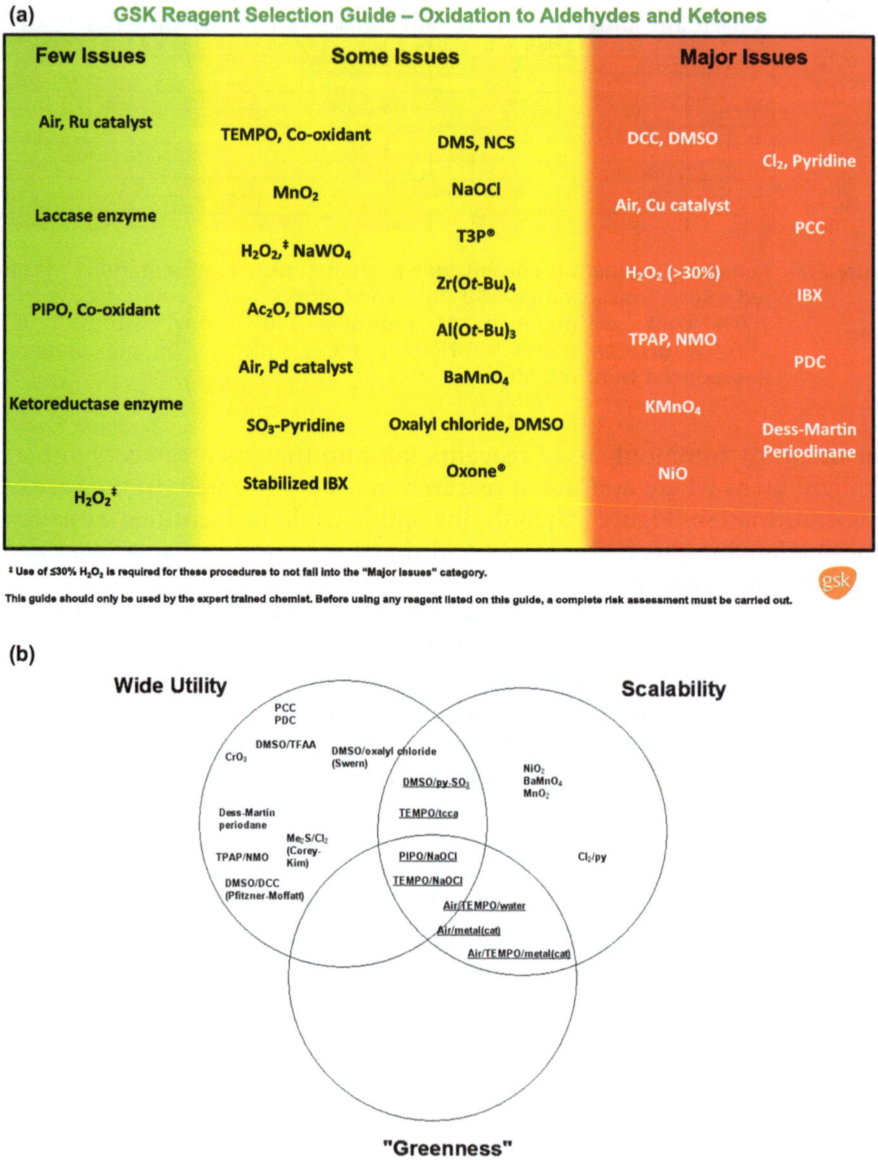


**Figure 3.6** Alcohol oxidation green reagent guides from GlaxoSmithKline[52] (top) and Pfizer[7] (bottom). The GSK guide shown has been updated since initial publication to incorporate newly available data. Pfizer's guide shown here specifically covers oxidation of primary alcohols to aldehydes.
Reproduced from ref. 7.

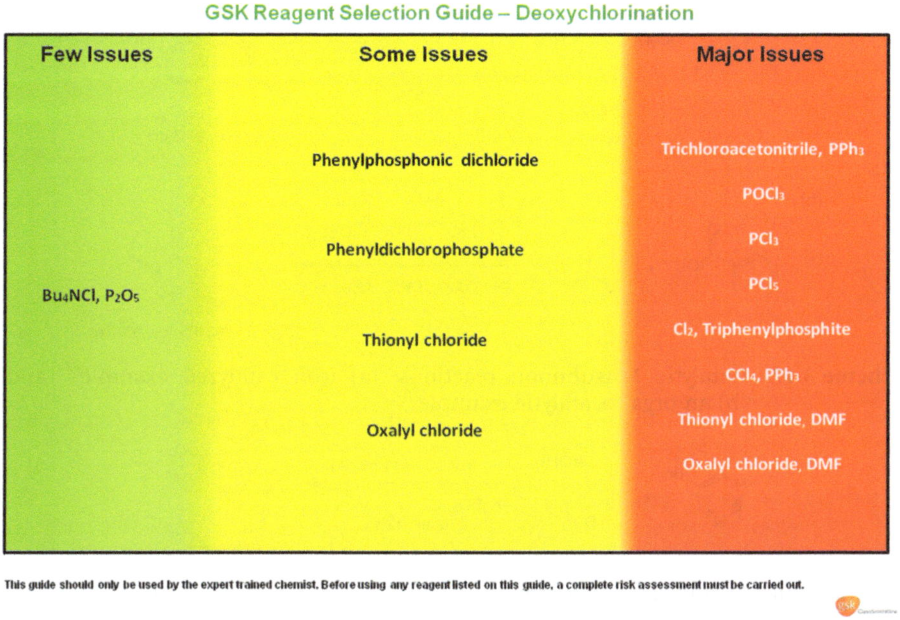


**Figure 3.7** Deoxychlorination green reagent guide from GlaxoSmithKline.[52] The GSK guide shown has been updated since initial publication to incorporate newly available data.

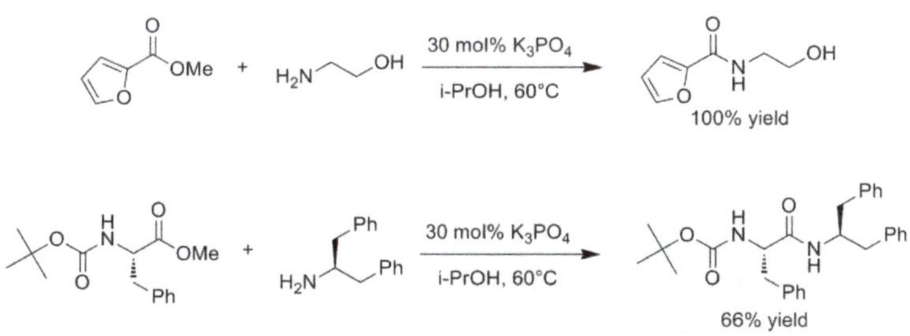


**Scheme 3.16** Catalytic amide formation from esters and amines.[59]

The Mitsunobu reaction allows for the efficient formation of ethers and amines. However, phosphines pose significant EHS concerns, and the presence of phosphine oxide byproducts often leads to complicated and solvent-intensive purification processes. Azodicarboxylate reagents are suspected carcinogens, carrying chronic human health risks, and are highly toxic to aquatic life. Multiple academic groups have investigated catalytic alternatives with varying success at making the overall transformation more sustainable. Taniguichi has published methods that retain the stoichiometric phosphine, but replace the azodicarboxylate with two catalytic reagents, an iron complex and a substituted hydrazine (Scheme 3.17a).[60] Another method uses azodicarboxylates catalytically along with a hypervalent iodine reagent as the stoichiometric oxidant (Scheme 3.17b).[61]

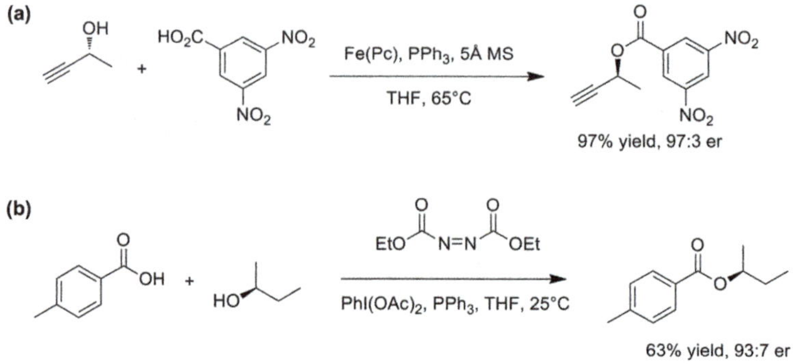

Scheme 3.17 Catalytic Mitsubnobu reactions: (a) iron catalyzed example[60] and (b) an organocatalytic example.[61]

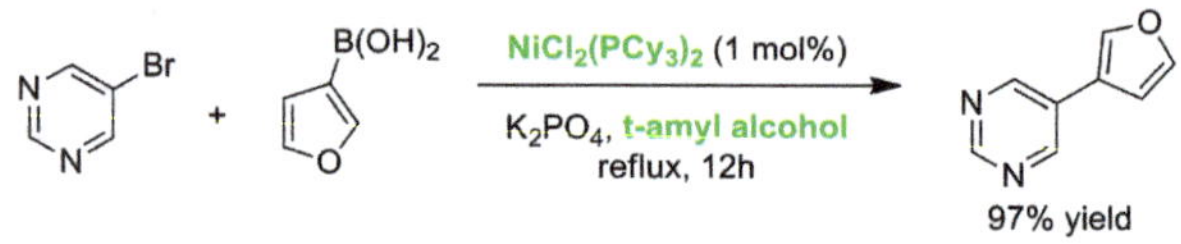

Scheme 3.18 Suzuki reaction example reported on gram scale by Garg *et al.*[35]

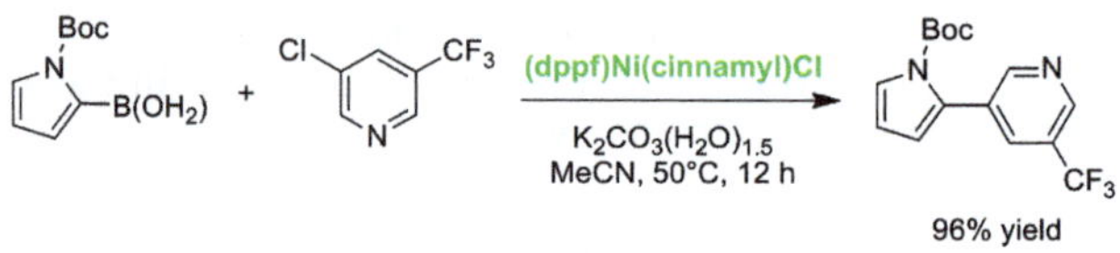

Scheme 3.19 Suzuki reaction example from Ge and Hartwig.[96]

Although stoichiometric phosphines are still required, these advances decrease the overall hazards and the total amount of reagents used. Research to further improve this transformation is warranted.

Cross-coupling chemistry has forever changed the way medicinal chemists design their routes. Currently, the Suzuki reaction is the most common form of C–C bond formation in pharmaceutical chemistry.[54] While these methods are used extensively, the increasing reliance on palladium poses a sustainability concern. Global reserves of platinum group elements (PGEs) are primarily found in four small mining regions located in the United States, Canada, Russia, and South Africa. Increasing demand and political considerations put the future availability of these metals in question. In addition, the methods for pure PGE recovery from mined ore are highly energy and water intensive, with one Montana area mine reporting that it processes 2400 tons of ore per day for a yield of only 0.6 oz of PGE per ton of ore.[62]

With these concerns in mind, there is increased interest in non-precious metal catalysts. Although nickel has long been known to effect Suzuki-type couplings,[63] only recently has it been studied in more depth and with wider scope. A range of high-yielding Ni-catalyzed Suzuki couplings in greener solvents has been demonstrated (Schemes 3.18 and 3.19). Other frequently

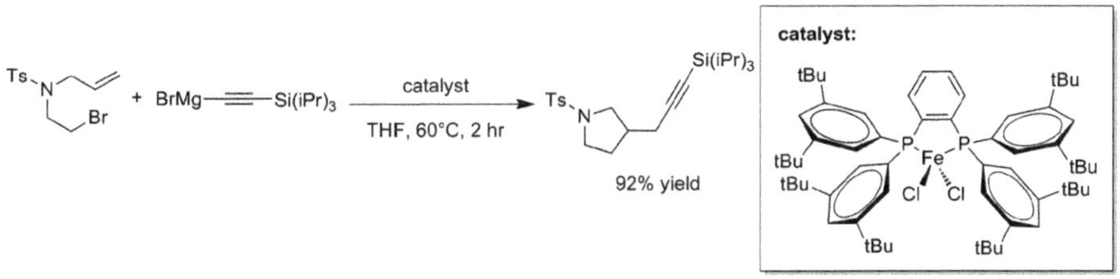


**Scheme 3.20** Sonogashira-like products obtained *via* iron catalyzed Grignard addition.[75]

employed Pd-catalyzed methods can utilize nickel catalysis, including Negishi,[64] Heck,[65] Kumada,[66] Stille,[67] and Sonagashira[68] couplings.

Iron, an abundant and inexpensive metal, is also attractive for catalytic processes. It has been known to catalyze Grignard additions since the early 1940s,[69] but early research was quite limited. As with nickel, the use of iron has now been demonstrated across a range of cross-coupling transformations,[70] including Suzuki,[71] Negishi,[72] Heck,[73] and C–H activation.[74] In addition, Sonogashira-like products were obtained *via* iron-catalyzed coupling of alkyl halides with alkynyl Grignard reagents (Scheme 3.20).[75] The known Suzuki examples at this time do not include biaryl couplings, as two published reports were later retracted, the original conclusions of which are believed to result from low level palladium contamination of iron reagents.[76]

Although catalysis is typically preferred, medicinal chemists should remain aware that there are situations in which it may be greener to use stoichiometric reagents. An example of this is illustrated by Moody *et al.* in their study of heteroaryl chloride amination.[77] This publication raises awareness that some Buchwald type substrates are sufficiently reactive to be suitable for $S_NAr$ processes using benign bases, thus eliminating the need for bulky ligands and precious metal catalysts. These substrate types are used extensively in pharmaceutical chemistry, demonstrating valuable opportunities to forgo precious metal catalysis.

Also central to the basic principles of green chemistry is the concept of atom economy. When considering multiple options, it is desirable to choose reagents and substrates with lower molecular weight. For example, to replace a phenolic hydroxyl group in a molecule, one choice is conversion into a sulfonate such as mesylate or tosylate. Among the sulfonate leaving groups, if reactivity is not an issue, the mesylate has the lowest molecular weight and would be preferred from an atom economy standpoint.

## 3.4 Reaction Setup

In line with the sixth principle of green chemistry, energy and water usage should be also considered when evaluating the sustainability of a chemical process.[3] Depending on scale, cooling or heating a reaction mixture requires significant energy. If all other parameters, including yields, are comparable, this energy usage renders the transformation less sustainable than using an ambient process. When a batch process requires thermal energy, heating the reaction mixture a few degrees below reflux temperature eliminates the use of chillers or continuously run water condensers. In fact, AstraZeneca scientists documented a six-fold increase in energy consumption running a reaction at reflux as opposed to 5 °C below the solvent's boiling point.[78] This small temperature difference should not result in a significant change in rate of conversion. Alternatively, using a higher-boiling solvent may be appropriate to eliminate the need for reflux conditions.

Literature assessments vary as to whether microwave or thermal heating is more energy efficient.[79] Factors such as reaction time and temperature should be considered in this assessment, as the higher pressures and the different method of energy transfer enabled by a microwave reactor may significantly decrease the time to reaction completion. Scaling up microwave conditions, however, can be difficult, as this typically requires special instrumentation or custom designed continuous processes to compensate for the poor penetration depth of microwaves.

Flow chemistry can save time and reduce safety hazards relative to batch processes, as reagents with special safety concerns are present in only very small amounts at any one time and are contained in a highly controlled environment.[80] Using systems of tubing as microreactors, high solubility of both reagents and products is necessary to avoid causing mechanical difficulties. The Thales Nano H-Cube® used for small-scale hydrogenations is one example. As additional such technology becomes more widely available, flow chemistry may prove an excellent option to avoid exposure to highly reactive chemicals or difficulties associated with transformations such as ozonolysis, generation of azides, lithiation, nitration, and carbonylation.[81] Further discussion on this topic may be found in Chapter 6 (Belecki and Gupton).

Electrochemistry employs electrical current instead of potentially hazardous and toxic reagents for certain oxidations and reductions; however, the special equipment necessary traditionally presents a high barrier against its adaption in medicinal chemistry. New technologies such as electrochemical microfluidics may make it easier for such techniques to be considered in drug discovery. A recent work in collaboration with both GlaxoSmithKline and Pfizer highlights TEMPO-mediated alcohol oxidations, although mostly on relatively simple substrates (Scheme 3.21).[82] The utility of

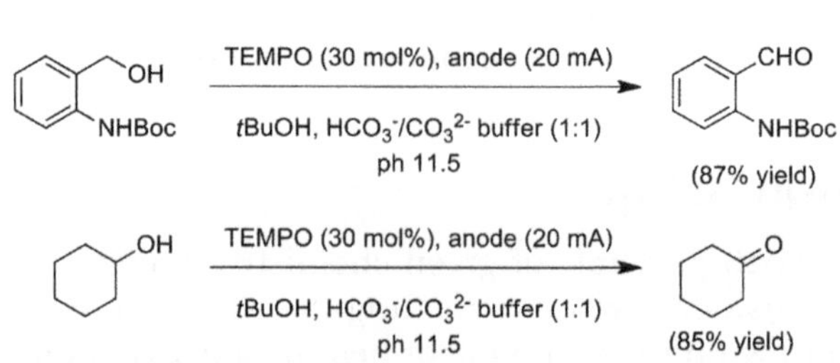


**Scheme 3.21** Electro-oxidation examples for primary and secondary alcohols.[84]

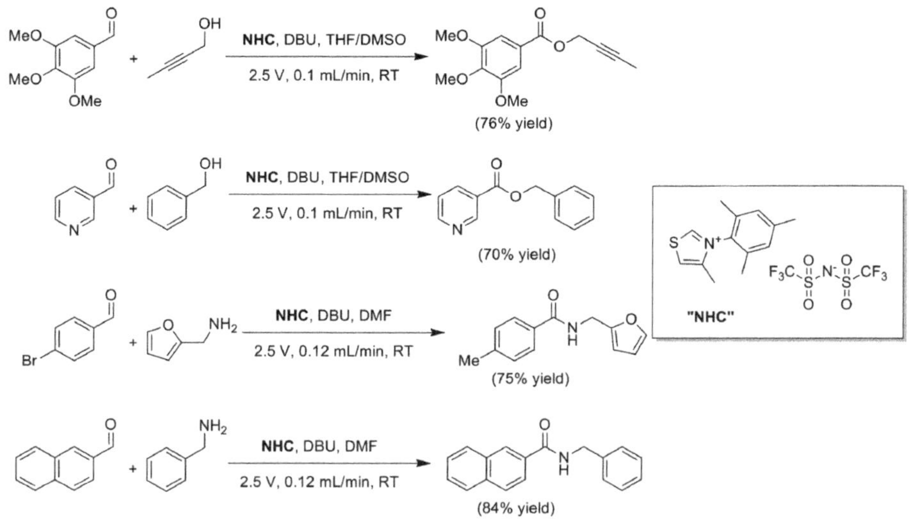


**Scheme 3.22** Electrosynthetic aldehyde conversion into esters or amides mediated by N-heterocyclic carbenes (NHCs).[85]

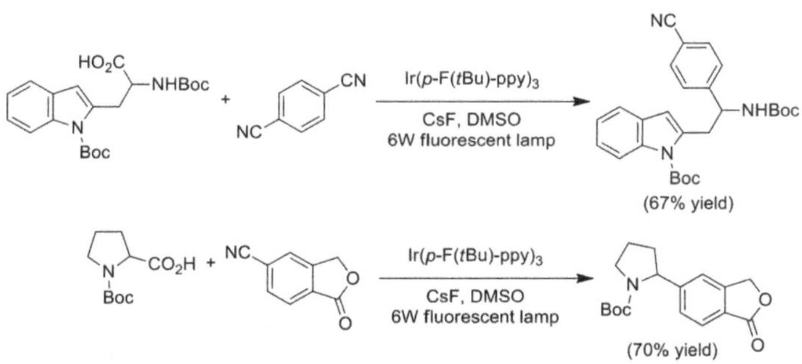


**Scheme 3.23** Decarboxylative arylation of α-amino acids.[86]

this setup has also been extended to the formation of esters or amides from aldehyde precursors, mediated by N-heterocyclic carbenes (Scheme 3.22).[83]

Photochemical reaction setups are also worth considering for some isomerization, polymerization, condensation, and photo-redox reactions, as the energy consumed is much less than what is required for maintaining a high temperature bath and possibly a condenser.[84] The use of a light source may reduce or eliminate the need for additional reagents and improve reaction selectivity. Reaction setups for scales up to a few grams simply require securing a lamp that emits the desired wavelength. While many traditional photochemistry techniques require high energy UV light sources, possibly posing safety and environmental concerns, newer reactions use catalysts that can utilize visible light sources.[85] The MacMillan group has developed multiple such methods, including decarboxylative arylation, direct benzylic ether arylation, and β-arylation of ketones and aldehydes. Schemes 3.23–3.25

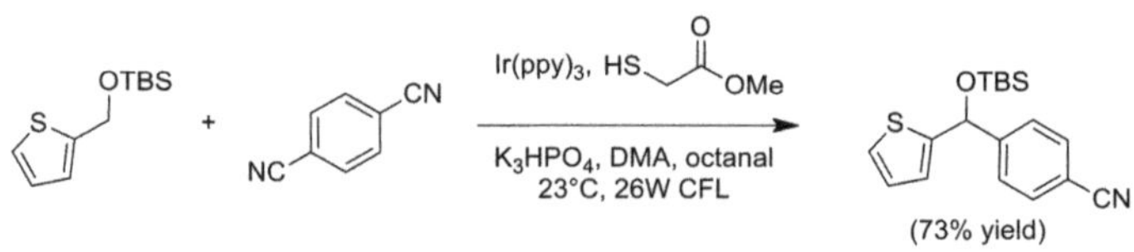

**Scheme 3.24** Direct arylation of benzylic ethers.[87]

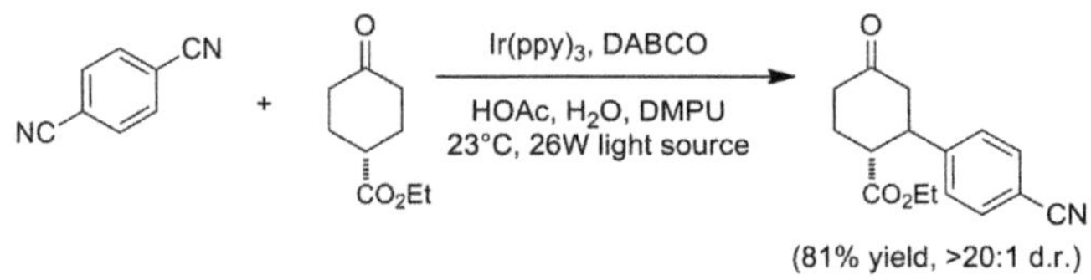

**Scheme 3.25** Direct β-arylation of ketones and aldehydes.[88]

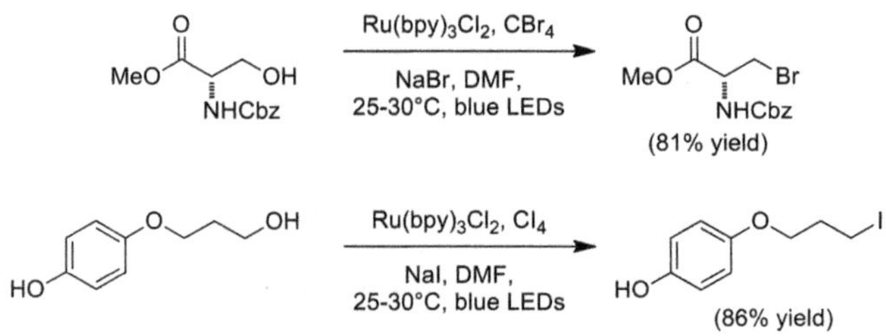

**Scheme 3.26** Visible light mediated deoxyhalogenation of alcohols.

highlight their synthesis of several drug-like substrates. Deoxyhalogenation, identified in Section 3.3.2 as an area needing additional research, may also employ photochemical methods, as shown by research from the Stephenson lab in Scheme 3.26. Challenges on a larger scale are similar to those of microwave chemistry, and are best addressed *via* continuous flow designs.

## 3.5 Reaction Workup

Once a reaction is complete, the desired product must be separated from the resulting mixture. The greenest process is isolation of an insoluble product by filtration. The product may also be precipitated by the addition of a co-solvent. In other cases it may be possible to precipitate only byproducts and impurities. Such procedures are generally less solvent intensive than extractive workups and avoid generation of contaminated aqueous waste; however, they are not always practical on small scales.

When considering an extractive workup, dichloromethane (DCM) is frequently chosen to facilitate substrate solubility and avoid potential hydrolysis of alternatives such as ethyl acetate. Eliminating DCM from an extractive workup has a large impact, as the amount of solvent used in workup far exceeds the original reaction volume. 2-MeTHF is a useful alternative when hydrolysis of the organic phase or aqueous solubility of the substrate is a concern. Unlike THF, 2-MeTHF is not miscible with water and has been shown to have a low propensity for emulsion formation.[34] The resulting organic layer can be dried using desiccants or azeotropic distillation, a more common method on production scale. Other greener solvents (Table 3.2) can also be used for the removal of water *via* azeotropic distillation.

Although ethyl acetate is suitable for most extractions, and has minimal green chemistry issues, isopropyl acetate, a commonly used process solvent, has even more advantages. Isopropyl acetate takes up only half as much water as ethyl acetate,[89] which facilitates more efficient drying procedures. In addition, isopropyl acetate has been shown to carry through lower levels of many commonly used dipolar aprotic solvents, which can improve chromatographic separations.[90]

**Table 3.2** Water azeotrope data for a selection of greener solvents, with non-preferred solvents toluene, dichloromethane (DCM), and tetrahydrofuran (THF) shown for comparison.

| Primary solvent | Solubility of water[a] (w/w) | Solubility in water[a] (w/w) | Azeotrope[b] (°C) | Azeotrope[b] (%water) |
|---|---|---|---|---|
| 1-Butanol | 20 | 7.7 | 92.7 | 42.5 |
| 2-Butanol | 36.3 | 15.4 | 87.0 | 26.8 |
| Cyclohexanol | 11.8 | 4.3 | 97.8[a] | 69.5[a] |
| Propyl acetate | 2.9 | 2.3 | 82.4 | 14 |
| Isopropyl acetate | 1.8 | 2.9 | 76.6 | 10.56 |
| Butyl acetate | 1.6 | 0.7 | 90.2 | 28.7 |
| Dimethyl carbonate | 2.6[c] | 12.7[c] | 77.5[d] | 11[d] |
| Cyclohexanone | 8 | 2.3 | 96.3 | 55 |
| Ethanol | Misc | Misc | 78.2 | 4 |
| 1-Propanol | Misc | Misc | 71.7 | 28.3 |
| 2-Propanol | Misc | Misc | 80.3 | 12.6 |
| Methyl acetate | 8.2 | 24.5 | 56.4[a] | 5.0[a] |
| Ethyl acetate | 3.3 | 7.7 | 70.4 | 8.5 |
| Cyclohexane | 0.01 | 0.006 | 69.5 | 8.4 |
| Heptane | 0.005 | 0.005 | 79.2 | 12.9 |
| 2-Methyl-THF | 4.1[e] | 14.4[e] | 71[e] | 10.6[e] |
| Methyl isobutyl ketone | 1.9 | 1.7 | 87.9 | 24.3 |
| Toluene | 0.033 | 0.052 | 84.61 | 13.5 |
| Dichloromethane | 2.3 | 1.7 | 38.1[a] | 1.5[a] |
| Tetrahydrofuran | Misc | Misc | 63.8[a] | 4.6[a] |

[a]From Smallwood, ref. 89.
[b]From ref. 91 unless otherwise noted.
[c]From ref. 94.
[d]From *Knovel Critical Tables*, ref. 92.
[e]From pennakem.com.[93]

## 3.6 Purification and Waste

Within drug discovery, purity requirements vary with each synthetic step. Some intermediates can be taken on after a simple workup, while final products and other intermediates typically require more extensive purification. Considering scale and intended use, it is difficult in medicinal chemistry to plan reactions to avoid the need for chromatography. Greener options to traditional chromatography methods are available and are continually being investigated.[95] Items that can easily be considered in drug discovery include use of preferred solvents, telescoping of multiple synthetic steps, and choosing the smallest reasonable column. Additional detail can be found in Chapter 4 (Miller and Peterson).

In addition to the impact of chemicals and consumables used in reaction setup, workup, and purification, one should also consider the waste generated by all these processes. Due to the complexity and required purity of submitted compounds, it is challenging within drug discovery to significantly reduce the amount of waste generated. However, changing the nature of the waste stream by using greener options can improve sustainability concerns. For example, a lower volume of chlorinated solvent waste decreases the need for use of scrubbers used to capture halogenated emissions upon incineration. There are also opportunities to upgrade the final fate of the waste stream by combining heat recovery with incineration, and even in some cases recycling. Additional information, including potential recycling of solvents and certain reagents is covered more extensively in Chapter 8 (Richter).

## 3.7 Conclusion

Options are available for the medicinal chemist to apply green chemistry principles at every stage of planning and carrying out a reaction. Conception of a synthetic route should take into account implications of the solvents and reagents used, including step and atom efficiency. In-house and literature resources can provide further detail on the scope, utility, and sustainability parameters of different alternatives, although wider research is needed in this area. Operational conditions, including energy and water consumption, should also be considered when setting up reactions. Careful evaluation of these factors in drug discovery can lessen the overall impact on worker health and safety, and on the environment. Early incorporation of green chemistry tools and principles may also save time and resources when further optimizing the medicinal chemistry routes to potential drug molecules.

## References

1. These reviews are published twice annually. For recent examples, see (a) R. Bandichhor, A. Bhattacharya, L. Diorazio, P. Dunn, K. Fraunhoffer,

F. Gallou, J. Hayler, M. Hickey, W. Hinkley, D. Hughes, L. Humphreys, B. Kaptein, S. Mathew, T. Rammeloo, P. Richardson and T. White, *Org. Process Res. Dev.*, 2013, **17**, 1394–1405; (b) R. Bandichhor, A. Bhattacharya, L. Diorazio, P. Dunn, K. Fraunhoffer, F. Gallou, J. Hayler, M. Hickey, W. Hinkley, D. Hughes, L. Humphreys, B. Kaptein, S. Mathew, T. Rammeloo, P. Richardson and T. White, *Org. Process Res. Dev.*, 2013, **17**, 615–626.

2. http://en.wikipedia.org/wiki/Green_chemistry_metrics; M. G. T. C. Ribeiro and A. A. S. C. Machado, *Green Chem. Lett. Rev.*, 2013, **6**(1), 1–18.
3. P. T. Anastas and J. C. Warner, *Green Chemistry: Theory and Practice*, Oxford University Press, New York, 1998, p. 30.
4. (a) R. W. Hoffmann, *Handbook of Green Chemistry, Part 7: Green Synthesis-Synthesis Without Protecting Groups*, Published Online: 15 APR 2012, DOI: 10.1002/9783527628698.hgc078; (b) P. S. Baran, T. J. Maimone and J. M. Richter, *Nature*, 2007, **446**, 404–408.
5. L. Zhou, J. Duquette and M. Zhang, *J. Org. Chem.*, 2003, **68**(9), 3729–3732.
6. R. K. Henderson, C. Jiménez-González, D. J. C. Constable, S. R. Alston, G. G. A. Inglis, G. Fisher, J. Sherwood, S. P. Binks and A. D. Curzons, *Green Chem.*, 2011, **13**, 854–862.
7. K. Alfonsi, J. Colberg, P. J. Dunn, T. Fevig, S. Jennings, T. A. Johnson, H. P. Kleine, C. Knight, M. A. Nagy and D. A. Perry, *Green Chem.*, 2008, **10**, 31–36.
8. D. Prat, O. Pardigon, H.-W. Flemming, S. Letestu, V. Ducandas, P. Isnard, E. Guntrum, T. Senac, S. Ruisseau, P. Cruciani and P. Hosek, *Org. Process Res. Dev.*, 2013, **17**, 1517–1525.
9. D. Prat, J. Hayler and A. Wells, *Green Chem.*, 2014, **16**, 4546–4551.
10. Manuscript in progress.
11. J. H. Clark, D. J. Macquarrie and J. Sherwood, *Green Chem.*, 2012, **14**, 90–93.
12. R. E. Mewshar, R. J. Edsall, Jr., C. Yang, E. S. Manas, Z. B. Xu, R. A. Henderson, J. C. Keith, Jr. and H. A. Harris, *J. Med. Chem.*, 2005, **48**, 3953–3979.
13. (a) S. Aparicio and R. Alcalde, *Green Chem.*, 2009, **11**, 65–78; (b) C. S. M. Pereira, V. M. T. M. Silva and A. E. Rodrigues, *Green Chem.*, 2011, **13**, 2658–2671.
14. M. Rose and R. Palkovits, *ChemSusChem*, 2012, **5**, 167–176.
15. (a) D. M. Alonso, S. G. Wettstein and J. A. Dumesic, *Green Chem.*, 2013, **15**, 584–595; (b) Z.-Q. Duan and F. Hu, *Green Chem.*, 2012, **14**, 1581–1583.
16. A. Perosa and F. Zecchini, *Methods and Reagents for Green Chemistry: An Introduction*, J. Wiley and Sons, Inc., Hoboken, 2007.
17. M. H. Tucker, R. Alamillo, A. J. Crisci, G. M. Gonzalez, S. L. Scott and J. A. Dumesic, *ACS Sustainable Chem. Eng.*, 2013, **1**, 554–560.
18. Y. Gu and F. Jérôme, *Green Chem.*, 2010, **12**, 1127–1138.
19. A. Wolfson and C. Dlugy, *Chem. Pap.*, 2007, **61**, 228–232.
20. Y. Gu, J. Barrault and F. Jérôme, *Adv. Synth. Catal.*, 2008, **350**, 2007–2012.
21. A. Wolfson, A. Snezhko, T. Meyouhas and D. Tavor, *Green Chem. Lett. Rev.*, 2012, **5**, 7–12.

22. A. E. Diaz-Álvarez, J. Francos, B. Lastras-Barreira, P. Crochet and V. Cadierno, *Chem. Commun.*, 2011, **47**, 6208–6227.
23. M. North, F. Pizzato and P. Villuendas, *ChemSusChem*, 2009, **2**, 862–865.
24. M. Seto, K. Aikawa, N. Miyamoto, Y. Aramaki, N. Kanazaki, K. Takashimia, Y. Kuze, Y. Iizawa, M. Baba and M. Shiraishi, *J. Med. Chem.*, 2006, **49**, 2037–2048.
25. D. J. Blythin, X. Chen, J. J. Piwinski, N.-Y. Shih, H.-J. Shue, J. C. Anthes and A. T. McPhail, *Bioorg. Med. Chem. Lett.*, 2002, **12**, 3161–3165.
26. R. J. Cvetovich, R. A. Reamer, L. DiMichele, J. Y. L. Chung and J. R. Chilenski, *J. Org. Chem.*, 2006, **71**, 8610–8613.
27. K. Watanabe, N. Yamagiwa and Y. Torisawa, *Org. Process Res. Dev.*, 2007, **11**, 251–258.
28. K. J. Filipski, A. Guzman-Perez, J. Bian, C. Perreault, G. E. Aspnes, M. T. Didiu, R. L. Dow, R. F. Hank, C. S. Jones, R. J. Maguire, M. Tu, D. Zeng, S. Liu, J. D. Knafels, J. Litchfield, K. Atkinson, D. R. Derksen, F. Bourbonais, K. S. Gajiwala, M. Hickey, T. O. Johnson, P. S. Humphries and J. A. Pfefferkorn, *Bioorg. Med. Chem. Lett.*, 2013, **23**, 4571–4578.
29. P. Liu, T. G. Hamill, M. Chioda, H. Chobanian, S. Fung, Y. Guo, L. Chang, R. Bakshi, Q. Hong, J. Dellureficio, L. S. Lin, C. Abbadie, J. Alexander, H. Jin, S. Mandala, L.-L. Shiao, W. Li, S. Sanabria, D. Williams, Z. Zeng, R. Hajdu, N. Jochnowitz, M. Rosenbach, B. Karanam, M. Madeira, G. Salituro, J. Powell, L. Xu, J. L. Terebetski, J. F. Leone, P. Miller, J. Cook, M. Holahan, A. Joshi, S. O'Malley, M. Purcell, D. Posavec, T.-B. Chen, K. Riffel, M. Williams, R. Hargeaves, K. A. Sullivan, R. P. Nargund and R. J. DeVita, *ACS Med. Chem. Lett.*, 2013, **4**, 509–513.
30. M. Krayer von Karuss and P. Harremoës, in *The Precautionary Principle in the 20th Century: Late Lessons from Early Warnings*, ed. P. Harremoës, D. Gee, M. MacGarvin, A. Stirling, J. Keys, B. Wynne and S. Guedes Vaz, Earthscan Publications Ltd., London, 2002, ch. 11, p. 113.
31. P. E. Maligres, S. W. Krska and P. G. Dormer, *J. Org. Chem.*, 2012, 77, 7646–7651.
32. S. Kobayashi, H. Kuroda, Y. Ohtsuka, T. Kashihara, A. Masuyama and K. Watanabe, *Tetrahedron*, 2013, **69**, 2251–2259.
33. (a) Z. Shiokawa, K. Hashimoto, B. Saito, Y. Oguro, H. Sumi, M. Yabuki, M. Yoshimatsu, Y. Kosugi, Y. Debori, N. Morishita, D. R. Dougan, G. P. Snell, S. Yoshida and T. Ishikawa, *Bioorg. Med. Chem.*, 2013, **21**, 7938–7954; (b) K. Hashimoto, B. Saito, N. Miaymoto, Y. Oguro, D. Tomita, Z. Shiokawa, M. Asano, H. Kakei, N. Taya, M. Kawasaki, H. Sumi, M. Yabuki, K. Iwai, S. Yoshida, M. Yoshimatsu, K. Aoyama, Y. Kosugi, T. Kojima, N. Morishita, D. R. Dougan, G. P. Snell, S. Imamura and T. Ishikawa, *J. Med. Chem.*, 2013, **56**, 1228–1246.
34. D. F. Aycock, *Org. Process Res. Dev.*, 2007, **11**, 156–159.
35. S. D. Ramgren, L. Hie, Y. Ye and N. K. Garg, *Org. Lett.*, 2013, **15**, 3950–3953.
36. J. C. Lorenz, C. A. Busacca, X. Feng, N. Grinberg, N. Haddad, J. Johnson, S. Kapadia, H. Lee, A. Saha, M. Sarvestani, E. M. Spinelli, R. Varsolona, X. Wei, X. Zeng and C. H. Senanayake, *J. Org. Chem.*, 2010, **75**, 1155–1161.

37. D. H. B. Ripin and M. Vetelino, *Synlett*, 2003, **15**, 2353.
38. V. Antonucci, J. Coleman, J. B. Ferry, N. Johnson, M. Mathe, J. P. Scott and J. Xu, *Org. Process Res. Dev.*, 2011, **15**, 939–941.
39. K. Capriotti and J. A. Capriotti, *J. Clin. Aesthet. Dermatol.*, 2012, **5**, 24–26.
40. U. Tilstam, *Org. Process Res. Dev.*, 2012, **16**, 1273–1278.
41. S. Bensaid, N. Laidaouri, D. El Abed, S. Kacimi and H. Doucet, *Tetrahedron Lett.*, 2011, **52**, 1383–1394.
42. R. Wagner, K. W. Mollison, L. Liu, C. L. Henry, T. A. Rosenberg, N. Bamaung, N. Tu, P. E. Wiedeman, Y. Or, J. R. Luly, B. C. Lane, J. Trevillyan, Y.-W. Chen, T. Fey, G. Hsieh, K. Marsh, M. Nuss, P. B. Jacobsen, D. Wilcox, R. P. Carlson, G. W. Carter and S. W. Djuric, *Bioorg. Med. Chem. Lett.*, 2005, **15**, 5340–5343.
43. P. L. Beaulieu, J. Gillard, D. Bykowski, C. Brochu, N. Dansereau, J.-S. Duceppe, B. Haché, A. Jakalian, L. Lagacé, S. LaPlante, G. McKercher, E. Moreau, S. Perreault, T. Stammers, L. Thauvette, J. Warrington and G. Kukolj, *Bioorg. Med. Chem. Lett.*, 2006, **16**, 4987–4993.
44. K. E. Andersen, J. Lau, B. F. Lundt, H. Petersen, P. O. Huusfeldt, P. D. Suzdak and M. D. B. Swedberg, *Bioorg. Med. Chem.*, 2001, **9**, 2773–2785.
45. B. Evers, G. Ruehter, M. Berg, J. A. Dodge, D. Hankotius, U. Hary, L. N. Jungheim, H.-J. Mest, E.-M. Martin de la Nava, M. Mohr, B. S. Muehl, S. Petersen, B. Sommer, G. Riedel-Herold, M. J. Tebbe, K. J. Thrasher and S. Voelkers, *Bioorg. Med. Chem.*, 2005, **13**, 6748–6762.
46. F. Baur, D. Beattie, D. Beer, D. Bentley, M. Bradley, I. Bruce, S. J. Charlton, B. Cuenoud, R. Ernst, R. A. Fairhurst, B. Faller, D. Farr, T. Keller, J. R. Fozard, J. Fullerton, S. Garman, J. Hatto, C. Hayden, H. He, C. Howes, D. Janus, Z. Jiang, C. Lewis, F. Loeuillet-Ritzler, H. Moser, J. Reilly, A. Steward, D. Sykes, L. Tedaldi, A. Trifilieff, M. Tweed, S. Watson, E. Wissler and D. Wyss, *J. Med. Chem.*, 2010, **53**, 3675–3684.
47. B. Hatano, S. Toyota and F. Toda, *Green Chem.*, 2001, **3**, 140–142.
48. M. Sutter, N. Sotto, Y. Raoul, E. Métay and M. Lemaire, *Green Chem.*, 2013, **15**, 347–352.
49. R. Pelagalli, I. Chiarotto, M. Feroci and S. Vecchio, *Green Chem.*, 2012, **14**, 2251–2255.
50. F. I. McGonagle, D. S. MacMillan, J. Murray, H. F. Sneddon, C. Jamieson and A. J. B. Watson, *Green Chem.*, 2013, **15**, 1159–1165.
51. D. S. MacMillan, J. Murray, H. F. Sneddon, C. Jamieson and A. J. B. Watson, *Green Chem.*, 2013, **15**, 596–600.
52. J. P. Adams, C. M. Alder, I. Andrews, A. M. Bullion, M. Campbell-Crawford, M. G. Darcy, J. D. Hayler, R. K. Henderson, C. A. Oare, I. Pendrak, A. M. Redman, L. E. Shuster, H. F. Sneddon and M. D. Walker, *Green Chem.*, 2013, **15**, 1542–1549.
53. https://www.osha.gov/dsg/hazcom/ghs.html, last accessed July 30, 2014.
54. (a) S. D. Roughley and A. M. Jordan, *J. Med. Chem.*, 2011, **54**, 3451–3479; (b) J. S. Carey, D. Laffan, C. Thomson and M. T. Williams, *Org. Biomol. Chem.*, 2006, **4**, 2337–2347.

55. H. Lundberg, F. Tinnis, N. Selander and H. Adolfsson, *Chem. Soc. Rev.*, 2014, **43**, 2714–2742.
56. R. M. Al-Zoubi, O. Marion and D. G. Hall, *Angew. Chem., Int. Ed.*, 2008, **47**, 2876–2879.
57. L. U. Nordstrøm, H. Vogt and R. J. Madsen, *J. Am. Chem. Soc.*, 2008, **130**, 17672–17673.
58. C. Gunathan, B.-D. Yehoshoa and D. Milstein, *Science*, 2007, **317**, 790–792.
59. N. Caldwell, C. Jamieson, I. Simpson and A. J. B. Watson, *ACS Sustainable Chem. Eng.*, 2013, **1**, 1339–1344.
60. D. Hirose, T. Taniguichi and H. Ishibashi, *Angew. Chem., Int. Ed.*, 2013, **52**, 4613–4617.
61. T. Y. S. But and P. H. Toy, *J. Am. Chem. Soc.*, 2006, **128**, 9636–9637.
62. http://www.stillwaterpalladium.com/concentrating.html, last accessed March 24, 2014.
63. (a) M. F. Semmelhack, P. M. Helquist and L. D. Jones, *J. Am. Chem. Soc.*, 1971, **93**, 5908–5910; (b) V. Percec, J.-Y. Bae and D. H. Hill, *J. Org. Chem.*, 1995, **60**, 1060–1065.
64. J. Shou and G. C. Fu, *J. Am. Chem. Soc.*, 2003, **125**, 14726–14727.
65. A. R. Ehle, Q. Zhou and M. P. Watson, *Org. Lett.*, 2012, **14**, 1202–1215.
66. (a) J. Terao, A. Ikumi, H. Kuniyasu and N. Kambe, *J. Am. Chem. Soc.*, 2003, **125**, 5646–5647; (b) J. Terao, H. Watanabe, H. Kuniyasu and N. Kambe, *J. Am. Chem. Soc.*, 2002, **124**, 4222–4223.
67. D. A. Powell, T. Maki and G. C. Fu, *J. Am. Chem. Soc.*, 2005, **127**, 510–511.
68. O. Vechorkin, D. Barmaz, V. Proust and X. Hu, *J. Am. Chem. Soc.*, 2009, **131**, 12078–12079.
69. M. Kumada and J. K. Kochi, *J. Am. Chem. Soc.*, 1941, **63**, 2316–2320.
70. W. M. Czaplik, M. Mayer, J. Cvengroš and A. J. von Wangelin, *ChemSusChem*, 2009, **2**, 396–417.
71. (a) T. Hatakeyama, T. Hashimoto, K. K. A. D. S. Kathriarachchi, T. Zenmyo, H. Seike and M. Nakamura, *Angew. Chem., Int. Ed.*, 2012, **51**, 8834–8837; (b) T. Hatakeyama, T. Hashimoto, Y. Kondo, Y. Fujiwara, H. Seike, H. Takaya, Y. Tamada, T. Ono and M. Nakamura, *J. Am. Chem. Soc.*, 2010, **132**, 10674–10676.
72. M. Nakamura, S. Ito, K. Matsuo and E. Nakamura, *Synlett*, 2005, 1794–1798.
73. R. Loska, C. M. Rao Volla and P. Vogel, *Adv. Synth. Catal.*, 2008, **350**, 2859–2864.
74. B. Moreau, J. Y. Wu and T. Ritter, *Org. Lett.*, 2009, **11**, 337–339.
75. T. Hatakeyama, Y. Okada, Y. Yoshimoto and M. Nakamura, *Angew. Chem., Int. Ed.*, 2011, **50**, 10973–10976.
76. R. B. Bedford, M. Nakamura, N. J. Gower, M. F. Haddow, M. A. Hall, M. Huwe, T. Hashimoto and R. A. Okopie, *Tetrahedron Lett.*, 2009, **50**, 6110–6111.
77. K. Walsh, H. F. Sneddon and C. J. Moody, *ChemSusChem*, 2013, **6**, 1455.
78. C. R. Hargreaves, presented in part at the 11th Annual Green Chemistry and Engineering Conference, Washington, D.C., June, 2007.

79. (a) J. D. Moseley and C. O. Kappe, *Green Chem.*, 2011, **13**, 794–806; (b) M. J. Gronnow, R. J. White, J. H. Clark and D. J. Macquarrie, *Org. Process Res. Dev.*, 2005, **9**, 516–518.
80. C. Wiles and P. Watts, *Future Med. Chem.*, 2009, **1**(9), 1593–1612.
81. http://www.thalesnano.com, last accessed May 5, 2014.
82. J. T. Hill-Cousins, J. Kuleshova, R. A. Green, P. R. Birkin, D. Pletcher, T. J. Underwood, S. G. Leach and R. C. D. Brown, *ChemSusChem*, 2012, **5**, 326–331.
83. R. A. Green, R. C. D. Brown, D. Pletcher, P. R. Birkin, and S. G. Leach, presented in part at the SCI/RSC Symposium on Continuous Processing and Flow Chemistry, West Sussex, September, 2013.
84. A. Albini and M. Fagnoni, *Green Chem.*, 2004, **6**, 1–6.
85. T. P. Yoon, M. A. Ischay and J. Du, *Nat. Chem.*, 2010, **2**, 527–532.
86. Z. Zuo and D. W. C. MacMillan, *J. Am. Chem. Soc.*, 2014, **136**, 5257–5260.
87. K. Qyortrup, D. A. Rankic and D. W. C. MacMillan, *J. Am. Chem. Soc.*, 2014, **136**, 626–629.
88. M. T. Pirnot, D. A. Rankic, D. B. C. Martin and D. W. C. MacMillan, *Science*, 2013, **339**, 1593–1596.
89. I. M. Smallwood, *Solvent Recovery Handbook*, Blackwell Science Ltd., Oxford, 2nd edn, 2002, ch. 7, p. 98.
90. L. Delhaye, A. Ceccato, P. Jacobs, C. Köttgen and A. Merschaert, *Org. Process Res. Dev.*, 2007, **11**, 160–164.
91. J. A. Dean, *Lange's Handbook of Chemistry*, McGraw-Hill, Inc., New York, 14th edn, 1992, ch. 5, pp. 5.57–5.61.
92. *Knovel Critical Tables*, 2nd edn, 2008; Online version available at: http://app.knovel.com/hotlink/toc/id:kpKCTE000X/knovel-critical-tables.
93. http://www.pennakem.com/pdfs/methf_tds01.pdf, last accessed 5 May 2014.
94. M. Góral, D. G. Shaw, A. Mączyński, B. Wiśniewska-Gocioswska and A. Jezierski, *J. Phys. Chem. Ref. Data*, 2009, **38**, 1093–1127.
95. (a) J. P. Taygerly, L. M. Miller, A. Yee and E. A. Peterson, *Green Chem.*, 2012, **14**, 3020–3025; (b) D. S. MacMillan, J. Murray, H. F. Sneddon, C. Jamieson and A. J. B. Watson, *Green Chem.*, 2012, **14**, 3016–3019.
96. S. Ge and J. F. Hartwig, *Angew. Chem. Int. Ed.*, 2012, **51**, 12837–12841.

CHAPTER 4

# *Greener Solvent Usage for Discovery Chemistry Analysis and Purification*

LARRY MILLER* AND EMILY A. PETERSON

Amgen, 360 Binney Street, Cambridge, MA 02142, USA
*Email: millerl@amgen.com

## 4.1 Introduction

During the synthesis of a small molecule it is critical to have analytical methods to determine purity as well as purification methods to generate high purity molecules. This is even more important in the drug discovery process where most molecules are synthesized only once and chemists may be unable to draw on knowledge/experiences from a previous synthesis. With the use of parallel chemistry and microwave synthesis, a large number of reactions can be performed in a short period of time, often resulting in analysis and purification becoming the rate limiting step to synthesize a new molecule. The use of analysis and purification tools that are phase appropriate as well as time and solvent efficient can greatly reduce the time and cost to synthesize a molecule for testing.

The concept of green chemistry was introduced approximately 15 years ago.[1] Early efforts in green chemistry targeted the largest opportunity areas, mainly chemical manufacturing and more recently process chemistry. Over the past few years green chemistry has been introduced to smaller volume activities such as medicinal and analytical chemistry. Green chemistry in medicinal chemistry synthesis is addressed in other chapters in this book.

RSC Drug Discovery Series No. 46
Green Chemistry Strategies for Drug Discovery
Edited by Emily A. Peterson and Julie B. Manley

Published by the Royal Society of Chemistry, www.rsc.org

This chapter will focus on the "greening" of analysis and purification activities in support of medicinal chemistry.

In addition to the 12 principles of green chemistry[1] there are three Rs commonly associated with green analytical chemistry (Reduce, Replace, and Recycle).[2] Greening of analysis and purification in support of medicinal chemistry has unique challenges due to the small volumes of materials being utilized and the wide assortment of molecules being analyzed/purified. These challenges have resulted in Reduce and Replace as the two main avenues for improving greenness of analysis/purification.

Topics of reduction covered in this chapter include use of smaller internal diameter (I.D.) columns for high-performance liquid chromatography (HPLC) analysis and transferring analysis and purification methods from HPLC to supercritical fluid chromatography (SFC). Topics of replacement covered in this chapter include substituting dichloromethane with greener solvents for flash chromatography and replacement of alkane based mobile phases for HPLC with carbon dioxide based mobile phase for SFC. Due to small volumes and the wide variety of solvent used in discovery purifications, recycling is often not a viable option in discovery chemistry and will not be addressed in this chapter.

Greening of analytical chemistry has been addressed previously in the literature but these publications are geared towards analytical chemistry in support of pharmaceutical development and manufacturing.[2,3] Greening of chromatographic purification has also been reviewed but covered mainly process scale chromatography.[4] The primary focus of this chapter is greening of analysis and smaller scale purifications encountered in medicinal chemistry support.

## 4.2 SFC *versus* HPLC for Chiral and Achiral Separations

### 4.2.1 SFC Introduction

For over 40 years, HPLC was the most frequently used technique for analysis and purification in support of pharmaceutical R&D. Over the past 15 years supercritical fluid chromatography (SFC) has become a viable alternative for the analysis and purification of small molecules during drug discovery.[5–7] This chapter provides a short introduction to SFC. If the reader is interested in an in-depth discussion on SFC they are directed to the literature.[8,9] With SFC, a majority of the solvent in the mobile phase has been replaced with supercritical $CO_2$. The critical point for $CO_2$ is a temperature of 31 °C and a pressure of 73 atm. Above the critical point, $CO_2$ exists as a supercritical fluid and has properties intermediate between a liquid and a gas. The low viscosity and high diffusivity of the SFC mobile phase allows higher flow rates relative to HPLC, resulting in shorter run times and increased efficiencies. Increased flow rates in SFC have significantly less impact on efficiency compared to HPLC. The use of carbon dioxide based mobile phases (SFC) allows linear velocities two to three times those seen with liquid mobile phases (HPLC) with approximately the same separation efficiency. In addition, the lower pressure drop of carbon dioxide based mobile phases in SFC allows higher linear velocities than possible with HPLC. This increase in flow rate without loss of efficiency results in shorter analysis time as well as higher purification productivities (material purified per unit time) relative to HPLC and other purification techniques. Shorter analysis time increases the number of analyses possible on an instrument; this reduces the number of instruments required in a laboratory, resulting in equipment cost savings as well as a reduction in laboratory space requirements and associated electrical and maintenance costs. The increased preparative productivity of SFC allows compounds to be purified in a shorter time frame, reducing the time required to generate pure compounds for pharmaceutical testing and accelerating the drug discovery process. This advantage is especially evident for medicinal chemistry purifications where the method development and purification cycle is performed thousands of times per year in a typical laboratory.

While supercritical $CO_2$ has a higher solvating power than gaseous $CO_2$, most pharmaceutical compounds are moderately polar and $CO_2$ alone is insufficient for elution from a chromatographic column. In nearly all cases a polar modifier must be added. While any solvent can theoretically be used as a modifier in SFC, most separations utilize methanol, ethanol, isopropanol, or acetonitrile. Addition of a modifier moves the mobile phase from a supercritical to a subcritical state. Even under subcritical conditions the mobile phase has many of the advantages described above. A major advantage of SFC *versus* HPLC is lower organic solvent usage. The lower solvent usage in SFC is achieved by replacing a majority of the mobile phase with $CO_2$. Solvent usage is reduced 2–10 fold compared to HPLC. This solvent reduction is especially critical in preparative purifications where solvent consumption can range from 1–5 L of solvent per gram of processed material. In preparative SFC, $CO_2$ is removed post chromatography by decreasing pressure, leaving only the modifier. This results in higher product concentrations, reducing the time and energy required for post purification solvent removal and product isolation. An additional advantage of SFC relative to reverse phase purification is the use of more volatile mobile phase solvents compared to aqueous based mobile phases typically used in reverse phase HPLC, saving time and energy during the distillation process. SFC is an environmentally preferred technology. $CO_2$ used in SFC is generally recovered as a by-product of manufacturing processes, resulting in no net increase in $CO_2$.[10]

While SFC has many advantages, it does have some disadvantages relative to HPLC. SFC equipment is one-and-a-half to two times more expensive than equivalent analytical and preparative HPLC instruments. Mobile phase delivery is more complex with SFC compared to HPLC. For HPLC mobile phase delivery is as simple as attaching an appropriately sized container to the pump inlet. For SFC a source of carbon dioxide is required. For small scale preparative (1 cm I.D. columns) and analytical equipment, $CO_2$ can be delivered from a cylinder. For larger scale preparative work using higher flow rates, the volume of a standard cylinder is insufficient. Alternative delivery systems such as dewars or tanks with a bulk delivery system must be purchased. Reductions in carbon dioxide requirements can be realized by incorporation of recycling systems into the preparative SFC system. SFC also has additional safety hazards relative to HPLC. The use of large volumes of carbon dioxide may lead to oxygen deficient environments. The use of appropriate safety controls such as ventilation and carbon dioxide monitors and alarms (especially for preparative systems) should be considered. An additional safety consideration is the use of carbon dioxide at high pressures. For analytical SFC, standard HPLC columns can be used. For preparative applications it is critical to use columns rated for pressures encountered under SFC conditions. In addition, the use of plastic or plastic coated bottles with appropriately sized vents for product collection is necessary to avoid over-pressurization and container rupture. The final limitation of SFC is the range of compounds for which it can be used. SFC is not as universal as HPLC, being limited mainly to non-polar to moderately polar compounds, although the addition of water to the modifier has recently been shown to extend the range of polar compounds that can be analyzed and purified by SFC.[11–14]

## 4.2.2 SFC for Chiral Analysis/Purification

While SFC has existed for more than 50 years, preparative enantioseparations have been the main technique utilizing SFC over the past 20 years.[5,7,15] The use of SFC has many advantages for the analysis and purification of racemates. The first step in the analysis or purification of a racemate is method development involving the screening of numerous chiral stationary phases and mobile phases. Using HPLC as the chromatographic technique requires 12–24 h analysis time and up to 1 L of solvent for each compound analyzed. The rapid flow-rates possible with SFC reduce time for screening to 1–2 h and solvent volumes to as low as 150 mL.[5,6] The solvent savings are amplified when these separations are scaled to preparative loadings. In a typical pharmaceutical purification laboratory where 2–20 racemates are resolved per day, the use of SFC drastically increases productivity as well as reducing solvent usage.

The advantages of SFC for preparative enantioseparations is illustrated in the resolution of a thiourea **1** (Figure 4.1). Details of the HPLC separation of this racemate have been previously published.[16] Owing to the short retention of the enantiomers under HPLC conditions (Chiralpak AD, 100% methanol), closed loop recycling was required to achieve baseline resolution of the enantiomers. This separation was also evaluated using SFC. Analytical screening showed the best enantioseparation using Chiralpak AD and 25% methanol. The preparative SFC separation of 350 mg of racemic **1** is shown in Figure 4.1. The short cycle time for the separation allowed an injection cycle time of 1.8 min, greatly increasing the productivity of the separation.

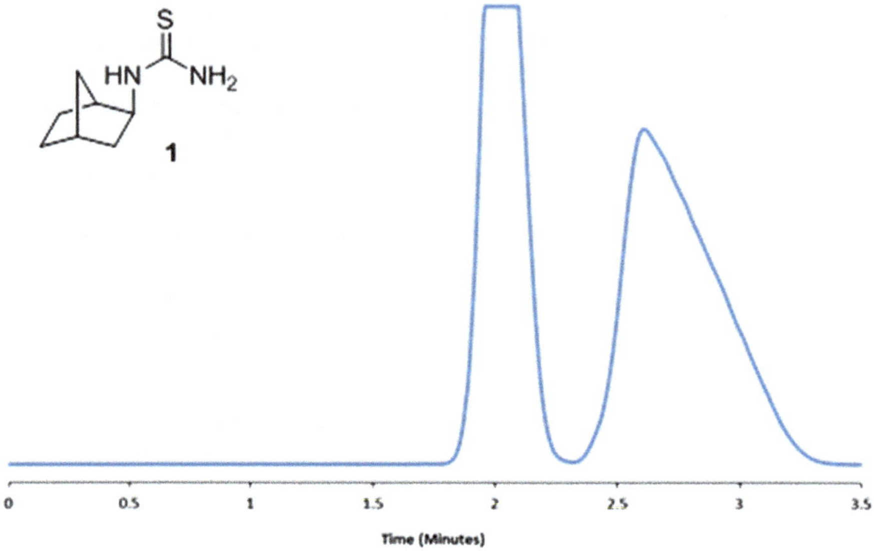


**Figure 4.1** Preparative SFC separation of thiourea **1**. Purification conducted on a Chiralpak AD-H column (150 mm × 50 mm I.D.) with a mobile phase of 25/75 methanol with 0.2% diethylamine/$CO_2$. A flow-rate of 350 mL min$^{-1}$, detection at 225 nm and a loading of 350 mg were used.

**Table 4.1** Productivity and solvent usage for HPLC and SFC separation of thiourea **1**.

| | Productivity (g racemate per kg CSP per day) | Solvent usage (L solvent per g racemate) |
|---|---|---|
| HPLC | 110 | 2.10 |
| SFC | 1600 | 0.44 |

Table 4.1 summarizes purification productivity and solvent usage for HPLC and SFC purification of thiourea **1**. Both purification techniques were optimized to generate product enantiomer (peak 2) at greater than 99% ee and greater than 95% yield. The advantages of SFC over HPLC are clearly demonstrated. Productivity increased approximately 15-fold while solvent usage was reduced more than four-fold.

### 4.2.3 SFC for Achiral Analysis/Purification

Over the past 20 years SFC has evolved to become the preferred method for separation of racemates at the mg to 100s of grams scale required in pharmaceutical discovery. Over the past five years the use of SFC for achiral purification has increased.[17–19] Most achiral SFC purifications utilize UV detection although recent equipment advances allow the use of mass spectrometers to direct collection based on mass.[20–22] Traditionally, flash chromatography or reverse phase chromatography is utilized for achiral purifications performed in support of medicinal chemistry. The limitations of flash chromatography will be discussed later in this chapter. While reverse phase chromatography has advantages of efficiency and relatively easy scale-up from analytical separations, it does have some disadvantages. The main disadvantage is the use of aqueous based mobile phases. Most pharmaceutical compounds have poor solubility in these mobile phases, leading to poor peak shape and/or poor loading. This impacts purity and/or yield of the product as well as the purification productivity. Another disadvantage of an aqueous based mobile phase is the increased time and energy required for distillation. The use of SFC for achiral purifications greatly reduces the amount of solvent required for purification as well as the energy required for distillation to recover purified materials. SFC is a normal phase type separation and offers orthogonal separations and selectivity relative to reverse phase HPLC.

The advantages of SFC for achiral purifications is illustrated in the following example. The target molecule (Compound **A**, confidential structure) was synthesized and purified using reverse phase mass directed preparative HPLC. The synthesis of the target molecule generated two positional isomers that were close eluting by HPLC (Figure 4.2). Purification using reverse phase HPLC generated a 30 : 70 mixture of the isomers.

Normal phase chromatography generally separates isomers better than reverse phase chromatography. As SFC is a normal phase separation process, it was explored for this separation. SFC method development was performed using a protocol previously described.[6] The best analytical separation is shown in Figure 4.3, Chromatogram A. The preparative resolution of 10 mg is shown in Figure 4.3, Chromatogram B. Due to the small sample size,

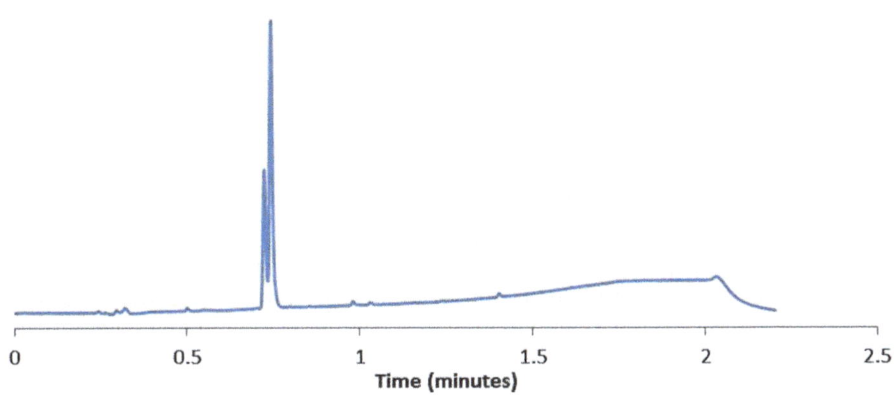


**Figure 4.2** Analytical HPLC separation of compound **A**. Analysis conducted on HALO C18, 2.7 µm, 3 mm I.D. × 50 mm, flow rate of 2 mL min$^{-1}$. Solvent A: water with 0.1% (v/v) trifluoroacetic acid; solvent B: acetonitrile with 0.1% (v/v) trifluoroacetic acid. Gradient from 5% to 95% B over 1.5 min, hold at 95% B for 0.3 min.

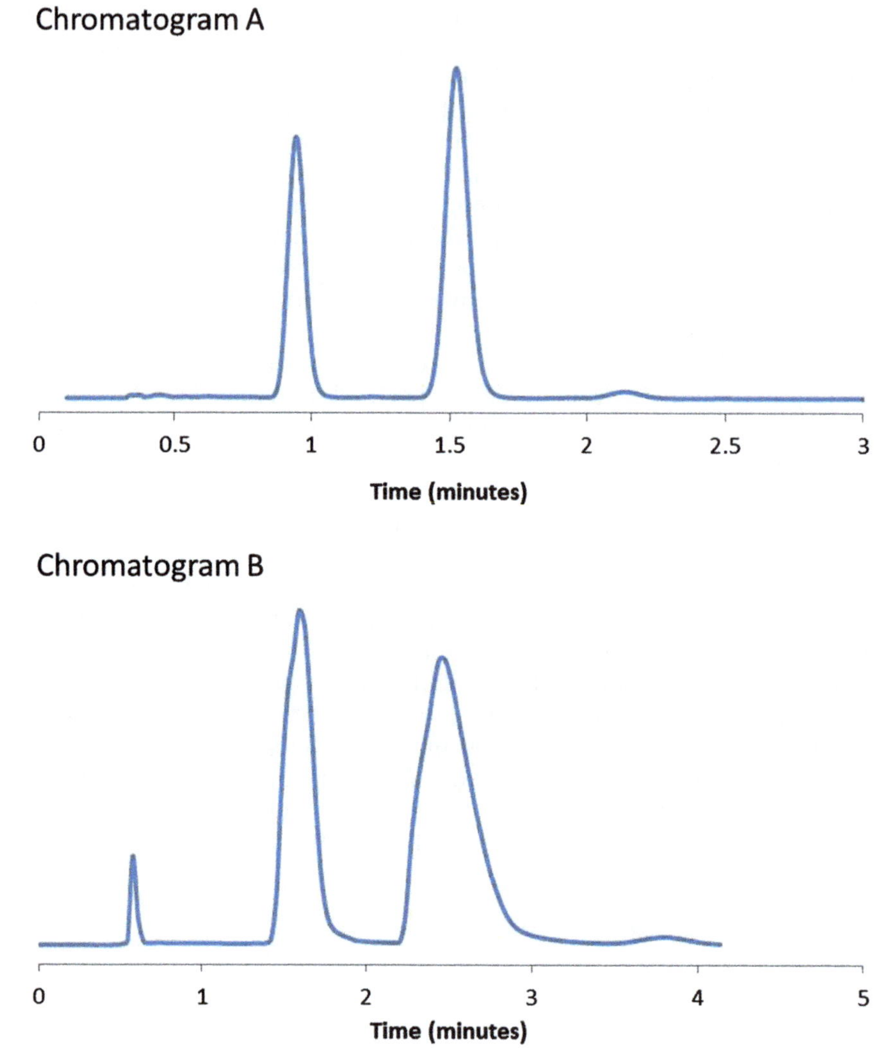


**Figure 4.3** Analytical (top) and preparative (bottom) SFC separation of compound **A**. Chromatogram A: analysis conducted on Chiralcel OJ-H (100 mm×4.6 mm I.D.) with a mobile phase of 55/45 methanol with 0.2% diethylamine/$CO_2$. A flow rate of 5 mL $min^{-1}$ was used. Chromatogram B: purification conducted on Chiralcel OJ-H (150 mm×20 mm I.D.) with a mobile phase of 55/45 methanol with 0.2% diethylamine/ $CO_2$. A flow rate of 80 mL $min^{-1}$, detection at 256 nm and a loading of 10 mg were used.

loading was not optimized for this separation. Even with the non-optimal loading of 10 mg per injection and a cycle time of 3 min, the separation was completed in approximately 30 min, generating both isomers at greater than 99% purity. Product fractions were isolated in less than 250 mL of methanol, requiring ~15 min for distillation. Total time for the purification from sample dissolution to generation of dry products was less than 2 h.

## 4.3 Optimization of Analytical LC/MS Conditions for Reaction Monitoring/Final QC

Purity determination is a critical step during the synthesis of pharmaceutical compounds. Most medicinal chemistry compounds require multiple synthetic steps, each step requiring analysis to provide information on reaction progress as well as purity. Over the course of synthesis of a single molecule anywhere from 5 to 50 separate analyses may be performed depending on the number of synthetic steps and purifications performed. While these analyses can be performed by thin-layer chromatography (TLC), over the past 20 years the advantages of HPLC (speed, chromatographic efficiency as well as molecular weight information that can be generated by coupling HPLC with a mass spectrometer (MS)) has resulted in HPLC (with or without a MS) becoming the preferred choice for analysis. In a modern medicinal chemistry laboratory it is not unusual to have one HPLC or LC/MS system for every five to ten chemists. While a single HPLC system does not generate large volumes of solvent ($\sim$1 L day$^{-1}$) these volumes are considerable when one considers a large pharmaceutical company has anywhere from 50 to 1000 systems.

In the late-1990s when HPLC became a routine analytical technique in medicinal chemistry laboratories, the state of the art for chromatographic columns was five micron particles packed into 4.6 mm I.D.$\times$150 or 250 mm length columns. While these columns provide efficiencies sufficient for this stage of pharmaceutical R&D, compared to today's technology they are inefficient in terms of analysis time, solvent usage, and waste generation. Over the past ten years there have been tremendous advances in column particle technology as well as instrument design that allow the use of smaller and/or shorter columns packed with smaller, more efficient particles that are operated at higher pressures and/or flow rates.[23]

Use of this new particle technology reduces solvent as well as the time required for a standard analysis. These advantages are shown in Figure 4.4. The three chromatograms show the separation of a standard test mixture utilized for column and system testing. The chromatographic conditions for all three methods have approximately the same chromatographic resolving power (peak capacity of 130 $\pm$ 10%). Excluded from this discussion are sub-two-micron columns. While these columns offer excellent efficiency, they operate at extremely high pressures and require equipment not routinely available in a medicinal chemistry laboratory. The reader is directed to the literature for additional information on sub-two-micron columns.[23]

Chromatogram A represents state of the art HPLC conditions from approximately 15 years ago. The method utilizes a five micron particle size and column dimensions of 4.6$\times$150 mm. These types of columns are still

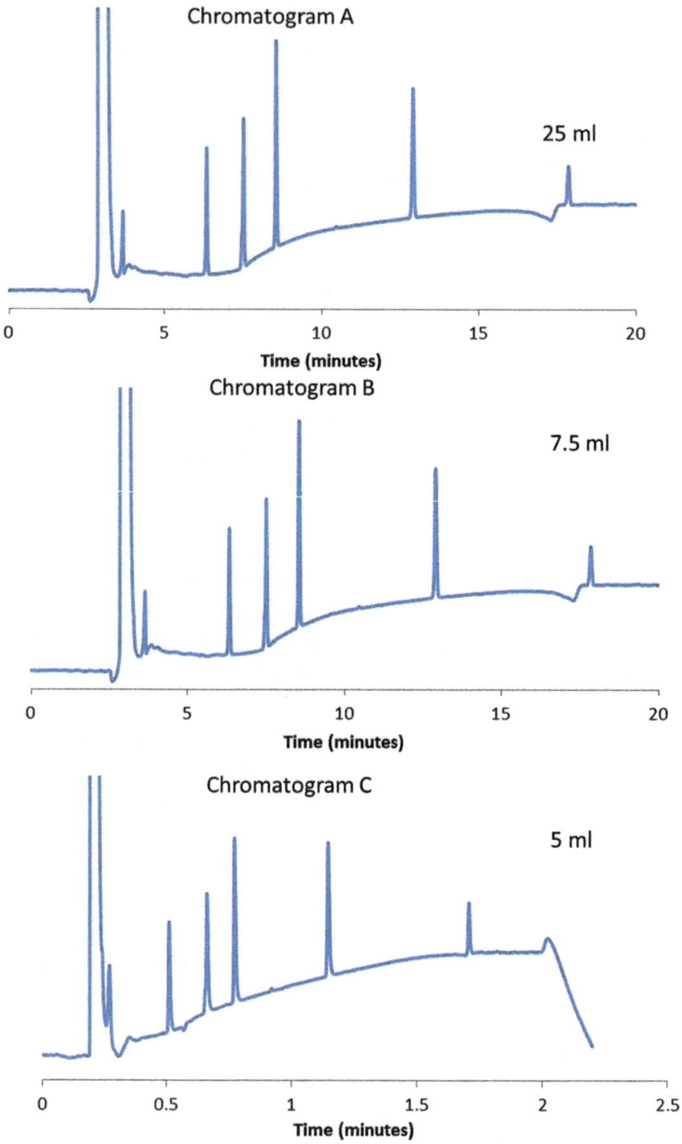


**Figure 4.4** Analytical HPLC separation of Amgen test mixture containing caffeine, glafenine HCl, labetalol HCl, chrysin, and di-*N*-pentyl phthalate. Solvent A: water with 0.1% (v/v) trifluoroacetic acid; solvent B: acetonitrile with 0.1% (v/v) trifluoroacetic acid. Chromatogram A: analysis conducted on Phenomenex Gemini, 5 micron, 4.6 mm I.D. × 150 mm, with a flow rate of 1 mL $min^{-1}$. Gradient from 5% to 95% B over 15 min, hold at 95% B for 5 min. Chromatogram B: analysis conducted on Zorbax SB-C18, 2.7 micron, 3 mm I.D. × 50 mm, with a flow rate of 1.5 mL $min^{-1}$. Gradient from 5% to 95% B over 3 min, hold at 95% B for 0.5 min. Chromatogram C: analysis conducted on HALO C18, 2.7 µm, 3 mm I.D. × 50 mm, with a flow rate of 2 mL $min^{-1}$. Gradient from 5% to 95% B over 1.5 min, hold at 95% B for 0.3 min.

routinely utilized in many pharmaceutical laboratories. Chromatogram B represents state of the art from approximately 8–10 years ago and utilizes a 3×50 mm column packed with 3.5 micron particles. The final chromatogram (Chromatogram C) represents current state of the art chromatographic efficiency (excluding sub-two-micron particles). This method utilizes a 3×50 mm column packed with 2.7 micron superficially porous particles. The 2.7 micron superficially porous particles are comparable in performance to sub-two-micron particles without the large pressure drop.[24] While each method is chromatographically equivalent they utilize drastically different amounts of solvents: Chromatogram A requires 25 mL, Chromatogram B requires 7.5 mL, and Chromatogram C just 5 mL of solvent (each volume includes solvent for column equilibration) for each analysis. The other advantage of smaller particle size columns is analysis time. Cycle time for the three methods is 25, 5, and 2.5 min, respectively. Switching from 5 micron to 3.5 micron columns reduces cycle time five-fold, with an additional two-fold reduction possible with superficially porous particles. The reduction in analysis time permits more analyses per hour, allowing, in turn, a reduction in the number of HPLC systems needed in a medicinal chemistry laboratory. Besides capital and maintenance savings the laboratory would also see a significant decrease in electrical requirements with a reduction in equipment numbers. Finally, an increase in efficiency may be realized due to the chemist not having to wait as long for data to proceed to the next activity.

The methods and chromatographic performance described above provide efficiencies suitable for final product analysis for discovery pharmaceuticals. Review of usage data shows the majority of HPLC analyses performed for medicinal chemistry support are not for final product analysis prior to biological testing but for reaction monitoring, fraction analysis of fractions from chromatographic purification, and analysis of raw material or intermediates. For these analyses the chromatographic performances demonstrated in Figure 4.4 may be more than is required. A lower efficiency method that utilizes even less solvent and affords shorter analysis times may be better suited for these types of analyses. A "Fast LC" method may be better suited for these analyses. One such method is identical to that described in Figure 4.4, Chromatogram B, but utilizes a 3×30 mm column *versus* a 3×50 mm column. This method provides a peak capacity of approximately 100, which is acceptable performance for these types of analyses. The main advantage of this method is a fast cycle time (~1.5 *versus* 2.5 min) while generating only 3.75 mL solvent waste per analysis. When developing an HPLC method it is critical to define the resolving power and separation efficiency required. A school of thought is that one can never have too much efficiency in a chromatographic method. Users need to remember increased efficiency is not free – there is a cost in terms of solvent used and discarded as well as analysis time.

Currently the vast majority of chromatographic technology utilized for analysis within medicinal chemistry laboratories is liquid chromatography

(LC) based. As discussed earlier in this chapter, use of this technology requires significantly higher solvent volumes relative to SFC. While SFC technology has begun to be located within medicinal chemistry laboratories, it is only currently being used in a handful of companies.[25] The main limitation with regard to replacing reverse phase HPLC is the restriction on the range of molecules that can be analyzed by SFC. Using reverse phase HPLC, molecules of a wide polarity and molecular weight range can be easily analyzed. SFC is limited to low to moderate polarity compounds although recent work utilizing water as an additive has increased the polarity range of molecules that can be successfully analyzed.[11,13] The other advantage of reverse phase HPLC is the existence of a universal stationary phase (C8 or C18) that can be used to analyze a wide variety of molecules. Currently, a universal stationary phase does not exist for SFC, often resulting in multiple phases to be explored for each separation. It is the authors' hope that current research in this field will allow increased use of SFC by medicinal chemists.

## 4.4 Optimizing Reverse Phase Purifications

Biological testing of medicinal chemistry compounds requires sufficient chemical purity to ensure observed biological activity is due to the compound of interest and not due to impurities. Purity requirements vary across the pharmaceutical industry but typical minimum purities of greater than 85–95% are necessary for initial testing. To reach these purity levels purification is often required. During the drug discovery phase of a project compounds are prepared on a small scale (<100 mg) and are often synthesized only once. It is not time or cost-effective to develop a crystallization process for each molecule. The standard purification process in pharmaceutical discovery is either flash or reverse phase purification.

In many pharmaceutical companies open-access reverse phase systems are available in the medicinal chemistry laboratories for compound purification. These instruments are co-located with chemists and the purifications are performed by medicinal chemists. Standard purification processes and purification tools are often developed to maximize purification success and minimize solvent requirements. Prior to purification, each sample is analyzed by LC/MS to confirm the amount of product present in the sample as well as retention time of the desired peak. The generic LC conditions utilize a 5–95% acetonitrile gradient to cover the wide polarity range of synthesized molecules (Figure 4.4). This type of gradient is not ideal for purification for several reasons. The largest potential issue is sample precipitation when injecting chemicals with low water solubility into a highly aqueous environment. Choosing appropriate starting gradient conditions can minimize risk of precipitation. Within Amgen medicinal chemistry an "analytical to prep prediction tool" has been developed (Figure 4.5) and is used throughout the medicinal chemistry laboratories. Using the analytical retention time of the desired product the chemist can choose a focused gradient that will minimize the risk of precipitation as well as provide

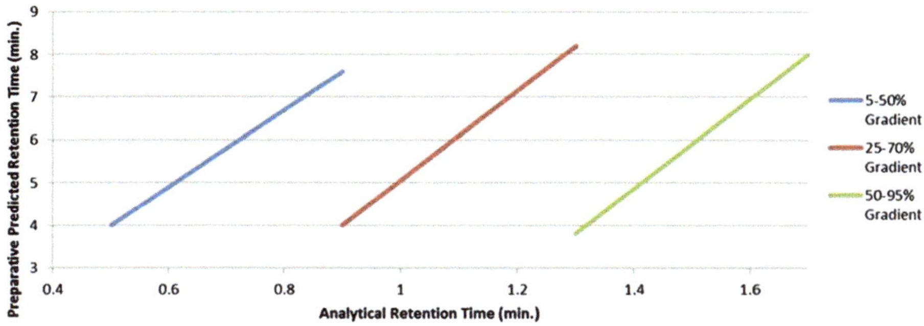


**Figure 4.5** Analytical to preparative prediction tool.

preparative retention sufficient to maximize separation efficiency.[26] For example, if the analytical retention time for the product of interest is 1.1 min, a preparative gradient of 25–70% acetonitrile with 0.1% (v/v) trifluoroacetic acid (TFA) in water with 0.1% (v/v) TFA would correlate to a preparative retention time of approximately 6 min. Use of this tool within Amgen has significantly reduced the incidence of sample precipitation as well as improved the first pass success rate for reverse phase open-access purifications. This is important not only for maintaining chemist efficiency by avoiding repeat purifications; it also reduces the amount of solvent waste generated during repeat purifications.

Sample dissolution in preparative reverse phase purification can significantly affect peak shape, impacting product purity and yield. For maximum efficiency it is desirable to dissolve samples in the HPLC mobile phase. As many pharmaceutical compounds have poor aqueous solubility, this is often not an option; instead, the standard procedure is to dissolve samples in DMSO or other polar solvent. Injection of a strong solvent into the chromatographic mobile phase can drastically impact peak shape, especially for early eluting peaks. Breakthrough of material can also occur, with part of the sample being non-retained, leading to significant impact on yield. This effect is illustrated in Figure 4.6, Chromatogram A. Concentrations were varied to maintain a constant amount of sample injected. With a 1 mL DMSO injection, suitable peak shape is observed. As the injection volume is increased, increases in peak width and poorly shaped peaks are seen until at 5 mL fronting peaks are observed for all peaks. Under these conditions it would be difficult to isolate material at high purity and recovery.

The impact of large volume injections can be decreased by increasing the column size or utilizing At-Column Dilution.[27] While At-Column Dilution works well, it requires the addition of a third pump as part of the sample injection process. Depending on the instrument being utilized, it may not be possible to control an additional pump through the system software. This was the case for Amgen's open access reverse phase purification equipment. To minimize the impact of large volume injections on chromatographic performance, a larger column (3 cm I.D. $\times$ 150 mm) was utilized.

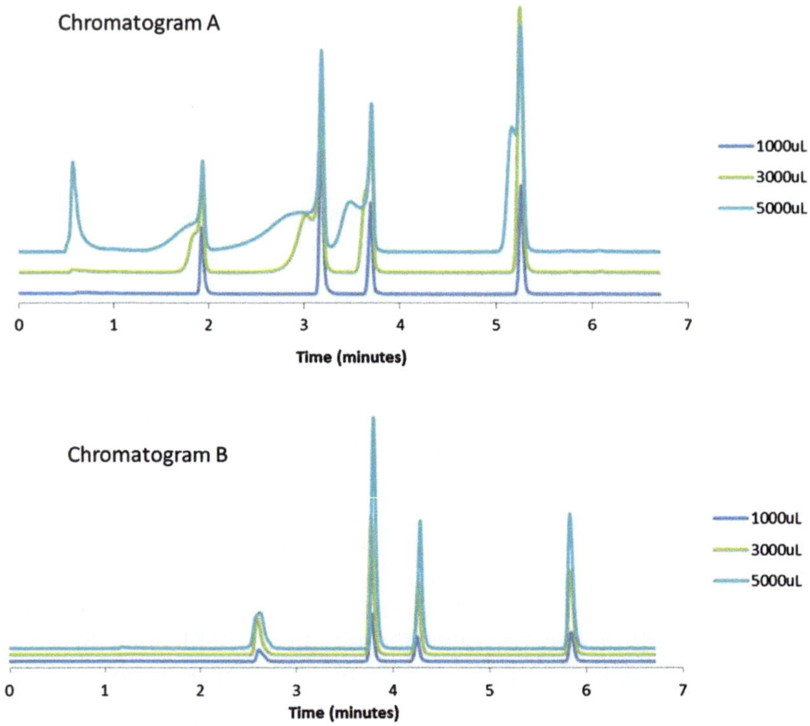


**Figure 4.6** Preparative HPLC separation of butyl 4-aminobenzoate (2 mg mL$^{-1}$), caffeine (1 mg mL$^{-1}$), carbamazepine (2 mg mL$^{-1}$), and 4-(phenylazo)-diphenylamine (1 mg mL$^{-1}$). All samples dissolved in DMSO. Solvent A: water with 0.1% (v/v) trifluoroacetic acid; solvent B: acetonitrile with 0.1% (v/v) trifluoroacetic acid Chromatogram A: purification conducted on XBridge C18, 10 micron, 19 mm I.D.×100 mm, flow rate of 40 mL min$^{-1}$. Gradient from 5 to 95% B over 7 min Chromatogram B: same conditions as chromatogram A except for the addition of a 25 mm I.D.×100 mm column packed with 50 micron silica gel prior to the C18 column.

At Amgen a technique called "In-Column Dilution" has been developed that reduces the effects seen in Figure 4.6, Chromatogram A without requiring a larger preparative column.[28] This technique involves insertion of a silica column prior to the reverse phase column. The silica column, packed with 50 micron particles, is approximately the same dimensions as the reverse phase column and provides the desired dilution of the DMSO injection. This effect can be seen in Figure 4.6, Chromatogram B. Even at 5 mL injection volumes, excellent peak shape is seen for all four peaks. Slight peak distortion is seen for early eluting peaks. Increased retention is seen due to extra system volume added by the silica column. Incorporation of this technique into standard open access reverse phase purification systems at Amgen has eliminated the need for multiple injections for poorly soluble samples as well as improved chromatographic purification efficiency.

**Table 4.2** Comparison of open access reverse phase purification methods.

| | Old method | New method |
|---|---|---|
| Column dimensions (I.D.×L, mm) | 30×150 | 19×100 |
| Column cost (US$)[a] | 5325 | 2110 |
| Flow rate (mL $min^{-1}$) | 40 | 40 |
| Cycle time (min) | 25 | 6.7 |
| Acetonitrile (mL) | 464 | 124 |
| Water (mL) | 536 | 144 |
| Waste generated (mL) | 1000 | 268 |
| Approximate cost/injection (US$)[a] | 26 | 9 |

[a]2011 costs.

A comparison of the improved purification process to the old method shows impressive solvent and cost savings (Table 4.2). Both methods produce approximately the same chromatographic performance. The use of a smaller column results in a reduction in column costs of approximately 65%. Significant solvent reductions were also realized (1 L per sample to 268 mL per sample). In addition, cycle time was reduced from 25 to 6.7 min due to smaller column size. This allowed the number of purification systems within Amgen to be reduced without impacting chemistry productivity, further reducing operating and ongoing maintenance costs. Purification costs for column and solvent were reduced approximately 65%. Finally, with the new method chromatographic peaks were sharper with products eluting in lower volumes. This reduces the volume of aqueous mobile phase to be evaporated during product recovery, resulting in lower energy costs.

Another technique that can be utilized to optimize reverse phase purification processes is mass directed purification.[29–31] Mass directed purification utilizes a mass spectrometer (MS) as a detector. The MS signal directs fraction collection such that only the mass of the desired product is collected. The main advantage of this approach is that the only material collected is the product of interest. Using UV directed purification, any peak above a set threshold will be collected, resulting in multiple fractions which must then be analyzed to determine which contain the desired product. The use of mass directed purification eliminates the need for LC/MS analysis (and the resulting solvent usage) to determine the peak of interest. In addition, collection of only one peak reduces the number of test tubes required for a purification by 2–10 fold.

The advantages of a mass directed purification approach are illustrated in Figure 4.7. Chromatogram A is the analytical HPLC analysis of the reaction mixture containing desired compound **B.** Analysis shows two major peaks as well as numerous minor peaks; MS data indicated the second major peak eluting at approximately 1.3 min to be the product of interest. This material was purified using reverse phase chromatography; the UV trace is shown in Figure 4.7, chromatogram B. If UV response was used to direct purification, 4–8 peaks would be collected, with each fraction requiring LC/MS analysis to determine the desired product. The use of mass directed purification greatly

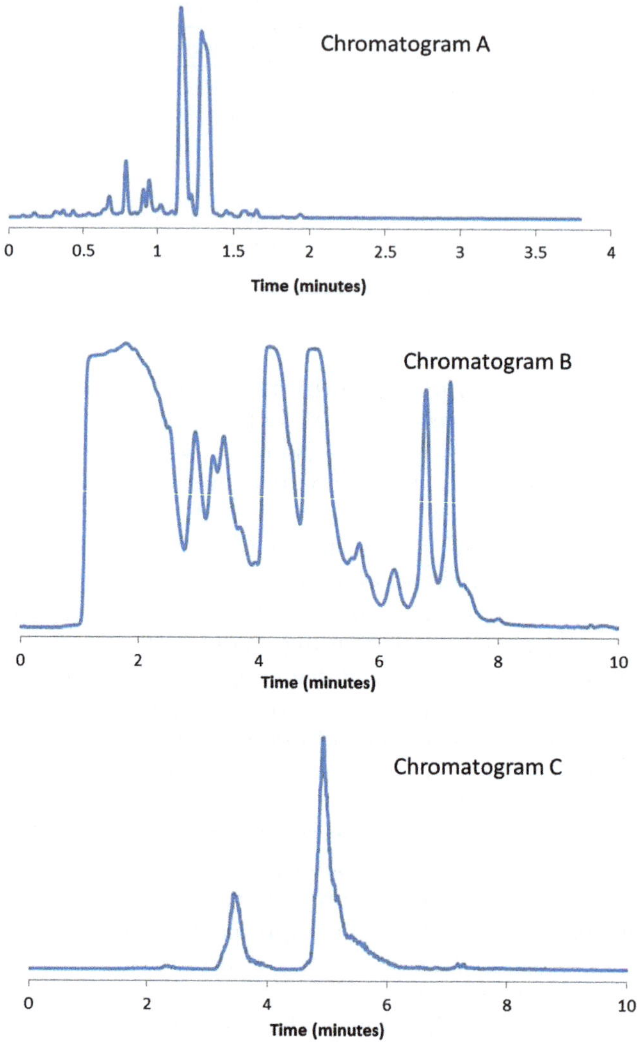


**Figure 4.7** Analysis and purification of Compound **B**. Solvent A: water with 0.1% (v/v) ammonium hydroxide; solvent B: acetonitrile with 0.1% (v/v) ammonium hydroxide. Chromatogram A: analysis conducted on XBridge C18, 2.7 micron, 3 mm I.D.×50 mm, with a flow rate of 1.5 mL min$^{-1}$. Gradient from 5% to 95% B over 3 min, hold at 95% B for 0.5 min. Chromatogram B: 200 mg sample dissolved in 2.5 mL DMSO, purification conducted on XBridge C18, 10 micron, 19 mm I.D.×100 mm, with a flow rate of 40 mL min$^{-1}$, detection at 215 nm. Gradient from 10% to 40% B over 10 min. Chromatogram C: ES negative, SIM = 492.03.

simplifies the collection process. The single ion trace (M + 492) is shown in Figure 4.7, chromatogram C. The trace shows two signals with the desired mass at 3.6 and 5 min. The smaller peak at 3.6 min is due to an isomeric

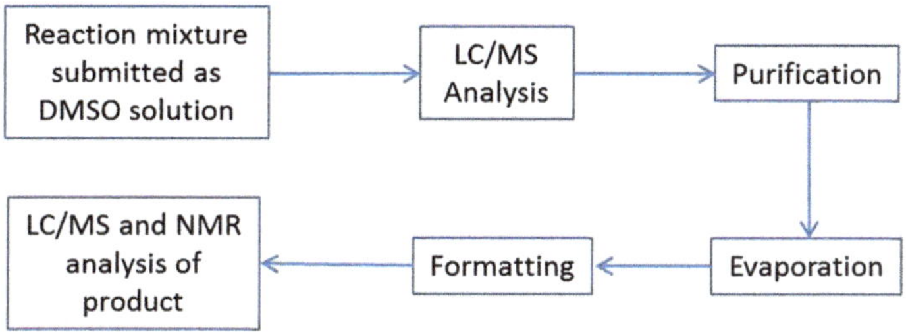


**Figure 4.8** Singleton process flowchart.

impurity in the reaction mixture. Mass directed purification resulted in only two fractions being collected. Based on response and retention time it was easily determined the desired product, compound **B**, eluted at 5 min. Use of mass based collection also allows run termination once the product is collected. Using run termination, once the mass of interest is collected, the solvent gradient ends and the column is flushed to elute any remaining material. Depending on product retention time this may allow up to 50% reduction in solvent usage per purification. Mass directed HPLC technology has been available for approximately 15 years. Only recently has mass directed SFC technology been introduced that will allow the advantages described above to be realized also for SFC purifications.[32]

To facilitate reverse-phase purification at Amgen, a process called the Singleton process has been implemented in the discovery laboratories in Cambridge, MA. This process was modelled after a purification workflow developed at Pfizer[33] and incorporates many of the techniques discussed above. The Singleton process utilizes mass directed purification as well as a high level of automation in the drying, formatting, and characterization steps to reduce solvent waste as well as minimize time for these steps. This process returns purified product, dried and ready for submission to biological testing, along with a complete package of characterization data required for compound submission to the sample bank. The process is summarized in Figure 4.8. Use of this approach can reduce scientist time for purification, drying, characterization, and formatting from 2–3 h to less than 30 min per sample. In addition to increased time efficiency, the use of this process reduces the amount of normal phase silica gel purifications (flash chromatography) performed by the chemists and thus contributes to reduction in organic solvent requirements.

## 4.5 Greener Solvent Alternatives and Solvent Reduction for Flash Purification

### 4.5.1 Flash Purification Introduction

During small molecule pharmaceutical drug discovery thousands of compounds are synthesized in an attempt to prepare a molecule with the desired biological and physical attributes for progression to human testing. A significant portion of this chapter has been dedicated to optimizing the reverse-phase and SFC purification and analysis of final compounds prepared in drug discovery. Each of these compounds requires multiple synthetic steps and purifications. Since it is usually not feasible to purify larger amounts of intermediates by reverse-phase HPLC (due to need to evaporate large volumes of aqueous mobile phase) or SFC (due to the time constraints of sending chemicals to the purifications group), the workhorse purification technique for discovery research is normal-phase flash chromatography. As a result of the frequent use of flash chromatography, it is the largest source of organic solvent waste in drug discovery laboratories. Despite the significant solvent waste generated, flash chromatography has numerous advantages. Flash chromatography separation conditions are easily explored *via* thin layer chromatography (TLC) and scale-up is easy and accurate from TLC plates to flash columns. The availability of relatively inexpensive, automated, reliable, and easy to use equipment coupled with disposable cartridges have made compound purification by silica effective and time efficient. The automated machines that are used most frequently in drug discovery research are medium-pressure liquid chromatography (MPLC) machines that have UV detection as the source of automatic peak collection (some newer systems also have mass-based detection, but those are less common at this time). The upside of these advances is a fast, easy process for chromatographic purification. The downside is this ease means less incentive exists to explore alternatives to chromatography that might lead to avoiding flash chromatography and thus less solvent waste production.

### 4.5.2 Reduction of Organic Solvent Use During Silica Gel Chromatography

The largest reduction of organic solvent use in purification would result from avoiding normal-phase flash chromatography by using techniques such as crystallization or reaction telescoping. The use of these methods can ultimately lead to time savings when intermediates are scaled up, but they are not discussed in this chapter to maintain focus on purification methods. The topic of avoiding flash chromatography through the use of crystallization or reaction telescoping has been previously reviewed.[34] Because automated flash chromatography is convenient, fast, and widely implemented in drug discovery research, it is unrealistic to expect chemists to avoid this method altogether for compound isolation. With that in mind, there are several easily implemented methods to reduce solvent use in silica flash chromatography such as the use of smaller columns packed with high efficiency silica gel, real-time monitoring of peak elution, and gradient optimization. While the smaller particle and/or spherical silica gel columns are generally more expensive, they can be reused without significant effort (see Chapter 8, Richter). The superior separation provided by these columns reduces solvent usage and required separation time. Furthermore, a report by Stevens and Hill indicates that pre-packed disposable columns do not require pre-equilibration with solvent, unless they are larger than 120 g size, which can also save on solvent.[35] To avoid unnecessary fraction tube use, automated MPLCs can be configured to collect only when peaks are detected by the UV detector. The chemist can assay the collected peaks in real-time to determine if the compound of interest has already eluted from the column and thus stop the run when the desired compound has been collected. This action will conserve fraction tubes, solvent, and chemist time.

Optimizing the gradient for compound elution is also a useful means to reduce solvent use, save time, and obtain the best separation for the desired compound. Often, chemists who have not assayed their reaction by TLC will run a default gradient from 100% non-polar eluent to 100% polar eluent (*e.g.* 100% heptanes → 100% ethyl acetate (EtOAc)) to ensure that the desired compound elutes. This is generally unnecessary and wastes solvent at both extremes of the gradient. A shallower gradient can often provide superior separation to a steep gradient that goes from one polarity extreme to the other. A simple formula given in ref. 35 provides an easy starting point for gradient development. Ideally, the desired compound has an $R_f$ between 0.1 and 0.5 in the chosen solvent mixture with $X$% of the polar solvent. The machine would be programmed to start the gradient at $X/4$, increasing the % of polar eluent to $2X$ over ten column volumes (CVs). The run would then finish with holding at $2X$ for 1 CV. For example, if a compound had an $R_f = 0.2$ in 20% EtOAc in heptanes, the gradient would start at 5% EtOAc and increase to 40% EtOAc over 10 CV, then hold at 40% EtOAc for 1 CV. Once efficient separation conditions have been developed for one analog in a series, the method can be saved on the machine for use on subsequent related analogs. Furthermore, recent advances in MPLC machine software packages include tools to analyze and optimize separation conditions on previously performed runs so that a more optimal method may be applied to subsequent separations.

### 4.5.3 Reducing Dichloromethane (DCM) Usage

Flash chromatography is one of the largest contributors to the use of DCM in the chemistry laboratory due to its property as a non-polar solvent with high dissolution power for many organic compounds. DCM is toxic to humans and the environment and is challenging to dispose of because its low flammability hinders incineration.[36,37] For these reasons, reduction of DCM use is a primary goal for many green chemistry programs. Amgen, Genentech, Glaxo-Smith-Kline, Pfizer, and Merck all have programs within Discovery Chemistry aimed at reducing DCM consumption.

DCM is commonly used in conjunction with methanol (MeOH) to purify heterocyclic compounds using flash chromatography, with or without acidic or basic modifiers. The routine use of DCM/MeOH for heterocycle purification presents an obstacle with regard to reducing DCM consumption. The main challenge in reducing DCM usage in chromatography is encouraging behavioral change (Chapter 2, Sneddon). Overcoming the inherent comfort level that chemists have developed with a solvent system such as DCM/MeOH that they have likely been using for many years is difficult. It is the opinion of the authors that once several chemists within a department become sufficiently familiarized with an alternative and greener solvent system, the adoption of that solvent system will increase as others see successful implementation of the system amongst their peers and teammates.

Toward achieving this behavioral change, solvent guides have been published that aim to help chemists reduce the use of DCM and replace it with greener solvent blends that will provide equal chromatography performance.[38,39] Figure 4.9 is a reproduction of an empirically derived solvent guide specifically tailored at converting a DCM/MeOH solvent system into greener alternative solvents.[38] The use of a blend of 3 : 1 EtOAc : EtOH as a polar eluent appears to be the most universally applicable to the range of polarity covered by DCM/MeOH. The 3 : 1 EtOAc : EtOH solvent blend can also be used with basic and acidic additives and is now commercially available from several suppliers including Aldrich and Burdick and Jackson. The alternatives mentioned above can provide adequate and at times superior alternatives to DCM. If solubility problems arise, a small amount of DCM can be mixed with the non-polar eluent, or, alternatively, many automated MPLC machines allow the use of a third additive solvent in addition to the polar/non-polar eluents used for a gradient. If DCM is required to solubilize the compound, using as little as 5% DCM as an additive can be sufficient to improve solubility. Even if a small percentage of DCM is added to improve compound solubilization, this technique still leads to a significant reduction in DCM use. To encourage adoption of greener solvent blends such as 3 : 1 EtOAc : EtOH, it is recommended that pre-mixed or purchased ready-to-use 4 L bottles are provided in research laboratories.

Another alternative to replacing DCM using normal-phase flash chromatography is to employ the use of reverse-phase MPLC. Many of the newer MPLC systems now employ pumps that can handle significantly higher pressures and can be readily switched between normal and reverse-phase modes with a simple solvent line flush. Additionally, reverse-phase columns now exist that can be used several times and then discarded without significant cost. There are both advantages and disadvantages to using the reverse-phase MPLC in the preparation of drug discovery compounds. The major advantage is the ability to purify very polar intermediates on a larger preparative scale. In general, reverse-phase HPLC will give superior separation to a reverse-phase MPLC, but will require multiple injections if a significant amount of material needs to be purified, as would be the case for a synthetic intermediate. Before the advent of reverse-phase MPLC, the

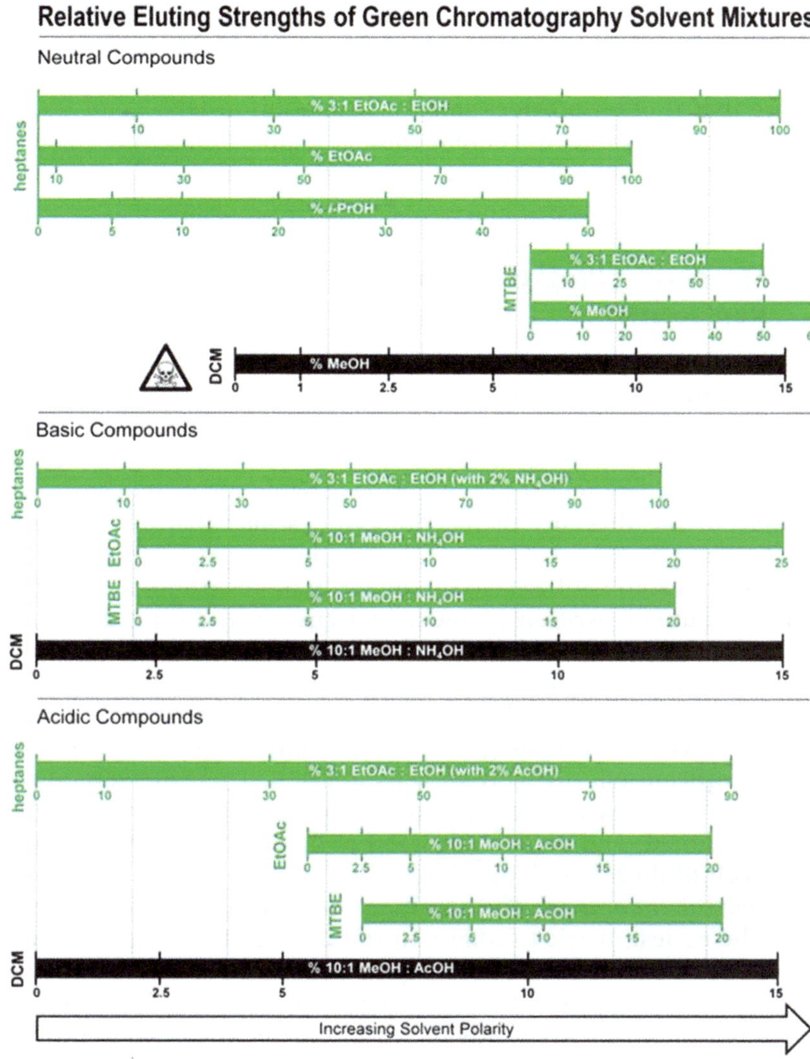


**Figure 4.9** TLC solvent guide to aid in replacing DCM in chromatography. Reprinted from ref. 38, with permission from the Royal Society of Chemistry.

chemist would have to decide between the choices of multiple HPLC injections, crystallization (still the best option from a Green Chemistry perspective), or a very long and solvent intensive normal-phase flash column likely employing the traditional DCM/MeOH solvent system with an acidic or basic modifier. Using the reverse-phase MPLC, a large amount of material can be loaded onto a column in a manner similar to normal-phase flash chromatography, in this case using Celite® instead of silica for any dry-loading or pre-column options. A range of solvents can be used as eluents, but standard HPLC solvent combinations (water, MeOH, or acetonitrile with TFA) work well in this application. The disadvantages of reverse-phase MPLC are that it often provides inferior resolution to HPLC and fractions obtained contain an aqueous component that requires extra time and energy for concentration compared to organic solvents.

### 4.5.4 Measuring Progress in Solvent Reduction and DCM Replacement

Metrics are the key to understanding if efforts to introduce greener practices in the laboratory have been effective. For chemists in manufacturing or process disciplines, this is more straightforward because of good manufacturing/laboratory practices (GMP/GLP) documentation requirements. There are few metrics that can be reliably measured in drug discovery due to the diversity of reactions and scales that are performed. Green chemistry metrics such as E-factor or Process Mass Intensity (PMI) are difficult to establish because the majority of electronic notebooks (or procedures routinely written by medicinal chemists) do not quantify the amount of solvent used in the chromatography step of the compound preparation. Since this is likely the largest contributor of waste for the synthesis of a compound, these metrics are inaccurate unless they capture the solvent used in chromatography. One metric that can be measured without significant effort is solvent consumption. There are several methods by which to measure solvent consumption: (1) solvent ordered, (2) solvent consumed, and (3) solvent waste. The advantages and disadvantages of each of these methods are described below.

*Measuring solvent ordered*: The advantage of measuring solvent ordered is that it is the easiest metric to collect since tracking of solvent both stocked and removed from a location is not required. For companies or laboratories that do not have an electronic inventory system in place, this is likely the best option. Although solvent ordered is an easy metric to collect, it is not often very accurate when considering that solvents are generally ordered in bulk and may remain in storage for lengthy periods before consumption. This means that the tracking of the ordering data would have to span a significant time period in order to have reasonable accuracy. This measurement method also leads to large fluctuations in the numbers, making it difficult to make timely comparisons between consumption of different solvents at any specific time.

*Measuring solvent consumed*: The most accurate means to measure solvent consumption is to track individual solvent bottles as they are removed for use from a common store room by the chemist. The biggest challenge to this method is determining how to capture what is replenished or taken from this stock room over a given period of time. The easiest way to overcome this challenge is through an electronic inventory that tracks when bottles are added and removed from a particular location. If an electronic system is not in place, a manual counting of the bottles added and removed would be required, which could provide a significant obstacle to this practice. Many companies and universities have switched to barcoded electronic inventory systems to maintain accurate inventory for safety, purchasing, and fire codes. Successful implementation of this practice requires chemist compliance in scanning out bottles from the stocking area to maintain an accurate inventory. The benefits of such a system are significant because the data are collected in real-time and can be analyzed in various ways because of the amount of data provided (chemist, location, amount, and date consumed).

*Measuring solvent waste*: This metric is likely the least useful because it provides little information about the identity of solvent disposed of as waste. Even if organic solvent wastes were segregated into halogenated and non-halogenated waste streams, this measurement would only provide a rough guide as to the solvent consumed. Although this method is consistent with the spirit of determining actual waste, the accuracy of this practice is questionable due to evaporation considerations as well as contamination with aqueous wastes, inorganic salts, and organic compounds. Additionally, the measurement would have to take into account the weight of the collection vessel, which may vary greatly and is unlikely to be tared before use. For these reasons, it is not recommended that solvent consumption be measured by outgoing waste.

As outlined in the discussion above, the best metric to collect is actual solvent consumed. In 2010, the Medicinal Chemistry department at the Amgen Massachusetts site began tracking solvent metrics for Green Chemistry using the electronic inventory system already in place at the site. Comparison of the years 2010 and 2013 revealed a reduction in absolute DCM consumption of 64%. Determination of the absolute consumption of DCM is a valuable metric to understand the scale of reduction to measure impact; in this case the scale of reduction is on the order of thousands of liters of DCM. However, there are drawbacks to relying solely on absolute consumption as a measurement of reduction. A valuable alternative metric is the measurement of DCM consumption as a percent of total solvent consumed. The reason for this is that personnel changes and productivity changes will not be captured by simply measuring the absolute consumption of DCM. If a reduction in force or chemist attrition occurs, there is no way to determine if the DCM reduction measured was as a result of Green Chemistry practices or simply fewer chemists working in the laboratories. Thus, tracking DCM consumption as a percent of total solvent gives an indication as to what choices chemists are making. Furthermore, tracking all of the solvent consumed using an electronic inventory system does not require many additional resources and can provide additional data for analysis.

As an illustration of the points above, Figure 4.10 shows the DCM consumption as a percent of total solvent over the same time period

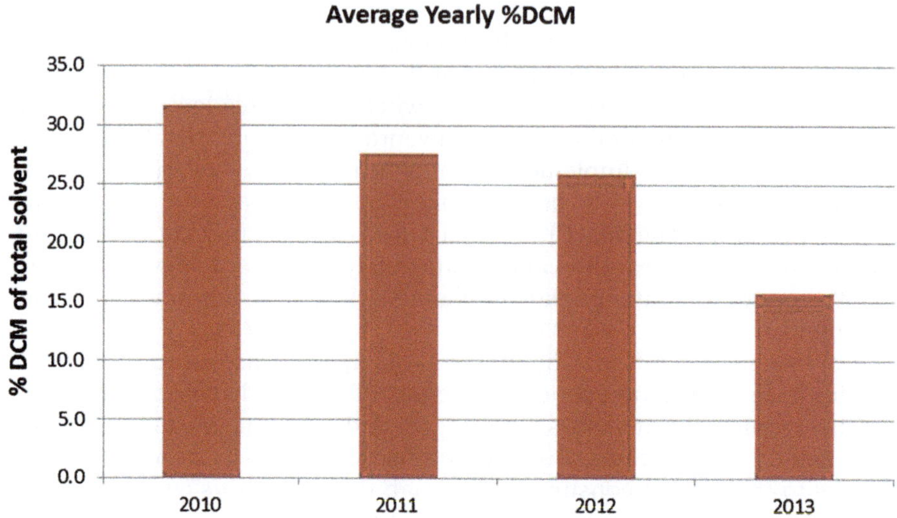


**Figure 4.10** Measurement of the percent DCM consumed compared to total solvent in the Drug Discovery laboratories at the Amgen Massachusetts site.

(2010–2013) for the Amgen Massachusetts Medicinal Chemistry Department. During this time period, there was a small decrease in the number of chemists within Medicinal Chemistry. For the year 2010, the amount of DCM consumed of total solvent was 32%. In the year 2013, this value was reduced to 16%, which means that the chemists used roughly 50% less DCM when measured as a percent of total solvent consumed. The numbers presented give an illustration of how different results can be obtained depending on how the metrics are analyzed. In looking at the graph in Figure 4.10, interestingly, a significant reduction occurred between the years 2012 and 2013. It was expected at the outset of the Green Chemistry program implemented in 2010 that the magnitude of DCM reduction would decrease each year as the easiest behavioral changes were implemented early. Upon reviewing these results shown in Figure 4.10, it was determined that the stocking of ready-made solutions of 3 : 1 EtOAc : EtOH in January of 2013 and the visible posting of the chromatography TLC guide shown in Figure 4.9 were responsible for this increased reduction of DCM usage.

## 4.6 Flash SFC as Alternative to LC Based Flash Purification

As discussed in Section 4.5, flash chromatography is the major source of solvent waste in a medicinal chemistry laboratory. While flash chromatography can be made greener by the solvent replacement strategies described above, the process still uses substantial amounts of solvent. SFC has been used to reduce solvent requirements for chiral and achiral purifications, but to date it has not been utilized as an alternative for flash LC purifications.

In 2010 a feasibility study comparing flash SFC to flash LC was performed in Amgen laboratories.[40] At the time of this work, most preparative SFC utilized small particle size spherical stationary phases (5–10 micron) compared to the larger irregular shaped particles typically used in flash chromatography (40–100 micron). In addition, little work had been performed in preparative SFC using the high loadings (sample : silica ratios of 1 : 10 to 1 : 100) routinely used in flash LC. The goal of this study was to develop purification conditions utilizing flash SFC and flash LC and compare the results in terms of purification productivities, purities, and solvent usage. Readers are directed to the article for complete details.[40] The results of this work show that flash SFC was a potential alterative to flash LC purifications performed within the medicinal chemistry laboratories. Comparative loadings were observed with flash SFC and flash LC. Flash SFC had advantages of greatly reduced solvent usage, approximately four-fold less compared to flash LC, and increased purification productivities due to higher chromatographic efficiencies seen at elevated flow rates. In addition, the products recovered from flash SFC are approximately four times more concentrated, reducing the time and energy requirements for isolation of dry products. Finally, the use of flash SFC has the opportunity to significantly reduce the amount of dichloromethane used in medicinal chemistry laboratories.

The advantages of flash SFC are demonstrated through the separation of a mixture of carbamezepine (**2**), flavone (**3**), and nortriptyline·HCl (**3**) (Figure 4.11). Sample injection was performed using solid injection technique.[41] The flash LC and flash SFC separation of 1200 mg of this mixture is shown in Figure 4.11. Comparison of the flash LC and flash SFC separations is shown in Table 4.3. Flash SFC displayed a two-fold increase in purification productivity and a greater than three-fold decrease in solvent usage. Finally, the products recovered from flash SFC purification were two- to five-fold more concentrated than those from flash LC, requiring significantly less time and energy for distillation.

While the advantages of flash SFC are evident relative to flash LC, there are still several technical issues that must be addressed before this technology becomes commonplace in the medicinal chemistry laboratory. The main issue involves the flash SFC equipment. Work to date in this field has utilized standard preparative SFC systems with pre-packed columns. Current SFC equipment is too expensive to replace flash LC systems. A significant reduction in cost is needed to compete with the relatively low cost of flash LC equipment. In addition, no technology currently exists for flash SFC cartridges that is as easy to use, has wide range of sizes, and is as versatile as flash LC cartridges. Furthermore, significant use of flash SFC would require a source and distribution network for $CO_2$ supply. Finally, most preparative SFC collection uses cyclone separators to separate

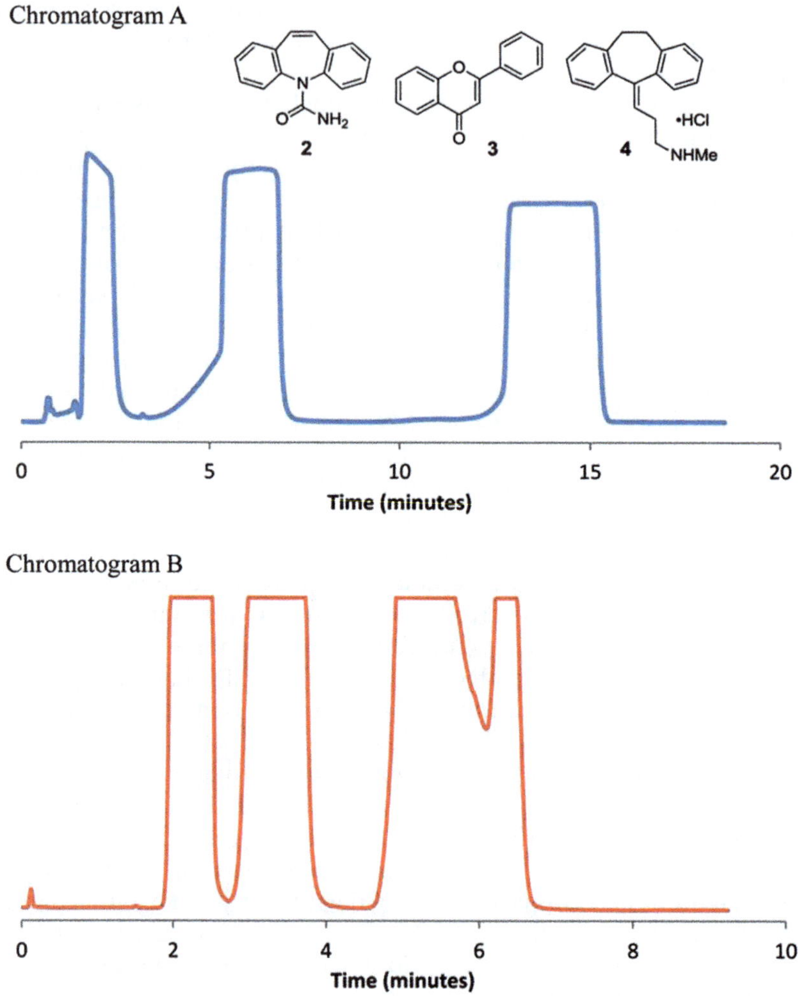


**Figure 4.11** Preparative separation of a 1200 mg mixture of carbamazepine (**2**), flavone (**3**), and nortriptyline · HCl (**4**). Chromatogram A: LC purification was conducted on Interchim 30 micron silica (150×19 mm I.D.) with a mobile phase gradient of methanol/dichloromethane/ammonium hydroxide (0.5/99.5/0.05 to 10/90/1) over 18.5 min. A flow rate of 40 mL $min^{-1}$ was used. Chromatogram B: SFC purification was conducted on Interchim 30 μm silica (150×19 mm I.D.) with a mobile phase gradient of 5–55% methanol (with 0.2% diethylamine) in $CO_2$ over 9 min, 100 bar. A flow rate of 80 mL $min^{-1}$ was used. Reprinted from ref. 40 with permission from Elsevier.

carbon dioxide from organic solvent post-purification. A design allowing collection into test tubes is mandatory for a flash SFC system. It is the authors' hope that SFC equipment vendors will begin to address these issues and introduce equipment suitable for flash SFC purification.

**Table 4.3** Flash LC/SFC comparison of preparative separation of carbamezepine (**2**), flavone (**3**), and nortriptyline · HCl (**4**).

| Technique | Load (mg) | Peak 1 volume (mL) | Peak 2 volume (mL) | Peak 3 volume (mL) | Solvent (L per g crude) | Productivity[a] ($g\ h^{-1}$) |
|---|---|---|---|---|---|---|
| LC[b] | 1200 | 43 | 100 | 108 | 0.62 | 2.52 |
| SFC | 1200 | 7.6 | 18 | 58 | 0.19 | 5.04 |

[a]Assume 10 min re-equilibration for LC, 5 min for SFC.
[b]Higher loadings may have been possible but not investigated due to limited size of dry pack LC cartridge.

## 4.7 Conclusion

Analysis and purification are critical steps in the synthesis of a molecule for drug discovery. These techniques often consume the bulk of solvent during synthesis of a small molecule at the low mg scale. Several approaches have been shown that can be used in analysis and purification that either reduce the amount of solvent utilized or replace them with more appropriate solvents. The use of these techniques can have a dramatic impact on drug discovery analysis and purification, reducing the amount of solvent utilized, the amount of waste generated, the cost of medicinal chemistry synthesis as well as the number of analytical/purification instruments required to support a drug discovery laboratory. In addition, the use of these techniques may result in increased time efficiency as well as improved compound purity.

## References

1. P. T. Anastas and J. C. Warner, *Green Chemistry: Theory and Practice*, Oxford University Press, New York, 1998.
2. C. J. Welch, N. Wu, M. Biba, R. Hartman, T. Brkovic, X. Gong, R. Helmy, W. Schafer, J. Cuff and Z. Pirzada, *TrAC, Trends Anal. Chem.*, 2010, **29**, 667.
3. P. Ferguson, M. Harding and J. Young, in *Green Techniques for Organic Synthesis and Medicinal Chemistry*, ed. W. Zhang and B. W. Cue Jr., John Wiley & Sons, 2012, ch. 25, pp. 659–683.
4. K. Mihlbachler and O. Dapremont, in *Green Techniques for Organic Synthesis and Medicinal Chemistry*, ed. W. Zhang and B. W. Cue Jr., John Wiley & Sons, 2012, ch. 22, pp. 589–611.
5. C. White, *J. Chromatogr. A*, 2005, **1074**, 163.
6. L. Miller and M. Potter, *J. Chromatogr. B: Anal. Technol. Biomed. Life Sci.*, 2008, **875**, 230.
7. L. Miller, *J. Chromatogr. A*, 2012, **1250**, 250.
8. T. A. Berger, *Packed Column SFC*, Royal Society of Chemistry, Cambridge, 1995.
9. G. K. Webster, *Supercritical Fluid Chromatography: Advances and Applications in Pharmaceutical Analysis*, CRC Press, 2014.

10. R. Pierantozzi, *Kirk-Othmer Enyclopedia of Chemical Technology*, John Wiley and Son, 2001, DOI: 10.1002/0471238961.0301180216090518.a01.pub2.
11. L. T. Taylor, *J. Chromatogr. A*, 2012, **1250**, 196.
12. M. A. Patel, F. Riley, M. Ashraf-Khorassani and L. T. Taylor, *J. Chromatogr. A*, 2012, **1233**, 85.
13. M. Ashraf-Khorassani, L. T. Taylor and E. Seest, *J. Chromatogr. A*, 2012, **1229**, 237.
14. J. Liu, E. L. Regalado, I. Mergelsberg and C. J. Welch, *Org. Biomol. Chem.*, 2013, **11**, 4925.
15. C. J. Welch, W. R. Leonard, Jr., J. O. Dasilva, M. Biba, J. Albanexe-Walker, D. W. Henderson, B. Laing and D. J. Mathre, *LCGC North Am.*, 2005, 23.
16. T. Q. Yan, C. Orihuela and D. Swanson, *Chirality*, 2008, **20**, 139.
17. M. L. de la Puente, P. L. Soto-Yarritu and C. Anta, *J. Chromatogr. A*, 2012, **1250**, 172.
18. M. L. de la Puente, P. Lopez Soto-Yarritu and J. Burnett, *J. Chromatogr. A*, 2011, **1218**, 8551.
19. V. Lazarescu, M. J. Mulvihill and L. Ma, *Supercritical Fluid Chromatography: Advances and Applications in Pharmaceutical Analysis*, 2014, p. 97.
20. Y. Zhang, L. Zeng, C. Pham and R. Xu, *J. Chromatogr. A*, 2014, **1324**, 86.
21. R. T. McClain, A. Dudkina, J. Barrow, G. Hartman and C. J. Welch, *J. Liq. Chromatogr. Relat. Technol.*, 2009, **32**, 483–499.
22. K. Ebinger, H. N. Weller, J. Kiplinger and P. Lefebvre, *JALA*, 2011, **16**, 241.
23. F. Gritti and G. Guiochon, *J. Chromatogr. A*, 2012, **1228**, 2.
24. J. J. DeStefano, S. A. Schuster, J. M. Lawhorn and J. J. Kirkland, *J. Chromatogr. A*, 2012, **1258**, 76.
25. C. M. Aurigemma, W. P. Farrell, J. Simpkins, M. Bayliss, M. Alimuddin and W. Wang, *J. Chromatogr. A*, 2012, **1229**, 260.
26. K. F. Blom, R. Sparks, J. Doughty, J. G. Everlof, T. Haque and A. P. Combs, *J. Comb. Chem.*, 2003, **5**, 670.
27. U. D. Neue, C. B. Mazza, J. Y. Cavanaugh, Z. Lu and T. E. Wheat, *Chromatographia Supplement*, 2003, **57**, S-121–S-127.
28. W. Goetzinger and S. Ayube, unpublished results.
29. L. Zeng, L. Burton, K. Yung, B. Shushan and D. Kassel, *J. Chromatogr. A*, 1998, **794**, 3.
30. J. Isbell, R. Xu, Z. Cai and D. Kassel, *J. Comb. Chem.*, 2002, **4**, 600.
31. D. Kassel, *Chem. Rev.*, 2001, **101**, 255.
32. R. T. McClain, A. Dudkina, J. Barrow, G. Hartman and C. J. Welch, *J. Liq. Chromatogr. Relat. Technol.*, 2009, **32**, 483.
33. B. A. Khunte and L. Philippe, *LC GC North Am.*, 2011, **29**, 170.
34. E. A. Peterson, B. Dillon, I. Raheem, P. Richardson, D. Richter, R. Schmidt and H. F. Sneddon, *Green Chem.*, 2014, **16**, 4060.
35. W. C. Stevens, Jr. and D. C. Hill, *Mol. Diversity*, 2009, **13**, 247.
36. A. f. T. S. a. D. Registry, Toxicological Profile for Methylene Chloride (Update). Draft for Public Comment., http://www.atsdr.cdc.gov/toxprofiles/tp14.pdf, accessed 16 December 2014.

37. U. S. E. P. Agency, Integrated Risk Information System (IRIS) on Methylene Chloride. http://cfpub.epa.gov/ncea/iris_drafts/recordisplay.cfm?deid=220583, accessed 12-16-2014.
38. J. P. Taygerly, L. M. Miller, A. Yee and E. A. Peterson, *Green Chem.*, 2012, **14**, 3020.
39. D. S. MacMillan, J. Murray, H. F. Sneddon, C. Jamieson and A. J. B. Watson, *Green Chem.*, 2012, **14**, 3016.
40. L. Miller and M. Mahoney, *J. Chromatogr. A*, 2012, **1250**, 264.
41. L. Miller, H. Bush and E. M. Derrico, *J. Chromatogr.*, 1989, **484**, 259.

CHAPTER 5

# Green Chemistry and High Throughput Screening

PAUL RICHARDSON

Pfizer Worldwide Research and Development, 10770 Science Center Drive, San Diego, California 92121, USA
Email: paul.f.richardson@pfizer.com

## 5.1 Introduction

This chapter describes the implementation and value that high throughput reaction screening can provide within a drug discovery environment. In addition, a particular focus is placed on its utility in making rapid and more informed decisions based on environmental aspects of the chemistry. The examples are drawn from internal projects in order to provide full details of background, learning, and concepts derived from each of these. The use of a reaction screening platform also enables one to rapidly evaluate the feasibility of potentially shorter synthetic routes to compounds of interest. Although the chemistry involved in the new synthetic approach may not be perceived to be green, the fact that the route is shorter, and hence likely to generate less waste, will lead to a gain in green metrics.[1] The examples within this chapter will focus on the use of reaction screening to discover new conditions for a specific transformation as opposed to its utilization in scouting for new synthetic routes. Notably, there are also numerous variations on how to set-up a screening platform from other companies/academic groups,[2] and many examples have been published demonstrating the diverse utility of this technology.[3]

RSC Drug Discovery Series No. 46
Green Chemistry Strategies for Drug Discovery
Edited by Emily A. Peterson and Julie B. Manley

Published by the Royal Society of Chemistry, www.rsc.org

## 5.2 Screening in Medicinal Chemistry

High throughput-screening or similar experimental approaches form a cornerstone of the drug discovery process.[4] Often, a project will be initiated with a high throughput screen of a large compound collection to identify suitable lead matter for subsequent optimization. In the optimization stage, the use of combinatorial chemistry to generate libraries of compounds takes advantage of automated liquid handling, and high throughput purification technology to enable the rapid generation of collections of compounds.[5] In this scenario, one carries out a chemical transformation on a diverse array of compounds under identical reaction conditions. Under this paradigm, the success of the reaction in terms of yield is less important with the onus being placed on the separation technology being able to identify and isolate the compounds of interest. As an extension of this, screening and reaction optimization looks at carrying out a reaction between the two same reactants while instead varying the reaction conditions. In this setting, the focus has been switched away from simple formation of the product onto identifying the best conditions to form the desired product in the highest yield. Whereas the value of this approach to the optimization of synthetic organic chemistry cannot be disputed, there are several barriers to its effective implementation in a discovery chemistry setting.

The first of these is in relation to having genuine standard samples for the desired product. Reaction screening is often carried out in chemical research and development as a project progresses to scale-up. However, in this desired setting, the product has previously been isolated and, as such, standard samples are in hand. Similarly, often, standards of undesired by-products from the reaction will also be available – thus extensive profiling of a reaction can be carried out to establish both robustness and the ideal conditions. Within discovery chemistry, in some cases, a sample of the desired product may be in hand, but in probably most cases a screen is initiated either because a reaction does not work or to evaluate a speculative approach to a lead molecule. As such, no standards exist for the desired product, and this puts the onus on developing an effective analytical method for evaluation of the reaction being screened. This as we will see is a key component of successful reaction screening.

Secondly, the quantity of material available for evaluating a reaction may be limited. This again is in stark contrast to process groups, and probably represents one of the biggest barriers for the uptake of a screening platform within medicinal chemistry departments. The trade-off of the use of such material either for a reaction screen or to potentially rapidly synthesize a series of analogues under sub-optimal conditions becomes a pivotal question. However, these two scenarios are not mutually exclusive. Use of material to provide optimum conditions for a specific transformation that can then be sequentially expanded over a series of analogues is of high value. A higher yielding transformation in a key diversification step can lead to a set amount of material being spread across a larger number of analogues, and can also ease the burden in terms of reaction purification if a cleaner reaction is observed, and by-product formation minimized. In addition, optimizing specific chemistries at an early stage can pay dividends throughout a project's life cycle. Naturally, the question of material use becomes somewhat of a moot point if the planned chemistry does not work under standard conditions. Herein, a reaction screen can be viewed as being a rapid means to see if the original synthetic approach has any merit, or if an alternative has to be evaluated. This can be a significant time saver particularly if any new route requires the preparation of new starting materials.

Finally, a further differentiator between screening in a discovery and in a process chemistry environment is the desired turnaround time for results. Although in process chemistry a rapid turnaround is required, often a logistical approach is taken to screening, focusing on specific parameters to make incremental improvements to the chemistry under investigation. However, in discovery the focus is very much on a fast turnaround. If the evaluation of a specific chemistry is likely to take in the order of weeks, then often the approach taken will be to look for a more established viable pathway to achieve the same transformation. However, implementing a screening workflow that enables a read-out in terms of viability of the chemistry in 2–3 days is a huge advantage in allowing rapid go/no-go decisions to be made both on speculative chemistry and established reactions applied to new drug-like templates. To achieve this, it is critical to evaluate all the stages of the implemented screening process to optimize in terms of throughput, and ensure that none of the various stages becomes a bottleneck.

While several distinctions have been made thus far between screening needs in process, and discovery, it is valuable to rank the order of importance of these parameters at these two stages of the drug discovery process, and this is summarized in Figure 5.1.

Clearly, this demonstrates that although yield is important to both there is a paradigm shift in other priorities between discovery and early development. However, the simple take-home message herein is that discovery chemists want answers both faster and through using less material when compared to early development.

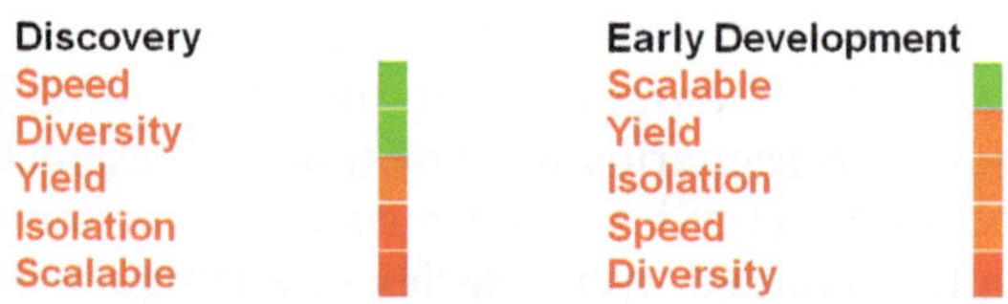


**Figure 5.1** Reaction screening – drivers for success.

## 5.3 Setting up a Reaction Screen

### 5.3.1 Workflow of Screening

Although several different issues can be addressed by high throughput reaction screening, they all consist of the same basic workflow. Initially, the project team presents a problem, leading to an experimental design being proposed followed by set-up and execution of the experiment. Upon completion, the results are analyzed and interpreted to provide a response to the team. This response can lead to a potential solution of the problem, proposals for further screening experiments to be carried out, or lead the team to the decision to focus resources on alternative approaches. The key to optimizing this workflow in terms of speed is to consider each step individually, and to attempt to utilize available resources and technology to eliminate the bottlenecks as they arise. An initial qualitative breakdown in the three key operational steps (analysis and interpretation combined) provides the scenario shown in Figure 5.2 in terms of time required for each step.

At first this may appear somewhat contradictory as one would expect the reaction set-up and execution to be the most labor and time intensive steps, though this is not the case. To initialize a screening platform, one commonly held belief is that a great deal of investment is required in terms of acquiring automated liquid handlers or weighing robots. However, in most reaction screens, this is simply not the case as the fastest way to execute reaction preparation is through liquid handling solutions using a multi-channeled pipette type set-up. Logistical planning of the screen set-up can expedite this to enable a minimum amount of manual operations in order to formulate the screen across the commonly utilized 96 well-plate format. Typically, the majority of screens are run using 2 mL glass HPLC type vials, and adoption of this plate-format allows easy direct transition to the desired LC-MS analytical platform employed.

### 5.3.2 How Low Can You Go? Material Requirements

As mentioned, one of the key limitations of reaction screening within a medicinal chemistry environment is restricted quantities of material

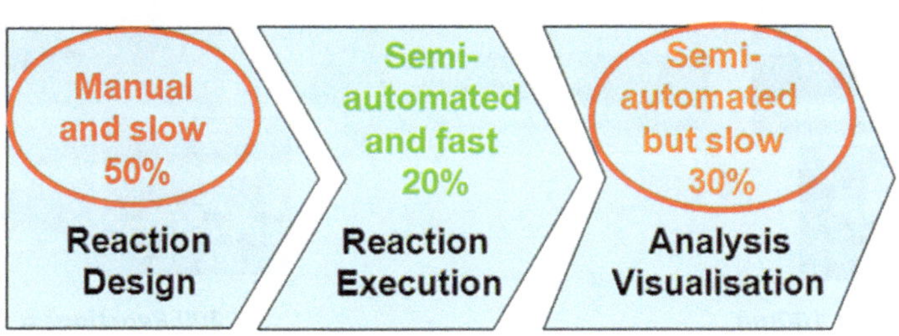


**Figure 5.2** Reaction screening workflow.

available for screening. Project teams often cannot afford >500 mg of material for a screen, and typically would like to reduce this amount as much as possible. A second parameter of note, which is discussed further in Section 5.3.3, is the desired extent of chemical screening space to be evaluated. Obviously, there is a direct correlation between increasing the amounts of material available to the amount of space that can be assayed. With this in mind, a goal within our screening platform has been driving towards minimizing the amount of material required per screen though without diluting the reactions. Over a 5 year period, this goal has been successfully achieved. Initially, screens looked at running reactions on a 5 mg scale in a 1.2 mL sample volume, which equates to 100 mg of material spread over 20 reactions. However, at the current time, 200 reactions can be run with the same amount of material, which is equivalent to a ten-fold decrease or 0.5 mg per 120 μL reaction volume (Figure 5.3).

The most important consideration throughout this more material-sparing approach was how to maintain the integrity of various catalytic reactions. These reactions are run with sub-mg quantities of substrate, and for the catalysts μg quantities are employed. This introduces not only challenges associated with aliquoting these amounts but also those associated with the air- and moisture-sensitivity observed with many of these materials. The breakthrough to achieve this was realized through setting up and running reaction screens under strictly inert and anhydrous conditions. This is most conveniently achieved through utilization of a glove-box, which not only provides the carefully controlled environment to ensure the integrity of the reactions, but also provides a mechanism to reduce the manual aspects of the set-up through the one-time preparation of stock solutions of ligands and catalysts, which can be used indefinitely. This has proved to be extremely productive in terms of making the workflows more efficient, and the long-term stability of these catalyst solutions has been demonstrated through LC-MS analysis of a series of air-sensitive phosphine ligands typically employed in palladium-mediated coupling processes. There is a critical caveat that should always be considered when running reaction screens under glove-box conditions: scaling successful conditions obtained from

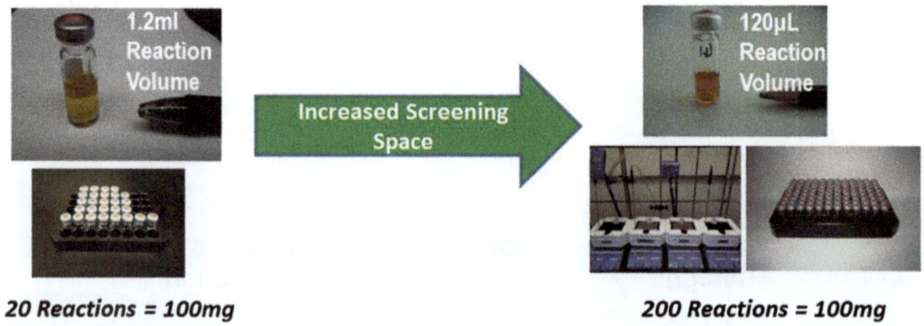


**Figure 5.3** Increasing chemical space through minimizing material requirements.

these screens is most likely to be carried out in a typical fume-hood-like setting. As such, if the scaled reactions fail to reproduce the success of a screening hit, oxygen and moisture sensitivity of the reaction components should always be a consideration for the lack of reproducibility. Of greater concern, though, is that numerous reactions are well known to be promoted by trace moisture or rely on the presence of air to promote an *in situ* oxidation within the system.[6] With this in mind, these reactions will show only low conversion or fail completely when screened under glove-box conditions. Given this, it is important to have an awareness of reactions that may benefit from traces of oxygen and moisture preferably at the design phase of a screen, but also if no positive results are observed within a reaction screen. In many cases, a reaction may be screened to improve on existing conditions, and in this scenario the standard conditions are typically embedded into the reaction screen template. If no conversion is observed herein across the screening plate, then the lack of air and moisture is more than likely the cause.

An added implicit advantage of a more material-sparing approach to reaction screening is the ability to conceptually cover a wider area of chemical space for a specific transformation.[7] This is important with regard to developing new chemistry, screening speculative transformations, and to exploit serendipitous discoveries.

### 5.3.3 Reaction Design – Coverage of Chemical Space

It goes without saying that the results of a reaction screen can only be as good as the quality of the initial design, and the importance of this is highlighted by this being the most time-intensive phase of the reaction screening workflow. A potentially alternative view for a specific transformation would be to just screen everything. It has already been stated that moving to a more material-sparing paradigm enables a screening of a wider area of chemical space, but in reality to screen all possible combinations of reagents, even for a simple transformation, is impractical. The variables described inherently within a reaction screen are typically split into "continuous" and "discontinuous" factors. Continuous factors refer to variables such as reagent stoichiometry, temperature, and reaction concentration, which can be selected from a range of possible settings for a given reaction, whereas discontinuous factors are variable such as individual reagents and reaction solvents. As the complexity of a screening design develops, the variation of both discontinuous and continuous variables can often be captured within the same screen. For example, a reaction screen specifically focused on evaluating a series of bases to mediate a specific transformation may match the reaction temperature under consideration with the strength of the base. As such, the screening template may feature reactions at three different temperatures to accommodate strong, medium, and weak-strength bases.

For the initial design of a specific screen, the first information that should be taken into consideration is obviously what is already known for the

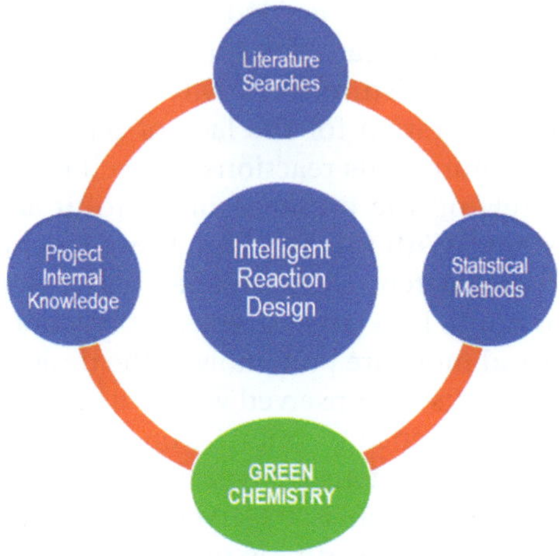


**Figure 5.4** Designing a reaction template.

transformation. Clearly, there is a plethora of sources of information available, which are summarized in Figure 5.4.

The primary source of information regarding a new reaction screen is what is already known within the project team regarding the specific reaction. This will include unsuccessful conditions tested for the substrates under consideration, successful exploitation of the desired methodology on closely related substrates as well as a long-term strategy of how to exploit the transformation if successful. This internally based knowledge is initially provided by direct communication with the project team/chemist. In addition, further specific details and related internal information can now be more efficiently accessed and collated through utilization of electronic laboratory notebooks and information repositories. Furthermore, input on internal preferred scale-up conditions can be accessed from CRD/Process resources, and incorporated within the screen. A second source of information comes through searching the literature using tools such as SciFinder® and Reaxys®, which enable facile structure-based searches to enable transformations on similar substrates to be accessed. Based on the nature of the substrate under evaluation, the structure similarity search can be broadened or narrowed to capture the most relevant conditions to be applied within the new screen.

There is a major disconnect with this literature/prior knowledge-based approach to designing a new reaction screen. By the very nature of the screen being initiated, it is assumed that the reaction under investigation has either not been performed before or has led to only poor results. Searching the literature on a per screen basis will indeed move us into the "correct" chemical space for similar reactions, but what if the chemistry does not work in this initial screen? At this juncture, the approach would either be abandoned or a significant amount of time would be spent searching or potentially settling at what could be a local maxima in terms of yield.

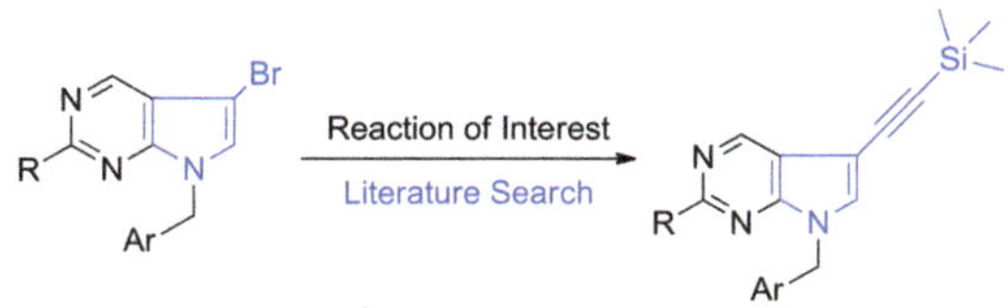


**Scheme 5.1** Sonagashira coupling – literature precedents.

This approach will exclude a large amount of unchartered chemical space and, as such, no new findings will ever be identified. This disconnect is illustrated by the following example. The project team wanted to optimize the Sonagashira reaction shown in Scheme 5.1, and the literature search shown was based on the similarity of the substrate under investigation.

The project team had already performed the reaction, and reported poor yields under standard Sonagashira reaction conditions. The literature search provided ∼200 examples, the vast majority of which utilized standard conditions employing $Pd[P(Ph)_3]Cl_2$ as the catalyst.[8] In this case, little has been learned from the literature other than the fact that there is a precedent for the reaction, as revealed by the ∼200 literature hits. Given that the literature conditions were tested and found to perform poorly, either an alternative synthetic approach to the molecule of interest was needed or a screen of wider chemical space was warranted. What if initially a more generic "one-size fits all" template was implemented for this reaction? Employing the discontinuous variables that are critical for success of a Sonagashira coupling, a screen template can be designed that dramatically increases the amount of chemical space covered while still minimizing the time involved to carry out the screen. Furthermore, the precedented literature conditions will be present within the evaluated screening space.

To illustrate this, it is valuable to re-visit the Sonagashira reaction considered previously.[9] In general, the important components for this reaction to proceed are the presence of a base, solvent, and Pd-based catalyst with or without a Cu-based co-catalyst. Typically, reactions are carried out either using DMF or an amine such as TEA as the solvent. Bearing in mind the components required for the reaction to proceed, a screen was designed in 96-well plate format evaluating six catalysts in eight solvents at two different temperatures, again highlighting the ability to vary both continuous and discontinuous factors within the same reaction screen. The results obtained are displayed in the Spotfire plot (Figure 5.5), which is the typical method to represent results obtained from a reaction screen. Herein, the two trellises represent the two reaction temperatures evaluated (left trellis 50 °C, right trellis 100 °C) with the *x*-axis showing the product present by extracted mass correlating with the *y*-axis, which shows the UV response. The colors show the catalyst employed, with the labels representing the solvent. The size of spot indicates the percentage of remaining starting material present in the reactions. From this analysis, ideally one is looking for a small spot present in the top right hand corner of the trellis to represent a promising hit

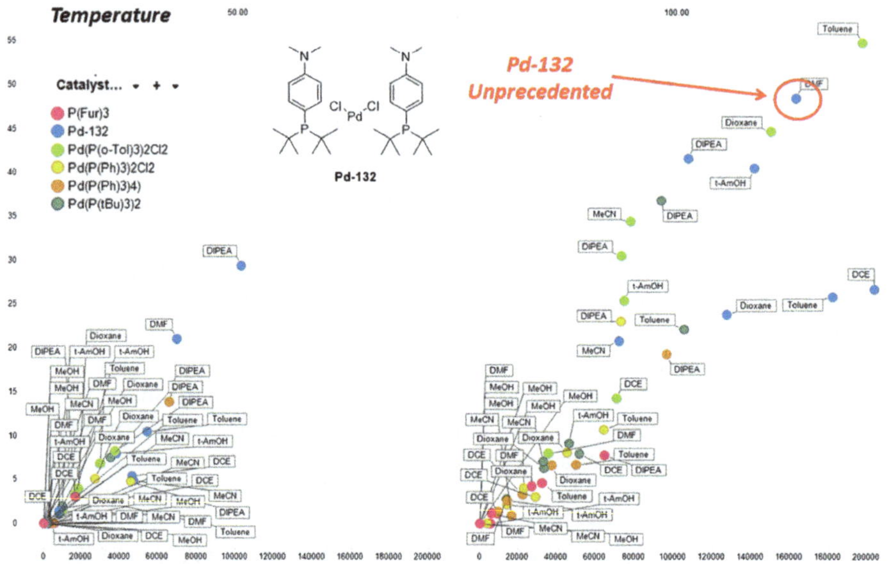


**Figure 5.5** Spotfire analysis of Sonagashira screen.

reaction for potential follow-up. Looking at the results on the 100 °C trellis, there are several promising reactions with one of the best highlighted on the plot. Although this still utilizes DMF as the reaction solvent, it is critical to note that this set of conditions employs Pd-132 as the catalyst for the transformation. Note that this catalyst system was not mentioned within the ~200 literature references searching similar transformations, and therefore represents unprecedented reaction conditions. From a more realistic perspective, the fact that these conditions have not been reported previously is not so surprising given that this is a relatively new ligand system. One further advantage provided by the Spotfire visualization is that, in this case, several potential reaction conditions are highlighted, which enables selections to be made based on factors such as (i) reaction profile, (ii) cost of catalyst, (iii) scalability, and (iv) greenness of the new conditions.

From the above simple example, the advantages of screening expanded chemical space can be appreciated, though as noted previously in the vast majority of cases it is impossible to screen every possible combination for a given reaction. Consider a Suzuki coupling for example: although one may be able to narrow down to a workable set of solvents and bases, there exists a wide-range of phosphine-based ligands that have been reported for the reaction and including all of these is simply impossible. With this in mind, statistical based methods based on principal component analyses have been developed to group such ligands (and other discontinuous variables such as solvents) in a map-format to enable selection of a sub-set with the most diverse array of properties so as to cover the maximum chemical space for a novel screen.[10] In addition, less sophisticated approaches can be adopted to access a sub-set of ligands simply considering the sterics or electronic properties of the ligands under consideration.

### 5.3.4 Strategies for Screening for Greener Solutions

Although the discussion so far has centered on maximizing the degree of chemical space covered by a reaction screen, from a green chemistry perspective it is desirable to design screens with a potential bias to identify greener reaction conditions. Considering the principles of green chemistry, there are a number of potential ways in which green concepts can be introduced through reaction screening. The most obvious perhaps is looking to replace non-ideal reagents, reactants, and solvents at the first pass with greener alternatives. Figure 5.6 represents this by showing a range of reagents included in a screen for amide bond formation with representative yields provided for the reaction under consideration.[11]

There is an important learning from these results presented herein. Although the results do indicate that common coupling reagents like HATU have performed well in this instance, it is important to note that greener alternatives such as T3P and CDI have also provided tangible levels of the desired product.[12] This is an important observation to make in that although

| Position | Reagent | Product Yield (%) |
|---|---|---|
| 1 | DPPA | 97.7 |
| 2 | TPTU | 95.5 |
| 3 | DECP | 89.4 |
| 4 | HSPyU.$PF_6$ | 85.9 |
| 5 | TSTU | 85.8 |
| 6 | HATU | 84.6 |
| 7 | DCC | 84.4 |
| 8 | HSTU | 76.5 |
| 9 | DIC | 74.0 |
| 10 | CMPI | 73.6 |
| | Selected "Green" Conditions | |
| 37 | T3P | 23.3 |
| 38 | CDMT | 22.8 |
| 39 | WSCDI | 20.4 |
| 40 | EEDQ | 11.7 |
| 41 | CDI | 9.4 |
| 43 | $Ac_2O$ | 7.5 |
| 47 | *i*-BuOCOCl | 5.7 |
| 48 | PivCl | 3.9 |

**Figure 5.6** Amide coupling – potential leads for greener conditions.

they may not be the reagents of choice if a one-off coupling to provide an analogue is required, they may warrant further investigation if one is interested in scaling up the desired product. It is important to note that in the screening manifold often all the reactions are set up under the same conditions, particularly with respect to order of addition. Often in amide bond formations, the reactions involve the formation of an activated intermediate and the rate of formation of this may be critical to obtaining the maximum yield. This intermediate will differ depending on the conditions employed, and the rate of formation may vary significantly. In a similar manner, varying other parameters such as temperature and concentration through a DoE (design of experiments) approach can expedite optimization of reaction conditions, which initially lead only to modest yields of the desired product.[13]

Similarly, utilizing reaction screens to investigate a range of different solvents for a desired transformation also represents an opportunity to identify alternative conditions right at the outset. Often with a classical transformation, a considerable body of literature evidence can rapidly accumulate indicating that certain specific solvents are the "solvents of choice" for a specific transformation, leading to a knee-jerk reaction in employing these often less than desirable solvents from an environmental perspective. For example, most amide bond formations reported in the literature employ either $CH_2Cl_2$ or DMF as the solvent, and although in some cases this may be attributable to substrate solubility, recent systematic studies from GSK have indicated that several other solvents can be successfully employed in this key transformation.[14] Likewise, Garg and co-workers at UCLA have carried out similar studies on solvents in nickel-catalyzed Suzuki–Miyaura couplings.[15] Over 30 solvents were evaluated in the model reaction between naphthyl sulfamate and phenylboronic acid with >95% conversion being observed in over half of these reactions. From these studies, the use of 2-MeTHF and *tert*-AmOH as greener solvent alternatives emerged. Despite these studies, it is possible to argue that investigating different solvents in an unprecedented transformation presents a risk; integrating these in a high throughput manner, though, mitigates the risk in terms of both the time and resource spent.

The concept of introducing templates for specific reaction screens has already been introduced, and the utilization of internal knowledge and literature precedent provide solid initial inputs into the design of these templates. There also exists several green chemistry-based information sources that can be utilized to influence this template design. Solvents have already been identified as important variables, and a number of tools have been developed and published classifying solvents based on properties pertaining to their greenness.[16] The literature also plays a role with journals focused on sustainability, green chemistry, and process chemistry enabling the latest developments in the field to be incorporated as well as capturing conditions that are amenable to scale-up, and as such likely to possess minimal green issues. In addition, several organizations have developed reagent guides,

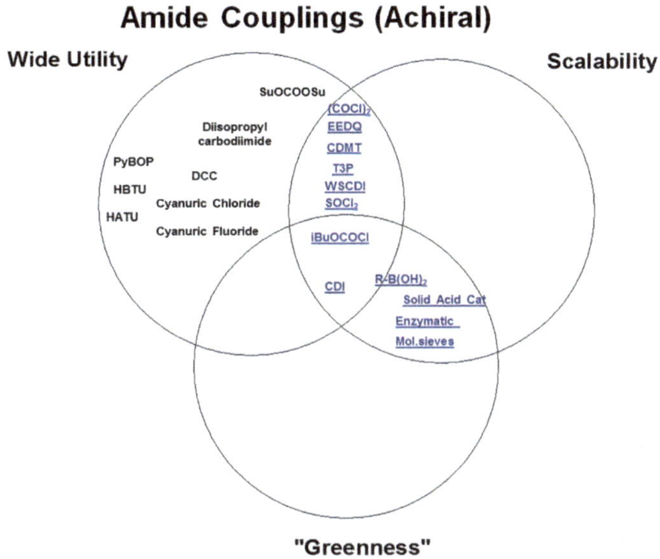


**Figure 5.7** Venn diagram reagent guide.

which look to classify reported conditions for a specific transformation based on a number of parameters.[17] An example of a grid from such a reagent guide is shown in Figure 5.7, in which the conditions provided have been partitioned based on their "scalability", "wide utility", and "greenness". The development of these guides is based on input from scientists from both discovery and process-based backgrounds utilizing literature and internal knowledge. If such a guide exists for the transformation of interest, this guide can form the foundation for the screen design.

It is important to consider that often it may not be possible to access a green chemistry reagent guide for a specific transformation, or one may not exist. Additionally, it should also be remembered that these guides are updated on a periodic basis, and as such may not capture the most recent methodology. There are several ways to facilitate accessing this information for incorporation into a screening template in addition to just current literature awareness. One such method is by tracking vendor newsletters to stay on top of the latest available reagents. This can be extremely useful in the Green Chemistry space with many vendors promoting a reagent based on its environmental attributes. It is important to keep an open mind when reviewing these resources as a vendor may tend to show bias to their reagents. In addition, there has been a growth in the development of academic–industrial relationships with a focus on developing more efficient solutions to well-established reactions, as well as tapping into novel methodology for organic synthesis.[18] Academic engagement in the former has been stimulated by the seminal publication highlighting the reactions, which are considered to be problematic from a "green" standpoint by industry scientists, and various publications in each of these areas is

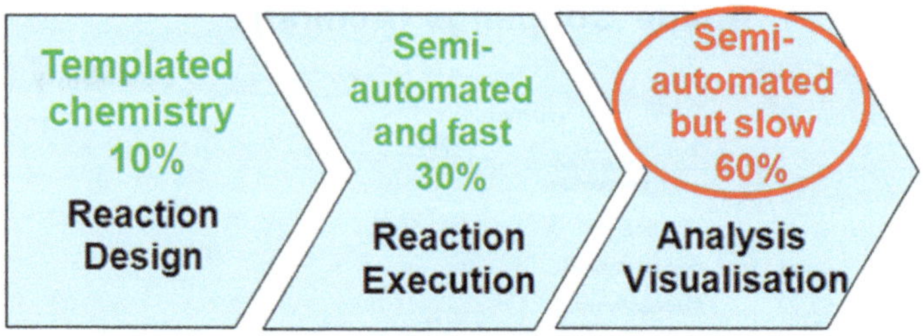


**Figure 5.8** Screening workflow for a templated reaction.

highlighted every 6 months in *Organic Process Research and Development* (OPRD).[19,20] For novel methodology, a range of initiatives has been instigated to enable academic partners to investigate new chemistry utilizing pharmaceutically relevant substrates.

With all these considerations taken into account, it is possible to design a generic template of screening conditions for the more commonly evaluated reactions, with the onus on covering a greater degree of chemical space rather than focusing on the specific substrates under consideration. This approach not only saves a significant amount of time in setting up the screen, but also enables novel chemistry to be accessed by covering uncharted territory as well as greener conditions to be found by including "green" chemical space. Within a discovery chemistry environment, there are a number of reactions that occur on a regular basis. These include for example palladium-mediated couplings (Buchwald, Suzuki, Heck), amide couplings, hydrogenations, Ullmann couplings, alkylations, and classical resolutions. It was found that by using generic templates designed for each of these reactions, 60% of the screens evaluated could be investigated *via* this approach. For such screens, this has a striking effect on speeding up the workflow (Figure 5.8). Generally when a screen for a template reaction is requested, the only up front work to do is to quickly check the literature to ensure that the expanded screen template covers any reported conditions, which it invariably does. However, although the time spent on reaction design has now been minimized, the onus is now on the analysis, visualization, and interpretation of the data, which is semi-automated. In addition, with a larger degree of chemical space covered, an increase in magnitude in analytical samples might be expected.

### 5.3.5 Data Capture, Analysis, and Visualization

Upon completion of the actual reaction segment, the key objective is to accumulate and analyze the data as rapidly as possible to enable discovery project teams to make expeditious decisions regarding the chemistry under investigation. The typical work flow is shown in Figure 5.9.[21]

Prior to embarking on the analysis depicted in the scheme, a visual inspection of the samples is important to determine their homogeneity. This is

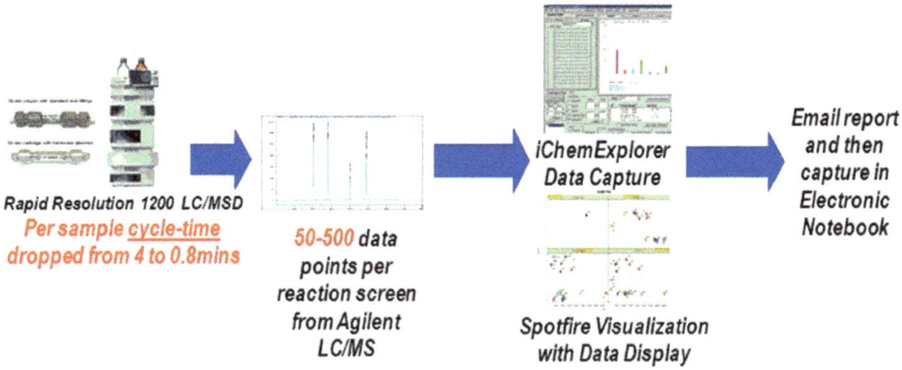


**Figure 5.9** Analysis, data capture, and visualization.

critical as insoluble materials can lead to clogging during the analytical run, and thus significantly delay the acquisition of data. In addition to the potential for such errors, insoluble materials in samples may also be either products or starting materials, and thus if not sampled will lead to an inaccurate representation of the data obtained. A dilution step with a solvent such as DMSO is usually sufficient to dissolve all the organics. If insoluble inorganic materials still remain, centrifugation and judicious setting of the needle sampling depth of the LC-MS autosampler can mitigate potential clogging errors.

Intrinsic to the successful evaluation of a screen is a suitable analytical method for the reactions under consideration. For asymmetric transformations, a further level of complexity is added with a chiral method being required. However, for analysis of a standard reaction, the focus is on implementing a general method, which enables adequate separation across a range of different reaction types. Herein, for a standard reverse phase gradient method, advances in column technology and the access to rapid resolution instrumentation has enabled cycle times per sample to be dropped from 4 to 0.8 min. Over a reaction screen consisting of for example 96 samples, this five-fold gain in efficiency is highly significant. In addition, running the screens in a 96-well plate format enables direct transition to the analytical platform. Upon completion of the data acquisition, the manual task of data analysis takes place. The first task herein is to evaluate the bulk data, and decide how best to envision this in terms of UV wavelength to observe in tandem with the mass trace. Often, as noted in the discovery space, no markers for the desired product exist, though of course one does have a marker for the starting materials, and evaluation of the absorption spectrum of this can aid in deciding the optimum wavelength for analysis. A further factor that can aid in this decision is to choose a wavelength in which to minimize effects of unrelated materials such as solvents. One of the first important tasks in carrying out the analysis is to utilize the mass spectroscopy information to determine if the product is formed in any of the reactions, and where it elutes. In addition, use of chemical intuition of the reaction under investigation is also important to determine what the possible byproducts of the reaction might be, and to identify these in the various reaction traces. In the final analysis, this allows factors to be determined in order to minimize these side reactions. Finally, when inspecting the various reaction traces, it is also possible to identify various peaks that may be solvent, ligand, or reagent-based in order to label and eliminate these from the final analysis of the reactions. Over these analytical steps, the instrument software enables labelling of the various identified peaks. The bulk data then can be processed through a data capture software (such as the iChemExplorer platform), and then visualized using a Spotfire data display.[22]

The Spotfire data visualization for the Sonagashira reaction was presented previously (Figure 5.5), and this represents a very powerful method to evaluate and manipulate the data. Looking at both the mass and UV response across the axes compensates for instances in which the desired product may co-elute with other products, and avoid having to develop a new and potentially longer acquisition method (with the corresponding overall increase in cycle time) to separate these. In addition, the Spotfire analysis rapidly allows the optimal conditions to be determined, as well as visualization of the most informative trends (solvent, ligand, reagent, temperature, *etc.*) for the reaction under consideration.

## 5.4 Case Studies in Reaction Screening

### 5.4.1 Case Study 1 – Amide Bond Formation

The reaction to couple an acid and amine to form an amide bond is one of the most prevalent in drug discovery chemistry with *ca.* 25% of all reactions estimated to involve this step.[23] Numerous methods and coupling reagents have been developed for this reaction. Despite the commonly held belief that amide bond formation is a solved problem in organic synthesis, the ACS GCIPR has listed the development of new methods for amide bond formation that avoid poor atom economy reagents as a priority.[19] In addition, as mentioned previously, the majority of such reactions are currently carried out in either $CH_2Cl_2$ or DMF, and these solvents were used in our original template described herein. Obviously, this template can evolve over time, and has now been expanded to include alternative greener solvents.[14]

Given the plethora of reagents available to mediate amide coupling, implementing a templated reaction screen to evaluate these in a high throughput format provides a powerful example of how this technology can impact discovery chemistry. From the perspective of green chemistry, again it possible to utilize a reagent guide developed for this transformation to ensure that the preferred reagents are included within our screening template. The amide bond screen template evaluates 36 coupling reagents in two solvents at two temperatures with and without an additive for a total of 144 reactions. Typically, such a screen will require between 150–300 mg of substrate, and take 0.5 full-time equivalent (FTE) days to set-up, execute, and analyze the data, with the whole process taking 2 days from start to finish. Without a reaction screen in place, and bearing in mind the time pressures involved in discovery chemistry, HATU is often the reagent of choice given the versatility of this reagent, and the belief that it offers the highest chance of success if only a single reaction is to be run. This decision though overlooks the drawbacks of this reagent in terms of its high MW, safety issues as well as a potentially difficult purification based on by-products such as TMU and the $PF_6^-$ anion, which can also be difficult to detect or visualize. The fact that HATU is included in our reaction screen enables us to benchmark the other coupling reagents against this, and make judicious choices based on numerous other factors rather than just simply whether the reaction worked or not to give product.

The four reactions shown in Scheme 5.2 were all subjected to the amide bond formation reaction screen utilizing the screen template, which we have previously described. Each of these reactions presents a different challenge. For example, reaction (1) highlights a potential selectivity issue between amide bond formation and SnAr displacement of the adjacent halide. For reaction (2), there exists a different selectivity issue in differentiating between the two reactive nitrogens. Reaction (3) features a sterically encumbered acid, whereas in reaction (4) a weakly reactive electron-deficient amine is utilized in the coupling reaction. For each reaction, the best performing coupling reagents/conditions are illustrated accompanied by the estimated yields from the screen.

The first thing to notice is that for none of these reactions has the screen identified HATU as the reagent of choice. In fact only for reaction (4) has HATU been selected as one of the stand out reagents. However, even here, superior conditions exist in terms of yield. In addition, the alternative

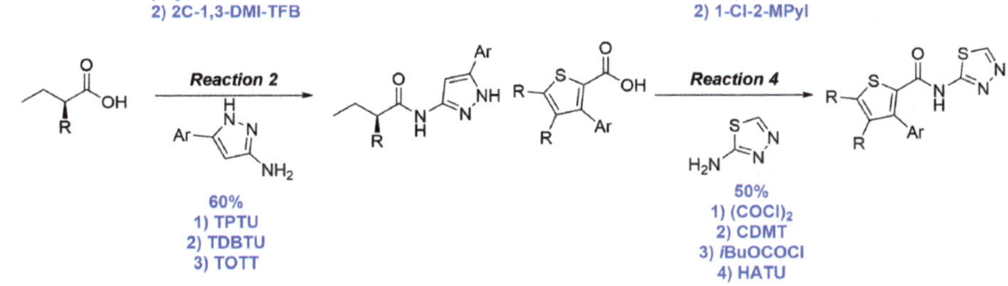


**Scheme 5.2** Examples of amide bond formations.

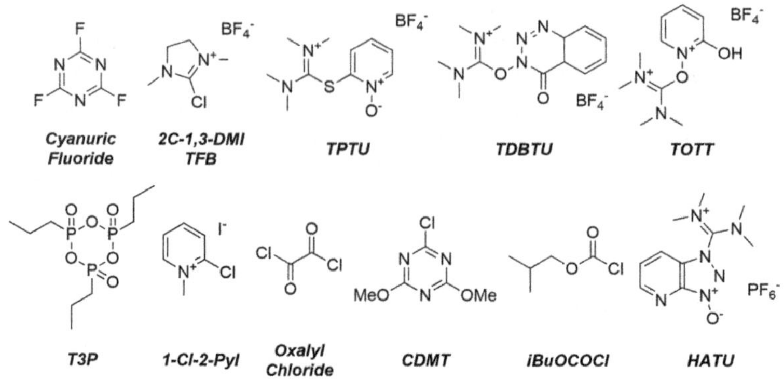


**Scheme 5.3** Hit reagents from amide bond formations.

conditions featuring *in situ* formation of either the acid chloride or, even better, a mixed anhydride are also preferable from a green chemistry perspective. In this vein as well, T3P, which has emerged as the best reagent identified for reaction (3), is also a favorable coupling reagent, and can easily be removed during work-up by a simple aqueous wash. The second striking result is the diversity in terms of the coupling reagents identified across these four reactions. Although one can argue that the oxalyl chloride conditions in reaction (4) and the use of T3P in reaction (3) represent well-established approaches for amide bond formation, the use of reagents such as TPTU (2) and generation of the acyl fluoride using cyanuric fluoride (1) would not typically be the first choice reagents unless an exact literature precedent existed for the substrate under investigation. This is unlikely particularly with elaborated substrates given that one of the key goals of discovery chemistry is to work and expand into novel chemical space. The final point to note, as alluded to previously, is that each of these screens provides a diverse range of possible reagents to mediate the coupling under investigation, enabling the chemist to make a strategic choice in terms of which to pursue (Scheme 5.3). This decision can be based on numerous factors such as scale, reagent availability, safety, ease of purification, and green considerations such as the ability to work in alternative solvents.

### 5.4.2 Case Study 2 – Suzuki Couplings

In a similar manner to amide bond formations, palladium-mediated C–C or C–N couplings form an integral part of most drug discovery research programs, and the growth in the exploitation of Suzuki, Stille, Heck, Negishi, Sonagashira, and Buchwald–Hartwig couplings has been exponential since their initial discovery.[24] In addition, new developments with regard to novel ligands, substrates, and coupling partners are constantly being disclosed. However, the disclosures often feature a narrow substrate scope, and often overlook pharmaceutically relevant substrates such as heterocycles. These reactions provide an ideal case for high throughput reaction screening though add an additional layer of complexity when compared to the amide bond formations. Herein, we will consider several examples of reaction screening applied to Suzuki coupling chemistry focusing on the decisions and learnings gained from these experiments.

A Suzuki coupling in its simplest form is a metal-mediated (generally palladium) reaction utilized to couple an organoboron species (which can be generated *in situ*) with an aryl halide (generally a bromide) in a solvent under base-mediated conditions. Numerous studies have been carried out on the mechanism, scope, and use of this reaction, but here we will present fairly basic examples of this chemistry. For the first example, we were interested in developing a robust synthetic route amenable to the scale-up of the three compounds shown in Scheme 5.4 for a program looking to identify Pi3K inhibitors.[25]

The original route to these compounds is shown in Scheme 5.5. As can be seen, this is an eight-step sequence for which the point of diversity (the aryl ring) is introduced right at the outset. In addition, although 2,4-dichloroacetophenone is commercially available, the other two acetophenones have to be synthesized, and finally a number of steps are somewhat capricious and not amenable to large-scale synthesis. For example, step 1

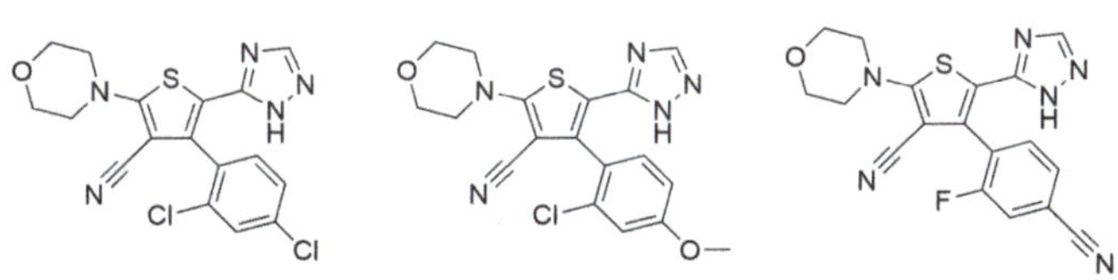

**Scheme 5.4** Pi3K lead compounds.

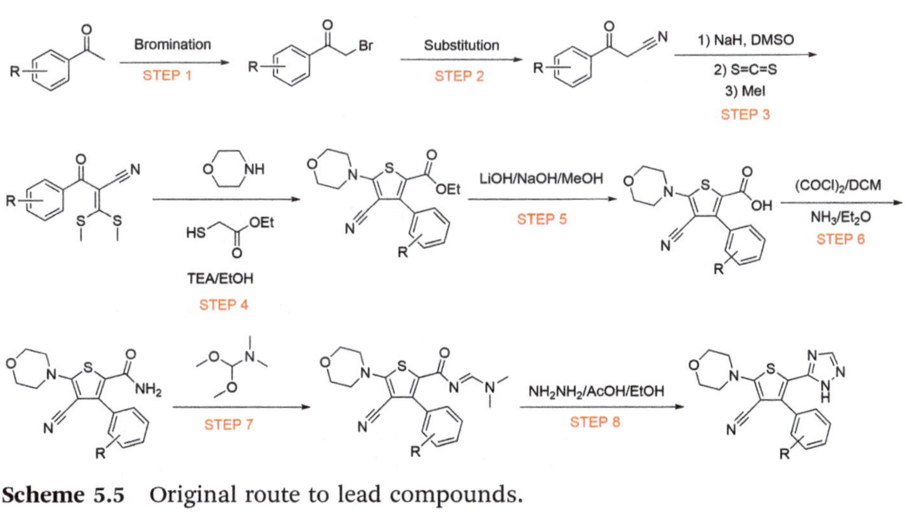


**Scheme 5.5** Original route to lead compounds.

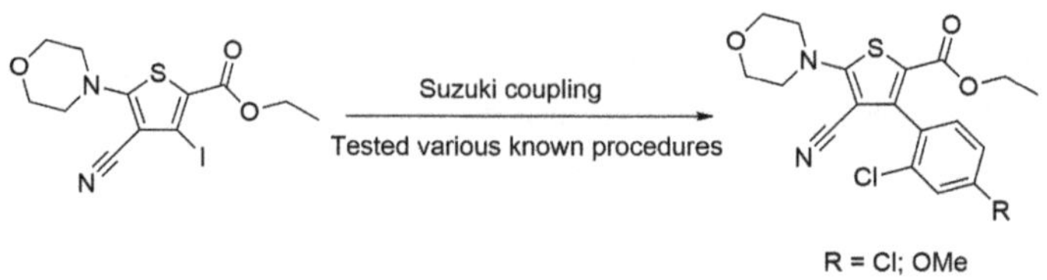


**Scheme 5.6** Projected Suzuki reaction for scale-up.

leads to rapid bis-halogenation, and the modified three-component Gewald reaction to form the thiophene in step 4 is messy, and difficult to purify.

An alternative plan would be to introduce the aryl fragment at a later stage in the synthesis, allowing both an opportunity to further probe the SAR of this position and also enabling an advanced common intermediate to be utilized in the subsequent scale-up chemistry. It was envisioned that Suzuki coupling of the iodide (of either the amide or the ester) would be an ideal reaction to introduce these, with the added advantage that all three of the desired boronic acids were commercially available (Scheme 5.6). With these pieces installed, then the amide or ester could be elaborated to the desired triazole in two or four steps, respectively.

A robust five-step route was developed to the ester, which could be subsequently easily converted into the amide. This would be an ideal substrate for the coupling in that it is closer to the end of the synthesis. However, efforts to utilize the amide in the desired Suzuki couplings were hampered by issues such as poor solubility, low conversion, and de-halogenation leading to tough purifications, and these findings were reinforced by an unsuccessful reaction screen of 96 possible conditions – none of which led to appreciable yields of the desired coupling product. As such, attention was switched to the more soluble ester. Initial attempts to utilize literature-based conditions generated significant amounts of des-iodo compound, and suffered from consistently low yields. In addition, the products had poor solubility in organic solvents, and displayed similar polarities to the starting iodide, thus presenting a challenge for purification. With this in mind, a reaction screen was designed not only to optimize the discrete reaction between each set of coupling partners, but also to identify generic conditions, which might be more widely applicable to this coupling in general. The screen evaluated three Pd-catalysts, four aqueous bases in four solvents for the ester, and two of the available boronic acids (2,4-dichlorophenylboronic and 2-chloro-4-methoxyphenylboronic acid). The results are displayed in the Spotfire plot shown in Figure 5.10. The reaction number is reproduced on the *x*-axis with the product total on the *y*-axis. The two trellises represent the different boronic acids assayed with the legend for color/shape reproduced on the plot.

The Spotfire plot shows that numerous reactions have performed well for both substrates, though in general the 2-chloro-4-methoxyphenylboronic acid is a better substrate. The three highlighted conditions in Figure 5.10

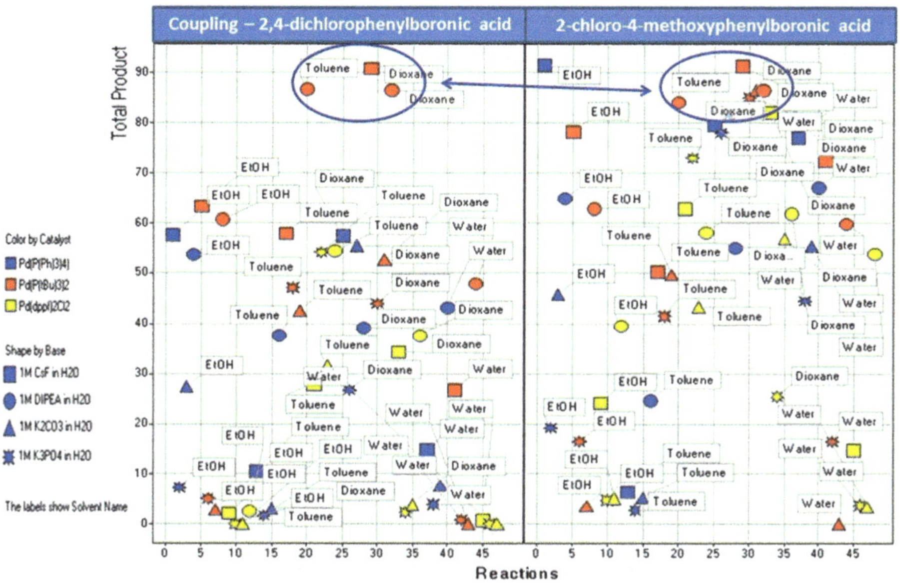


**Figure 5.10** Spotfire analysis of Suzuki couplings.

worked extremely well for both substrates (>80% yield), all of which utilize $Pd(P^tBu_3)_2$ as the catalyst. In this case, a scale-up of these compounds was desired, and with this in mind it was important to do a reality check prior to scaling. There are several potential disconnects between a screening hit and actually reproducing the reaction on a practical scale. The difficulty of using standard conditions to reproduce a screen that is run under strictly anhydrous and oxygen-free conditions in a glove-box has already been discussed. Although this is not the case for this example, other potential issues may arise, which may hinder the proposed path forward. The first of these is a concentration effect. At this discovery level, reactions in the screening paradigm tend to be run at concentrations between 0.01 and 0.05 M. Although to produce mg quantities of an analogue it may be practical to directly scale the screening conditions, for gram quantities, as in this case, an adjustment in concentration has to be made. This can lead to an irreproducible outcome to the reaction under investigation, and one has to be aware of this. If this is observed, more advanced DoE studies can be carried out to examine the most important variables (temperature, concentration, *etc.*) to establish a successful reaction.[13] Secondly, to facilitate the setting up of screens utilizing pipetting techniques, many catalysts/ligands or reagents are introduced as stock solutions, which are made up in a single solvent. Given that solvent is one of the variables, which is often evaluated in a reaction screen, the use of such stock solutions introduces an additional solvent into the reaction mixture. Although this is likely to only amount to *ca.* 2–10% of the total solvent volume, one cannot underestimate that it may affect the outcome on scaling up or reproducing the reaction. Usually, on running the reaction after the screen, the tendency is to introduce the reagents neat, and so the solvent forming the stock solution is simply omitted. Finally, perhaps the biggest disconnect in reproducing the result obtained in a reaction screen is a mass transfer effect.[26] As noted, these screens are typically run in small HPLC type glass vials, which have a small stirrer bar. Particularly for biphasic, heterogeneous, and gas-phase reactions, the effects observed from mixing in the small vial are going to be very different even when scaling the reaction into conventional glassware. Process engineers are familiar with these phenomena on transferring reactions into the plant, and can model such effects. However, this is often a concept that is alien to many discovery chemists, but is probably the most likely to cause issues with irreproducibility of data from reaction screens.

With the current reaction from Scheme 5.6, mass transfer is likely to be an issue for some of the reaction conditions that are biphasic. In addition, the concentration must be increased for scale up. Returning to the three sets of reaction conditions established from the Suzuki screen, each of these was reproduced on a 1 g scale. From this, it was observed that the two conditions that utilize Hunig's base lead to lower yields and more des-iodination possibly due to Pd-aggregation and precipitation on the larger scale. The remaining conditions using CsF as the base in dioxane/water provided both analogues investigated in the screen in greater than 98% conversion, and were also shown to be successful for the other compound of interest. Palladium removal on small scale was achieved using a Si-thiol resin, with the final products being obtained after aqueous wash and crystallization.[27] The key lead compound was scaled to 250 g using minor modifications of these conditions (lower catalyst loading, ten-fold increase in concentration). Extraction into toluene on scale followed by cooling and crystallization led to direct isolation of the compound in 90% yield and >98% purity without the need for a Si-thiol treatment (Scheme 5.7). One thing to note with this example is that the solvent utilized to make up the stock solutions (in this case toluene) was retained throughout the scale-up studies, thus illustrating the potential to find a more environmentally-benign solvent system for this reaction.[28]

The following example looks at a further advantage of high throughput reaction screening in the discovery space with the rapid optimization of reaction conditions being required to expedite reaction time for an initial scale-up of material (~100 s of mg), preserve precious starting material, and

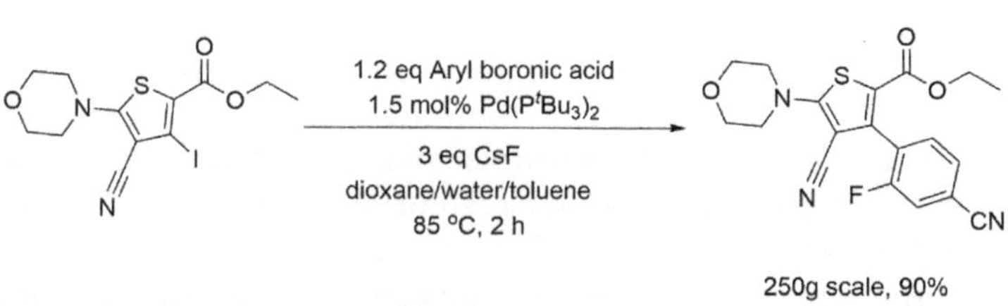


**Scheme 5.7** Scale-up of lead Pi3K compound.

facilitate the purification sequence. The reaction is shown in Scheme 5.8. The lead conditions utilize DME/water with Pd(dppf)$Cl_2$ as the catalyst and $NaHCO_3$ as the base leading to a 50% yield after two days reaction time at 80 °C, with an arduous separation of the desired product and the des-brominated starting material. In addition, both the starting materials are custom pieces, and as such, in short supply.

A reaction screen was designed to examine this coupling looking at 96 reactions using 113 mg of starting material with each reaction being run on a 0.003 mmol scale. The screen design focused on six catalyst/ligand combinations, and four solvents and bases, and the reactions were run for 18 h in the glove-box at 80 °C. From the conditions examined, only one out of 96 gave the desired product in >80% *in situ* yield (Figure 5.11).

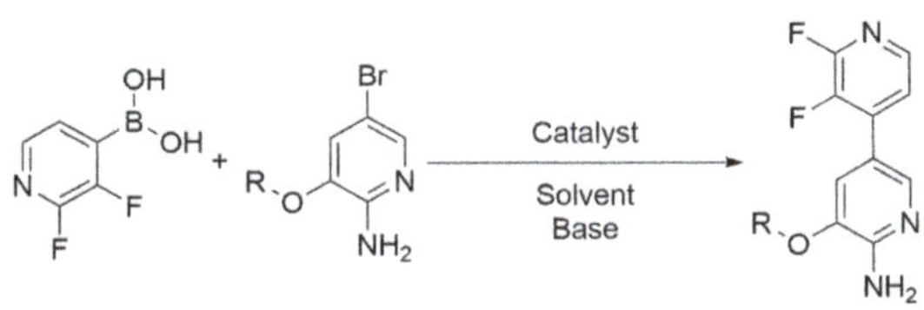


**Scheme 5.8** Novel Suzuki coupling.

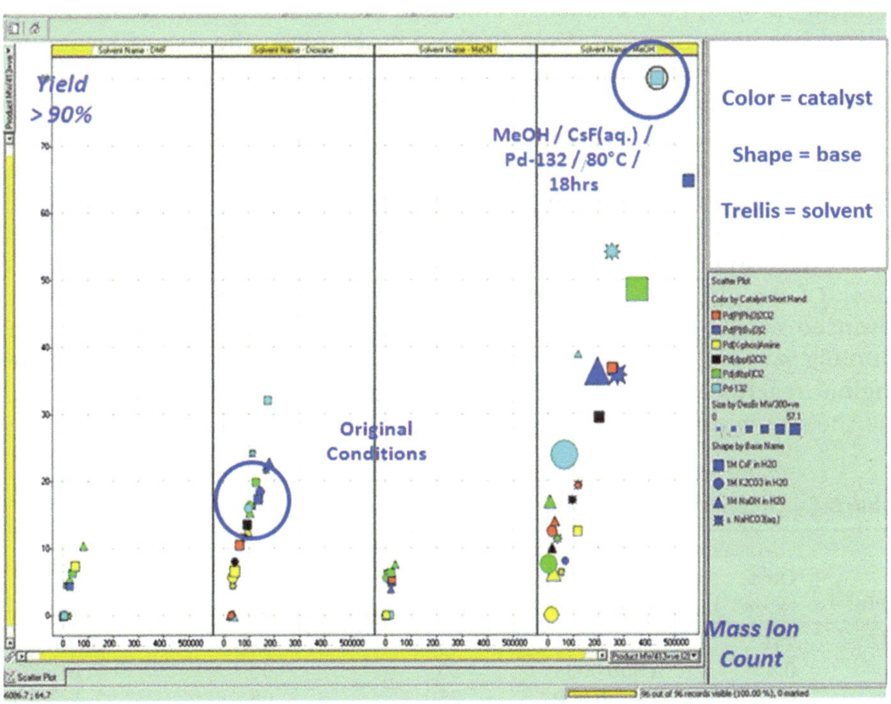


**Figure 5.11** Spotfire analysis reveals novel conditions.

As shown, the conditions utilize MeOH, CsF, and Pd-132 indicating the crucial synergistic interplay between these three factors. Such interactions are extremely difficult to find, and it is unlikely that an optimization focused on optimizing one factor at a time would have worked here. Equally impressively, the screen results are turned around in a day with the compound then scaled to 200 mg under these conditions with a 90% isolated yield.

The concept of using screening as a tool to provide green conditions for a specific transformation has been briefly covered, and it has also been revealed that an alternative solvent for a reaction under investigation can perhaps provide the quickest and most tangible impact in this respect. However, the Suzuki couplings so far considered illustrate another real problem within organic synthesis today: the prevalence of Pd-mediated reactions in discovery chemistry. Due to their efficiency, versatility, and overall utility, these reactions are often inherited into a development process if a compound progresses. In particular, for Suzuki reactions, only Pd has been employed within process at the current time.[29] There is a growing fundamental concern with the use of Pd, and there is an on-going drive to replace this precious metal in these processes with base metals such as Ni, Cu, and Fe for cost, toxicity, supply, and environmental reasons. A comparison of these metals is shown in Table 5.1.[30]

There have been reports of base-metal catalyzed Suzuki reactions in the chemical literature, but implementing them directly into a discovery chemistry program is a challenge unless an identical substrate to that being utilized has been reported. Even in these cases, given the risk of having to venture into chemically uncharted territory, the discovery chemist will still likely explore the well-established Pd-route. Reaction screening though offers an opportunity to explore the potential for utilization of an alternative metal in a coupling process with low risk but a potential high reward. The cost to attempt this is minimal in terms of material (the material-sparing approach of glove-box screening has been discussed) and time (given the ease of reaction set-up using stock solutions and liquid handling techniques). Of the alternative base-metals employed, Ni appears to be the most advanced at the current time, and, with this in mind, a project Suzuki coupling was evaluated using both Pd and Ni sources side-by-side.[14,31] The original reaction returned a 20–40% yield under Pd-mediated conditions, and the screen examined six Ni and six Pd sources with a range of bases in a

**Table 5.1** Greener alternatives to Pd.

| Metal | Cost ($\$$ $oz^{-1}$) | Annual production (tons) | Oral exposure limits (ppm) | Natural abundance (ppm) | Carbon footprint |
|---|---|---|---|---|---|
| Pd | 746 | 24 | 10 | 0.015 | 6649 |
| Ni | 0.44 | 1 350 000 | 25 | 90 | 7.1 |
| Cu | 0.23 | 15 000 000 | 250 | 68 | 1.2–3.4 |
| Fe | 0.005 | 1 200 000 000 | 1300 | 56 300 | 1 |

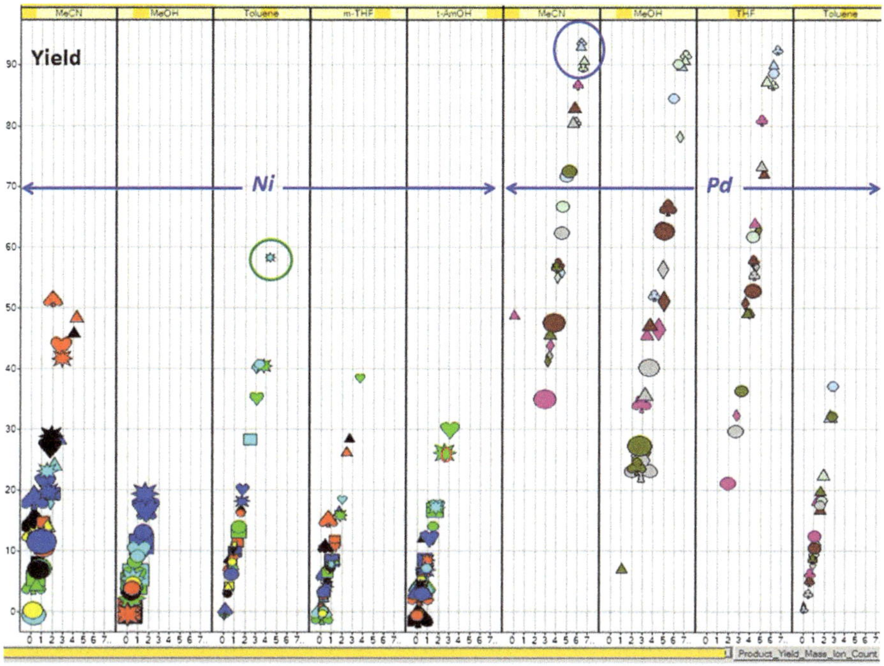


**Figure 5.12** Comparison of Ni to Pd catalysis.

series of five organic solvents. In total *ca.* 300 reactions were run utilizing ~150 mg of material. The results are displayed in Figure 5.12 with the trellises sorted by metal-source utilized.

Although the screen has identified several Pd-mediated conditions that give >90% *in situ* yield, which represent a significant improvement over the original reaction conditions, one can observe that a more modest hit for the reaction has been identified (50–60% yield) using a Ni-based system. Although the team may choose to persist with a Pd-mediated reaction at the given time, this is still useful information to have in hand if the compound is to progress with the opportunity to optimize this hit. In addition, it is critical to remember the development in Ni-based coupling chemistry is still in its infancy (particularly when compared to Pd), and as such future developments with regards to both ligands and catalysts are only going to enhance its scope and utility.

## 5.5 Biotransformations

The significant growth in the application of biotransformations in drug discovery over the past decade has been no doubt linked to the realization that often these can be carried out in typical laboratory glassware using readily available enzyme reagents. In addition, numerous screening kits of classes of enzymes have become commercially available, enabling a chemist to adopt a high throughput screening approach to assay the viability of this particular route in an expeditious manner at an early stage of a program. Kits of numerous esterases, hydrolases, ketoreductases, and transaminases are available from a range of suppliers with detailed instructions on how to perform the screen for which the enzymes are intended. Major advantages of the use of enzymes have been well documented in terms of often working in an aqueous environment, and the ability to obtain highly enantioselective transformations. Drawbacks that are often cited with this approach are the inherent bias of many enzymes to give a particular enantiomer and limitations in terms of substrate scope. Many of these transformations are often resolutions, which do enable access to both enantiomers (though loss of 50% of material is an issue unless a recycle or *in situ* racemization can be performed). It is also important to note that often a sub-optimal enzyme hit can be further optimized for a specific substrate and transformation through a series of site-directed mutagenesis. There are numerous subtleties involved in scaling biotransformations even to gram scale (co-factor recycling, pH control, co-solvents for optimal concentrations), though these are becoming increasingly well documented to enable widespread uptake of the methodology in an efficient manner.[32]

An example of a biotransformation identified through screening is given in Scheme 5.9, showing the synthesis of the chiral alcohol used in the synthesis of crizotinib. Initial screening of the asymmetric reduction using chiral borane reagents failed to give appreciable levels of enantioselectivity. Looking at a resolution approach led to the early identification and implementation of a PLE-mediated resolution of the racemic acetate, which is highly efficient in terms of conversion and enantioselectivity.[33] Moreover, it is possible to convert the undesired enantiomer into the correct stereochemistry through a Mitsunobu reaction. The availability of gram scale quantities of the chiral alcohol allowed the medchem team to rapidly advance the program through optimization of the right-hand side of the compound. Upon development, this resolution approach was switched to

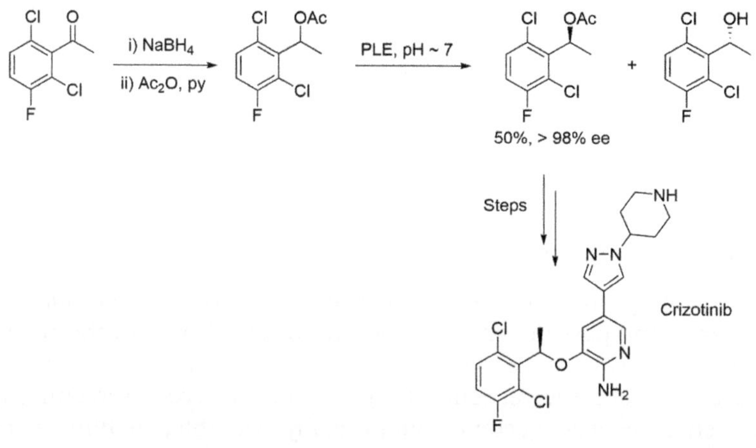


**Scheme 5.9** Pig liver esterase (PLE)-mediated resolution in crizotinib synthesis.

utilize a modified ketoreductase for the reduction of the acetophenone.[34] Notably, the results here were somewhat fortuitous in that the ketoreductases evaluated in the screens at the time only gave the desired (*S*)-enantiomer, and no system was identified to give the (*R*)-enantiomer.

## 5.6 Screening Gaseous/Heterogeneous Reactions

With regard to setting up reaction screens, typically there is often no requirement for specialist equipment, and to implement a program from the ground up the initial investment and focus should be placed on having a robust analytical platform in place to evaluate the reactions in as expeditious a manner as possible. Small glass vials, and multi-channel pipettes, are commonly available, and although using a glove-box is an advantage in many cases, in particular in terms of minimizing scale, this is by no means a necessity. Expanding the scope of reactions that one is able to screen to encompass gas-mediated reactions does usually require specialized equipment, and one must consider the benefit of such an investment. In addition, for reactions such as hydrogenation, heterogeneous catalysis is often the method of choice, and as such use of stock solutions is not an option. In this case, access to an automated weighing robot presents a major advantage as the alternatives are either inaccurate or extremely labor intensive. The other two areas in which such equipment can provide an advantage is in weighing immobilized enzymes for biotransformation screening to provide customized plates as well as in preparing plates of chiral acids and bases for resolution screening (*vide infra*).

The screening of gas reactions in a 96-well plate/vial format can be accomplished using a HEL Cat96 (Figure 5.13), which features a chamber in which the plate can be placed, and then pressurized with the gas of interest,

**Figure 5.13** Cat96 for screening gaseous reactions.

whilst the constituent vials are heated and agitated. The software for these systems allows automated venting, and heating and cooling cycles, and it is important to remember to pierce the septa of the vials to allow gas in-flow. Although this platform is used for carbonylation and other gaseous reactions, its most extensive use is in evaluating reductions using hydrogen. Although there are valid safety concerns with use of hydrogen in discovery, the cleanliness of the reactions, ease of work-up, and access to specialized facilities makes this a very attractive transformation within process chemistry.

The following represents an example of a heterogeneous catalyst hydrogenation reaction screen, and highlights factors to be aware of when running this chemistry. There was interest in determining a potentially scalable route to 1-bicyclo[1.1.1]pentylamine, and it was found that the azido iodide shown in Scheme 5.10 could be accessed in two steps from a solution of propellane. Reduction of this material with concomitant des-halogenation would lead to the desired amine.[35]

Several attempts to previously reduce this material had proven unsuccessful, and in order to quickly investigate potential reduction conditions, a high throughput hydrogenation screen was initiated in a plate format using the Cat96. Each reaction was run using 0.01 mmol of substrate at 0.045 M concentration. Overall, the screen evaluated *ca.* 40 catalysts, three solvents, two additives, and two different temperatures with the reactions being run at 4 bar hydrogen pressure for 15 h. In total, over 240 different combinations were evaluated (Figure 5.14).

From this screen, $Pd(OH)_2$ with HCl in MeOH emerged as the most promising conditions to mediate this transformation, appearing to give ~50–60% yield (based on comparison with an authentic standard of known concentration). On attempting to scale these conditions, only a 16% isolated yield was obtained with substantial amounts of methanol displacement being observed. This is believed to be a reflection of the difference in the aspects of mass transfer as discussed previously when transferring a reaction from a vial to a hydrogenation vessel, which features an impeller and is better able to agitate the solution. This example does emphasize a further advantage of reaction screening in that one is able to evaluate an approach using a diverse array of reagents/conditions and rapidly make a go/no go decision on the validity of the methodology and look for an alternative solution if necessary, as in this example.[36]

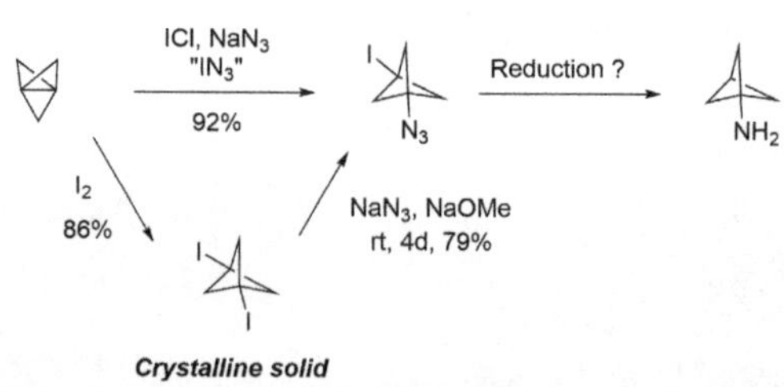


**Scheme 5.10** Proposed synthesis of 1-bicyclo[1.1.1]pentylamine.

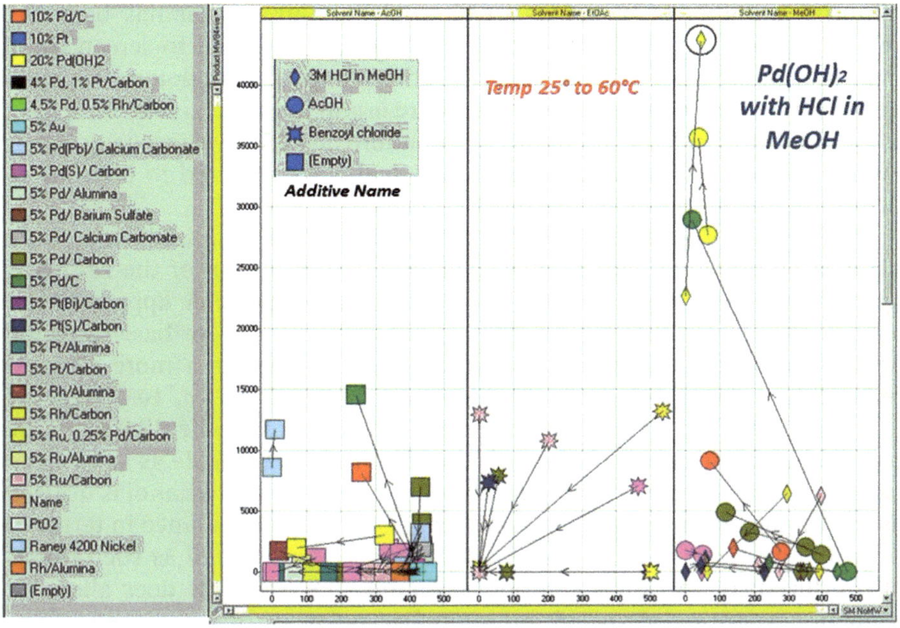


**Figure 5.14** Spotfire analysis on hydrogenation screen.

## 5.7 Resolutions/Purifications and Purges

Classical resolutions are a well-established method for the formation of enantiopure compounds.[37] The technique relies on formation of a pair of diastereomeric salts, which display different solubilities enabling their separation by techniques such as recrystallization. Obviously, the greater the difference in the solubilities, the more effective is the resolution. Classical resolutions are commonly carried out within process chemistry, and are usually placed at the early stage of a synthesis. The key reason for this is that 50% of the material (the undesired enantiomer) is discarded, and it is judicious to do this at an early stage of the synthesis. Their widespread use in process illustrates the utility and scalability of a classical resolution. The key to investigating this technique is the presence of a functional handle within the molecule of interest, for example a carboxylic acid or a basic amine that can react with a chiral base or acid, respectively, to generate the desired diastereomeric salt.

There are numerous sources of chiral acids and bases and, over time, it is possible to acquire extensive collections (75–100 compounds) of both chiral acids and chiral bases. Given the number of possible combinations of substrate/resolving agent/solvent, this to some degree represents the classical scenario of searching for a needle in a haystack, and as such classical resolutions are an ideal transformation to rapidly evaluate using a high throughput screen employing a semi-automated workflow and utilizing minimal amounts of material. An alternative solution that has been developed is to group the resolving agents into families, and to screen these families against a specific substrate.[38] Although this methodology has been demonstrated to be highly successful, there is still both a degree of screening to be done to find the optimal combination of substrate/resolving agent family/solution and also additional analytical work required to be done to determine the structure of the resolved complex.

The key part of a screen is the "salt preparation," which initially requires a plate to be prepared with a known amount/concentration of the resolving agent dispensed to each discrete well. There are two possible approaches to achieve this. Given that the majority of either chiral acid- or base-resolving agents are solids, using an automated solid-dispenser is a more efficient method to prepare the plate, allowing accurate weighing (*ca.* 10% error) of milligram amounts of solids to the vials, which can then be diluted to the required concentration. The reagents, which are not available as solids, can easily be added as pre-prepared stock solutions. Typically, ethanol is utilized as the initial screening solvent as this represents a good balance in terms of polarity for solubility of potential salts that are formed, and as such this is also utilized for the stock solutions. Notably, the workflow does allow for sequential variation of the solvent (more/less polar) without the need for the preparation of a new plate. After the plate is made up, a visual inspection of the wells is critical to aid the potential identification of false positives. Once completed, an ethanolic solution of the compound under investigation is added with the wells again inspected for precipitation. The plate is then typically processed through heat cycle, and then slowly cooled, allowing any precipitates formed to settle. The plate is again inspected for precipitates. At this point, the key component for analysis is having a chiral method in place to analyze the enantiopurity of samples. This is usually not an issue, as it is likely that the material has previously been separated or analyzed in house. From the vials, which display precipitates, samples of the mother liquor are withdrawn to assay the enantiopurity. If enantioenrichment is observed, then the solids are isolated, and also analyzed to determine the ee. If no precipitates are observed in the initial cycle, then either a slow evaporation or replacement of the solvent with a less polar solvent followed by repeating this cycle can be attempted to promote the formation of precipitates.

Once hit(s) are identified through isolation of a moderate to high ee solid, optimization studies can be carried out through variation of the solvent (switching between alcohols), concentration, recrystallization, *etc.*, to obtain enantiopure material. At this juncture, it is also critical to assess the recovery from the process, as obviously one wishes this to be as close to 50% as possible. If multiple hits are obtained, then selection of the optimal conditions will depend not only on the ee upgrade/recovery of the process, but also on the cost and availability of the resolving agent being employed, as well as its ease of recovery from the salt break process. The workflow detailed here is iterative in nature, simple to carry out and implement, and allows rapid identification of possible classical resolution systems for scale-up. In

70–80% of the cases studied, this approach has enabled identification of a suitable classical resolution.

The following illustrates an example of this approach. This represents an ideal case to evaluate a classical resolution approach. The project was sufficiently advanced within medicinal chemistry so as that bulk quantities of the racemate were available. In addition, the molecule possessed multiple salt handles with a highly basic center close to the chiral center, and general synthetic approaches to access this class of compounds are non-trivial. The initial approaches to access these compounds utilized SFC separation. Although this was highly efficient, the optimal separation conditions required derivatization of the molecule as the Cbz derivative. As such, two further steps were required in addition to the chiral separation to obtain the desired enantiopure fragment. Given this, a resolution screen was initiated. Following the process outlined above, several possible resolving agents were identified. Dibenzoyl-L-tartaric acid was selected as the resolving agent for optimization due to it being readily available in both enantiomeric forms (thus allowing access to both enantiomers of the substrate), and through minimal optimization of the solvent system a scalable highly efficient resolution system was identified (Scheme 5.11).

This example illustrates the resolution of an amine, though an identical method is equally applicable to acid resolution using a collection of chiral amines to form the salts. Of further interest is the applicability of reaction screening in development of purification procedures based either on crystallizations or purges of impurities from solid materials. Herein, one can envision similar processes being implemented with variations on the work flow described for a classical resolution.

For example, in the case of a crystallization, one can either look at formation of salts (replacing the chiral acids and bases described above) or simply dispense crude material, and look for precipitation after a heat–cool sequence in a range of different solvents. Initial analysis focuses on whether solids have been formed, and an inspection of their physical appearance (ease of filtration). A subsequent scaling will look at both purity and optimizing the recovery of desired material. In the case of investigating a purge of impurities from solid material, again the crude solid can be dispensed across a series of vials with different solvents being added. After a heat–cool cycle, analysis of the supernatant will enable a determination to be made (based on the known crude standard) on both whether impurities have been purged and how much of the desired parent has been lost to the solvent. Both are critical parameters in determining if an effective purge through a slurry type process is logistically viable for the material under investigation.

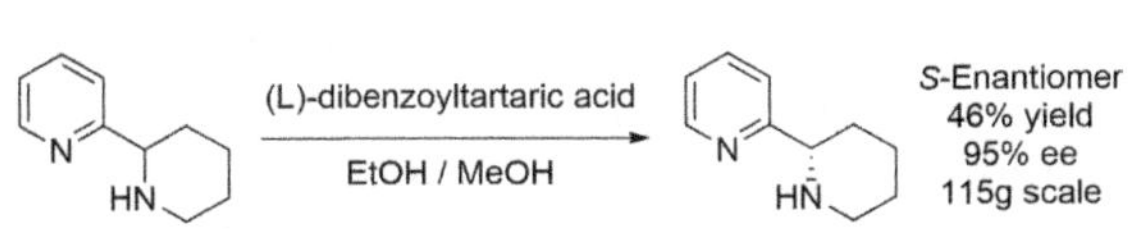


**Scheme 5.11** Scale-up of a classical resolution identified through screening.

## 5.8 Capturing Information/Conclusions

It is hoped that this chapter has briefly illustrated the value of implementing a high throughput reaction screening platform within discovery chemistry, and its applications in exploring greener alternatives to conventional reaction conditions and purifications. One final concept that is very important is the acquisition of data over time that such a platform will acquire, and is important to capture. Although this chapter has focused on how the workflows have been expedited by screening, it is still important to realize that screening represents a technology that should be employed judiciously based on the information available to the team. Although direct examples of the types of substrates being worked upon are not often available in the chemical literature, by building up an internal database of screening reactions, results on similar substrates will be acquired, and will represent a rich source of information to make a rapid informed choice of initial conditions to assay in new projects.

## Acknowledgements

The author would like to thank Neal Sach (Pfizer), who introduced and streamlined the reaction screening platform to the Discovery group in La Jolla, CA. The content of this chapter is based on discussions and collaborations with Neal.

## References

1. P. T. Anastas and J. C. Warner, *Green Chemistry Theory and Practice*, Oxford University Press, 1998.
2. (a) J. R. Schmink, A. Bellomo and S. Berritt, *Aldrichimica Acta*, 2013, **46**, 71; (b) A. I. Morrell, *New Synthetic Methodologies in Medicinal Chemistry*, RSC Publishing, Cambridge, 2012.
3. See for example, J. G. de Vries and A. H. M. de Vries, *Eur. J. Org. Chem.*, 2003, 799.
4. R. Macarron, M. N. Banks, D. Bojanic, D. J. Burns, D. A. Cirovic, T. Garyantes, D. V. S. Green, R. P. Hertzberg, W. P. Janzen, J. W. Paslay, U. Schopfer and G. S. Sittampalam, *Nat. Rev. Drug Discov.*, 2011, **10**, 188.
5. E. M. Gordon, J. F. Kerwin, Jr., W. J. Dower, S. P. A. Fodor and M. A. Gallop, *J. Med. Chem.*, 1994, **37**, 1385.
6. S. Ribe and P. Wipf, *Chem. Commun.*, 2001, 299.
7. P. M. Murray, S. N. G. Tyler and J. D. Moseley, *Org. Process Res. Dev.*, 2013, **17**, 40.
8. See for example, H.-D. Xu, R.-W. Zhang, X. Li, S. Huang, W. Tang and W.-H. Hu, *Org. Lett.*, 2013, **15**, 840.
9. R. Chinchilla and C. Nájera, *Chem. Soc. Rev.*, 2011, **40**, 5084.
10. (a) J. D. Moseley and P. M. Murray, *J. Chem. Technol. Biotechnol.*, 2014, **89**, 623; (b) N. Fey, J. N. Harvey, G. C. Lloyd-Jones, P. Murray, A. G. Orpen,

R. Osborne and M. Purdie, *Organometallics*, 2008, **27**, 1372; (c) J. Jover, N. Fey, J. N. Harvey, G. C. Lloyd-Jones, A. G. Orpen, G. J. J. Owen-Smith, P. Murray, D. R. J. Hose, R. Osborne and M. Purdie, *Organometallics*, 2010, **29**, 6245; (d) J. Jover, N. Fey, J. N. Harvey, G. C. Lloyd-Jones, A. G. Orpen, G. J. J. Owen-Smith, P. Murray, D. R. J. Hose, R. Osborne and M. Purdie, *Organometallics*, 2012, **31**, 5302.

11. (a) E. Valeur and M. Bradley, *Chem. Soc. Rev.*, 2009, **38**, 606; (b) A. El-Faham and F. Albericio, *Chem. Rev.*, 2011, **111**, 6557.
12. Basavaprabhu, T. M. Vishwanatha, N. R. Panguluri and V. V. Sureshbabu, *Synthesis*, 2013, **45**, 1569.
13. H. Tye, *Drug Discov. Today*, 2004, **9**, 485.
14. D. S. MacMillan, J. Murray, H. F. Sneddon, C. Jamieson and A. J. B. Watson, *Green Chem.*, 2013, **15**, 596.
15. S. D. Ramgren, L. Hie, Y. Ye and N. K. Garg, *Org. Lett.*, 2013, **15**, 3950.
16. See for example, R. K. Henderson, C. Jiménez-González, D. J. C. Constable, S. R. Alston, G. G. A. Inglis, G. Fisher, J. Sherwood, S. P. Binks and A. D. Curzons, *Green Chem.*, 2011, **13**, 854.
17. See for example, K. Alfonsi, J. Colberg, P. J. Dunn, T. Fevig, S. Jennings, T. A. Johnson, H. P. Kleine, C. Knight, M. A. Nagy, D. A. Perry and M. Stefaniak, *Green Chem.*, 2008, **10**, 31.
18. See for example, A. Kadam, M. Nguyen, M. Kopach, P. Richardson, F. Gallou, Z.-K. Wan and W. Zhang, *Green Chem.*, 2013, **15**, 1880.
19. D. J. C. Constable, P. J. Dunn, J. D. Hayler, G. R. Humphrey, J. L. Leazer, Jr., R. J. Linderman, K. Lorenz, J. Manley, B. A. Pearlman, A. Wells, A. Zaks and T. Y. Zhang, *Green Chem.*, 2007, **9**, 411.
20. See for example, R. Bandichhor, A. Bhattacharya, L. Diorazio, P. Dunn, K. Fraunhoffer, F. Gallou, J. Hayler, M. Hickey, B. Hinkley, D. Hughes, L. Humphreys, B. Kaptein, S. Mathew, L. Oh, P. Richardson, T. White and S. Wuyts, *Org. Process Res. Dev.*, 2013, **17**, 1394.
21. (a) P. D. Higginson and N. W. Sach, *Org. Process Res. Dev.*, 2001, **5**, 331; (b) P. D. Higginson and N. W. Sach, *Org. Process Res. Dev.*, 2004, **8**, 1009.
22. For an example of the iChemExplorer platform in use, see L. M. Stencel and N. E. Leadbeater, *New J. Chem.*, 2014, **38**, 242.
23. The prevalence of amide bond formation reactions varies across analyses ranging from 15 to as high as 50%. For a typical analysis, see S. D. Roughley and A. M. Jordan, *J. Med. Chem.*, 2011, **54**, 3451.
24. C. C. C. Johansson Seechurn, M. O. Kitching, T. J. Colacot and V. Snieckus, *Angew. Chem., Int. Ed.*, 2012, **51**, 5062.
25. K. K.-C. Liu, J. Zhu, G. L. Smith, M.-J. Yin, S. Bailey, J. H. Chen, Q. Hu, Q. Huang, C. Li, Q. J. Li, M. A. Marx, G. Paderes, P. F. Richardson, N. W. Sach, M. Walls, P. A. Wells and A. Zou, *ACS Med. Chem. Lett.*, 2011, **2**, 809.
26. See for example, R. Klaewkla, M. Arend and W. F. Hoelderich, *Mass Transfer – Advanced Aspects*, Intech, 2011.
27. E. J. Flahive, B. L. Ewanicki, N. W. Sach, S. A. O'Neill-Slawecki, N. S. Stankovic, S. Yu, S. M. Guinness and J. Dunn, *Org. Process Res. Dev.*, 2008, **12**, 637.

28. Q. Huang, P. F. Richardson, N. W. Sach, J. Zhu, K. K.-C. Liu, G. L. Smith and D. M. Bowles, *Org. Process Res. Dev.*, 2011, **15**, 556.
29. J. Magano and J. R. Dunetz, *Chem. Rev.*, 2011, **111**, 2177.
30. (a) Metal prices refer to commodity prices. All given prices ($/troy Oz) from 29 August 2013 were accessed from www.metalprices.com/dailyexchangedata/Index; (b) Annual 2006 production accessed 28 August 2013 at www.worldofwallstreet.us/metals-annual-production-.html; (c) Specific limits for residues of metals accessed 29 August 2013 at www.ema.europa.eu/docs/en_GB/document_library/Scientific_guideline/2009/09/WC500003586.pdf; (d) The carbon footprint has historically defined as the "total set of greenhouse gas (GHG) emissions caused by an organization, event, product or person".
31. See, for example, S. Ge and J. F. Hartwig, *Angew. Chem., Int. Ed.*, 2012, **51**, 12837.
32. *Practical Methods for Biocatalysis and Biotransformations 2*, ed. J. Whitton and P. Sutton, John Wiley & Sons, Inc., Hoboken, New Jersey, 2012.
33. C. A. Martinez, E. Keller, R. Meijer, G. Metselaar, G. Kruithof, C. Moore and P. P. Kung, *Tetrahedron: Asymmetry*, 2010, **21**, 2408.
34. P. D. de Koning, D. McAndrew, R. Moore, I. B. Moses, D. C. Boyles, K. Kissick, C. L. Stanchina, T. Cuthbertson, A. Kamatani, L. Rahman, R. Rodriguez, A. Urbina, A. Sandoval (nee Accacia) and P. R. Rose, *Org. Process Res. Dev.*, 2011, **15**, 1018.
35. K. D. Bunker, N. W. Sach, Q. Huang and P. F. Richardson, *Org. Lett.*, 2011, **13**, 4746.
36. Recently, a successful reduction of this compound has been published. See, Y. L. Goh, E. K. Tam, P. H. Bernardo, C. B. Cheong, C. W. Johannes, A. D. William and V. A. Adsool, *Org. Lett.*, 2014, **16**, 1884.
37. R. M. Kellogg and M. Leeman, *Compr. Chirality*, 2012, **9**, 367.
38. T. R. Vries, H. Wynberg, E. van Echten, J. Koek, W. ten Hoeve, R. M. Kellogg, Q. B. Broxterman, A. Minnaard, B. Kaptein, S. van-der Sluis, L. A. Hulshof and J. Kooistra, *Angew. Chem., Int. Ed.*, 1998, **37**, 2349.

CHAPTER 6

# Continuous Processing in Drug Discovery

KATHERINE BELECKI AND B. FRANK GUPTON*

Department of Chemical and Life Science Engineering and
Department of Chemistry, Virginia Commonwealth University, Richmond, VA 23284-9068, USA
*Email: bfgupton@vcu.edu

## 6.1 Introduction

The advancement of continuous processing methods in pharmaceutical applications has been identified as one of the highest research priorities of the American Chemical Society Green Chemistry Institute Pharmaceutical Roundtable.[1] This chapter aims to provide an overview of the concepts, perspectives, techniques, and equipment that make continuous chemistry an enabling technology for improving the efficiency and environmental impact of the drug discovery process. The untapped reaction space that could be made accessible by flow chemistry provides a compelling argument for incorporating continuous methods into medicinal chemistry programs. Moreover, the potential to "build in" a continuous process strategy from the early stages of drug discovery represents an opportunity not only to gain efficiencies in the identification of new biologically active molecules but also to streamline and integrate the downstream processes that lead to commercial implementation. For this reason, we have elected to expand the scope of this chapter to include issues and best practices for continuous larger scale active pharmaceutical ingredient (API) production in order to

RSC Drug Discovery Series No. 46
Green Chemistry Strategies for Drug Discovery
Edited by Emily A. Peterson and Julie B. Manley

Published by the Royal Society of Chemistry, www.rsc.org

provide a more holistic perspective related to the challenges and opportunities for implementing flow chemistry into drug discovery.

## 6.2 Green Process Strategies

### 6.2.1 Process Intensification

The principles of green chemistry and the objectives of process intensification share a common goal, which is to increase the overall efficiency and selectivity of chemical processing.[2] To achieve this goal, process intensification methods often take advantage of more extreme reaction conditions by employing higher temperatures and pressures. Most batch processes in pharmaceutical operations are carried out under atmospheric pressure, which limits the ability to achieve reaction temperatures in excess of the selected solvent's boiling point. In contrast, most continuous reactor systems possess the capability to operate at significantly higher temperatures and pressures to expand the window of process operability (Figure 6.1). Extremely fast reactions with large exotherms can be effectively controlled in continuous reactor systems by leveraging the enhanced heat and mass transfer properties of flow reactors. Slower reactions can be accelerated at higher temperatures and pressures. Byproduct formation can be limited by adjusting flow rates to effectively control reaction times.

Although there may be an energy cost to access higher reaction temperatures and pressures, successful process intensification efforts channel more material down the desired reaction pathway, resulting in higher yields and less byproduct formation. Moreover, the improved purity profile of the product stream often enables a simpler isolation strategy; the reduction in waste that accompanies a more efficient purification protocol tends to decisively outweigh the energy costs associated with extreme operating windows.

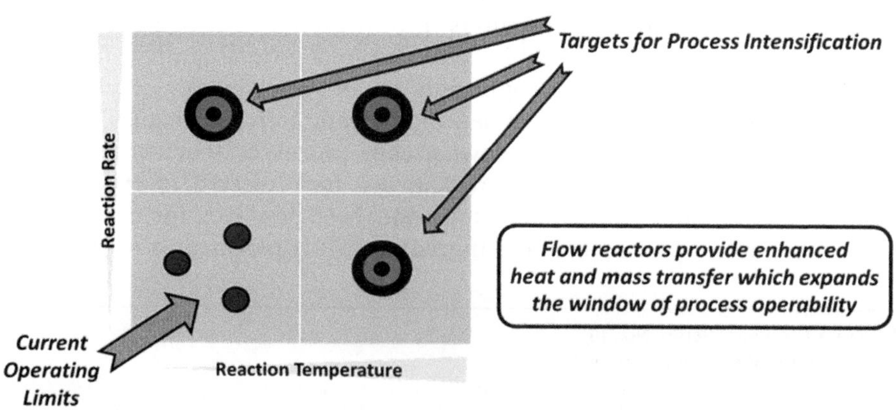


**Figure 6.1** Operating ranges for batch and continuous reactor systems.

**Table 6.1** *E* factors for chemical sectors.

| Industrial segment | *E*-factor |
|---|---|
| Bulk chemicals | <1–5 |
| Fine chemicals | 5–50 |
| Pharmaceutical chemicals | 25–100 |

### 6.2.2 Telescoping Operations

A comparison of the kilograms of waste produced per kilogram of product (*E* factor) from bulk chemicals, fine chemicals, and pharmaceuticals shows the dramatic difference in these three market sectors (Table 6.1).[3] There are several compelling explanations that can be used to rationalize this metric; however, it is still crucial that steps are taken to reduce the amount of waste associated with API production. The development of more sustainable pharmaceutical processes is not only environmentally responsible, but economically sensible. Fortunately, continuous chemistry has the potential to make a real and noticeable impact at every stage of the pipeline, from discovery to manufacturing.

Large *E*-factors often reflect the higher level of complexity associated with multi-step processes. In most cases, each individual step requires isolation and purification, which contribute to elevated *E* factors as well. The major source of waste associated with these processes is the solvent that is discarded during the separation process. The purpose of these isolation steps is to purge byproducts and impurities from the system and to allow for the use of different solvents in subsequent steps. Individual steps can be "telescoped" by eliminating these intermediate isolations, provided the yields are significantly high for each reaction step to preclude accumulation of impurities. In addition, a common solvent must be used throughout the telescoped steps. Flow reactors can be readily spliced together in linear fashion to perform this function in high throughput applications for drug discovery as well as in commercial operations. In both cases the reduction in solvent consumption can be quite significant.

## 6.3 Reaction Classes Enabled by Continuous Chemistry

Some of the most frequently cited advantages of continuous chemical operations include the expanded heat transfer capabilities, improved command over mixing, and increased access to alternative reaction pathways and/or high energy intermediates that would be inaccessible in batch operations. From a process development perspective, many of these features explain the appeal of continuous chemistry. Moreover, these same elements make flow processes inherently well-suited to a number of specific classes of chemical processes and transformations, as described below. Beyond these reaction categories, continuous processing can offer improvements even in cases where traditional round-bottom flask chemistry has been adequate. For example, in many cases flow chemistry provides enhanced reproducibility and a more accessible path to automation. Despite all of these potential advantages, continuous processing should be viewed as a complement to, rather than a supplement for, traditional batch methodology.[4–6]

### 6.3.1 Catalysis

The use of catalysis, and particularly catalysis that imparts selectivity, is one of the main tenets of the green chemistry movement. In the context of flow chemistry, heterogeneous or solid-supported catalysts can be packed into a cartridge while a dissolved substrate is passed over the immobilized catalyst. This approach has an obvious benefit to product isolation, as no discrete operation is necessary to remove the catalyst from the product stream. Recovery of the catalyst itself is straightforward as well, which can facilitate recyclability. Moreover, continuous heterogeneous catalytic reactions enable chemists to leverage the reaction stoichiometry in a way that would be (economically) prohibitive in the analogous batch setting. Packed-bed flow reactors can create a much higher effective concentration of catalyst, resulting in lower catalyst loading and overall improved efficiency. Although this approach has been used extensively in commodity chemical manufacturing operations, its acceptance in fine chemical and pharmaceutical operations has been rather limited.

Metal-catalyzed reactions such as hydrogenations and cross-coupling reactions are widely used by medicinal chemists, but these reactions pose challenges upon scale up during drug development. Homogenous metal catalysts are particularly problematic, often resulting in heavy-metal contamination of the final product. Alcazar and coworkers have recently demonstrated the application of flow technology to aryl–alkyl Negishi cross-coupling reactions using a commercially available palladium catalyst on a proprietary silica-based support.[7] By immobilizing the catalyst in a packed-bed reactor, they were able to leverage stoichiometry to overcome the kinetic disadvantages of heterogeneous catalysis (Figure 6.2). In addition to illustrating the versatility of this continuous process by coupling various aryl halide substrates and organozinc reagents, the authors were able to demonstrate excellent functional group compatibility and scalability. Furthermore, the catalyst cartridge could be re-used many times without compromising yield; the ability to recycle the catalyst greatly improves the "green" profile of Negishi cross-coupling reactions. In related work, Mennecke and Kirschning were able to perform Suzuki–Miyaura and Heck–Mizoroki cross-coupling reactions in flow with palladium nanoparticles supported on functionalized Raschig-rings.[8]

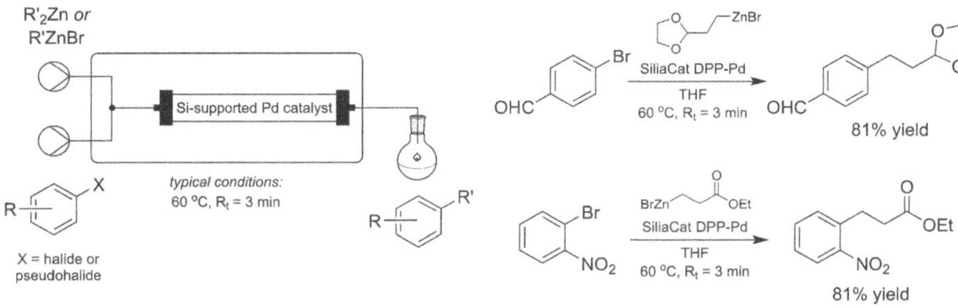


**Figure 6.2** Alkyl–aryl Negishi cross-coupling reactions in flow using a heterogeneous catalyst cartridge.

Another application of continuous metal-catalyzed heterogeneous catalysis is the oxidation of primary and secondary alcohols to aldehydes and ketones.[9] Using a fixed-bed continuous reactor fitted with a component that allows for the introduction of a compressed gas, molecular oxygen was exploited as the hydrogen acceptor for the oxidation of alcohols in flow and the oxidative aromatization of indolines to indoles. These reactions, catalyzed by a commercially available gold catalyst packed into a closed cartridge, have several "green" elements, including isolation and recyclability of the catalyst cartridge, a very clean byproduct profile (only water), and excellent overall atom economy.

Non-metal heterogeneous catalysis has also been demonstrated in flow. McQuade and coworkers functionalized a commercially available resin with TEMPO, thereby creating a heterogeneous oxidation catalyst that could be employed in a continuous platform.[10] By packing the immobilized catalyst into an improvised packed-bed microreactor, they were able to convert primary and secondary alcohols into the corresponding aldehydes and ketones in high yield. Furthermore, the catalyst was continuously regenerated during the reaction, and maintained excellent activity over many cycles.

Chemists have developed strategies for using homogeneous catalysts in flow chemistry settings as well. Johnson, May, and coworkers designed a continuous hydrogenation reactor for the homogeneous asymmetric reduction of a tetra substituted double bond at pilot scale, which required hydrogen pressure up to 70 bar (Figure 6.3).[11] Stirred-tank reactors capable of meeting this pressure requirement were not available at their manufacturing site, and the significant capital input that would have been necessary to realize the same chemical output using a batch approach was avoided by implementing a continuous strategy. Separation of the homogeneous catalyst and isolation of the product were achieved by a continuous liquid–liquid extraction followed by a continuous crystallization. This fully continuous process generated the same weekly throughput using laboratory infrastructure that would be expected from a batch process approach using 400 L reactors.

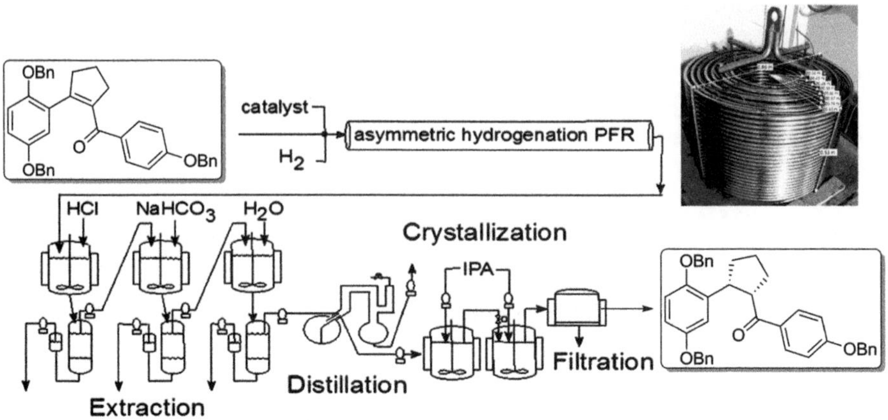


**Figure 6.3** Continuous process for homogeneous alkene hydrogenation, isolation, and purification, achieving plant-scale throughput with pilot-scale equipment.
Adapted with permission from ref. 11. Copyright (2012) American Chemical Society.

## 6.3.2 Kinetic *versus* Thermodynamic Reactions

Continuous chemistry effectively expands the scope of potential reaction space to include transformations that are hindered or prohibited by the inherent constraints of traditional round-bottom flask chemistry. For example, flow reactors can be operated under pressure, enabling chemists to safely exceed the atmospheric boiling point of the reaction solvent. Flow microreactors also offer exquisite control over residence time under reaction conditions at a molecular level. Combined with the spatial regulation of temperature, ease of reagent introduction, and enhanced mass and heat transfer, flow technology can provide chemists with functional access to reactive intermediates and/or enable researchers to selectively direct reactions down pathways that are inaccessible or disfavored in a traditional batch setting.

In its simplest version, this feature of continuous chemistry can mean optimal quenching of a reaction where the desired product is prone to over-reaction or degradation under continued exposure to the reaction conditions. Jamison and coworkers have exploited flow chemistry to provide a useful platform for DIBAL-H in the partial reduction of esters to aldehydes.[12] Under batch conditions, the practicality of DIBAL-H is compromised by problems of over-reduction and low reproducibility, and scale-up is often disfavored due to the cryogenic conditions required for reagent addition. Using a fabricated flow system, the selective reduction of several different esters to the corresponding aldehydes was achieved in high yields (>95%) with very little observed over-reduction. The rapid optimization of reaction parameters (45 reactions in less than 5 h) not only allowed the Jamison team to identify robust operating conditions for each particular substrate, but also revealed which parameter(s) are critical to controlling the reaction outcome.

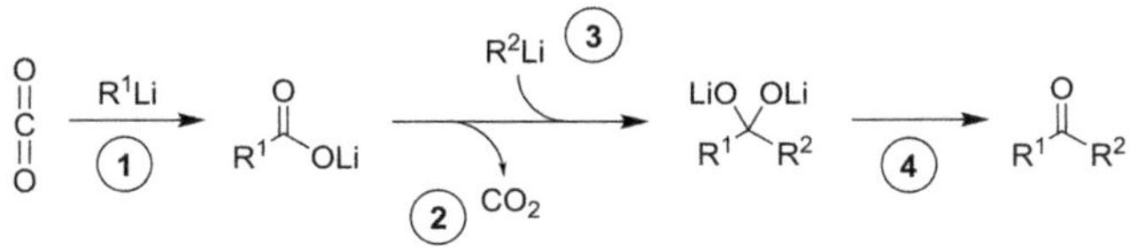


| | | Batch: | Flow: |
|---|---|---|---|
| 1 | carboxylation: | **low yield; generation of $R^1_2C{=}O$ byproducts** | **increased selectivity due to exceedingly high surface-to-volume ratio** |
| 2 | removal of excess $CO_2$: | **laborious and time-consuming** | **straightforward or unnecessary** |
| 3 | formation of dilithium salts: | **long reaction time** | **improved reactivity due to better mass/heat transfer** |
| 4 | quenching: | **low yield; generation of tertiary alcohol byproducts** | **high yield; alcohol byproducts suppressed by controlled quench** |

**Figure 6.4** Preparation of unsymmetric ketones in flow.
Adapted with permission from ref. 13. © 2014 Wiley-VCH Verlag GmbH, Weinheim.

Jamison, Hatton, and coworkers also recently illustrated the inherent benefits of flow chemistry in the preparation of unsymmetric ketones, a fundamental building block category for the synthesis of a wide variety of compounds.[13] A one-pot process had been reported by the Breitmaier group for the construction of these ketones from organolithium reagents and carbon dioxide, a "green" $C_1$ feedstock.[14] This batch approach was hampered by competing reaction pathways and operational challenges to scale-up, but Jamison and coworkers recognized the opportunity to harness the kinetic control available through continuous chemistry to address these limitations (Figure 6.4). The increased contact between gas and liquid phases afforded by a flow system, along with enhanced control over $CO_2$ stoichiometry in a flow setting, enabled the researchers to improve the selectivity of the carboxylation reaction and avoid formation of the symmetric ketone byproduct. Furthermore, the reaction time necessary for addition of the second organolithium reagent into the carboxylate intermediate was greatly reduced in flow, and tertiary alcohol byproducts that formed upon quenching the batch reaction could be easily prevented in the flow system. The continuous process was extended to make use of Grignard reagents in addition to organolithium substrates. Moreover, the starting organometallic reagents could be generated in flow and integrated into the overall process, resulting in a three-step continuous synthesis of ketones from simple precursors.

The application of flow chemistry to control extremely fast reactions and highly reactive intermediates has been explored in depth by Yoshida and coworkers, who have coined the term "flash chemistry" to describe this burgeoning field.[15,16] Yoshida effectively demonstrates how flow reactors offer precise command over the residence time to establish regio-control

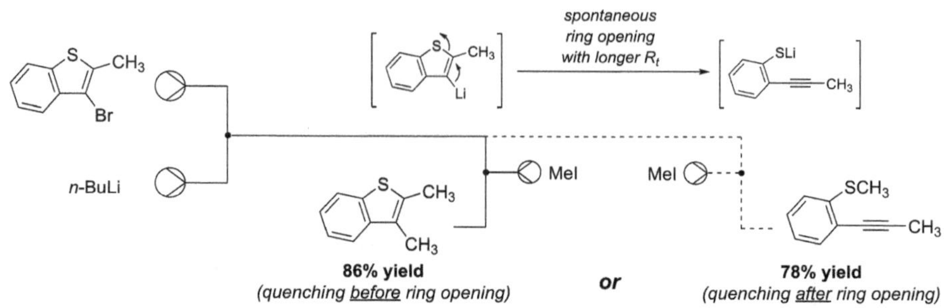


**Figure 6.5** Selective control of reaction outcome by adjusting residence time in the flow reactor.

with highly functionalized substrates. By adjusting the flow rate and/or the length of the flow reactor, it is possible to sufficiently restrict residence times on the millisecond time scale. This powerful technology allows chemists to rationally control reaction pathways of short-lived reactive intermediates by introducing a quenching agent at the appropriate time to maximize throughput and minimize degradation and/or side reactions.

The Yoshida group has applied flash chemistry to successfully use organolithium species bearing halides or electrophilic functional groups.[17] These types of reagents would normally be prone to rapid β-elimination or intermolecular side reactions, but flow technology allows researchers to generate unstable organolithiates *in situ* and selectively intercept these short-lived intermediates with a desired electrophile before alternative reaction pathways can ensue. These same concepts were applied to a protecting-group-free formal synthesis of pauciflorol F, demonstrating the potential of continuous flow chemistry to promote a more environmentally benign approach to total synthesis.[18]

A compelling illustration of the flexibility afforded by high-resolution control of reaction conditions in flow microreactors was offered by the Yoshida group as well.[19] Halogen–lithium exchange of 3-bromo-2-methylbenzothiophene gave an organolithium intermediate that could undergo a ring-opening rearrangement to a thiophenolate derivative (Figure 6.5). The organolithium intermediate could be selectively quenched with an electrophile either before or after this ring-opening rearrangement. Using the same substrate and reagents, chemists were able to generate either 2,3-dimethylbenzothiophene in 86% yield or 1-methylsulfanyl-2-(prop-1-ynyl)benzene in 78% yield by adjusting only the residence time and temperature. This report also exemplifies the capability of flow chemistry to access kinetic rather than thermodynamic reaction pathways.

### 6.3.3 Photochemistry

Photochemistry and flow technology are a natural pairing for various reasons. The inability to transmit ultraviolet light through a traditional batch reactor represents a major barrier to the use of this important synthetic tool in drug discovery and development. In contrast, the high surface-area-to-volume ratio in a typical flow reactor provides a more efficient option for irradiation of the reaction solution. Moreover, a uniform exposure level can be achieved with flow chemistry since reactor design ensures that substrates undergo a single, directional pass through the illuminated section of the reactor. In a traditional batch reactor, individual molecules may make several passes by the light source or none at all. Finally, continuous chemistry enables simple yet precise control over reaction time by adjusting residence time; this aspect of flow reactors can be essential in photochemical reactions, as under- and over-exposure to irradiation are typically both detrimental to optimizing conversion into the desired product.

There are several more subtle benefits to the utilization of continuous chemistry principles in photochemistry as well. Continuous flow reactions are amenable to the dilute reaction conditions that are sometimes necessary to avoid competing intermolecular side reactions following photoactivation, especially when paired with solvent recycling *via* continuous evaporation. Additionally, the improved control over temperature can be important in photochemical reactions, as the light source is often a heat source as well. Given that the lamp could also be viewed as a potential ignition source, there is an associated safety benefit to doing photochemistry in flow. Under continuous conditions, the overall quantity of potentially flammable material near the light source is limited.

The potential benefits of photochemistry in flow were recently illustrated in the context of a timely and relevant public health issue. Lévesque and Seeberger reported a multi-step continuous process that uses a photochemical ene reaction in the conversion of dihydroartemisinic acid (DHAA) into the natural product artemisinin (Figure 6.6).[20] Artemisinin is one of the most effective treatments for malaria and is often used as part of a

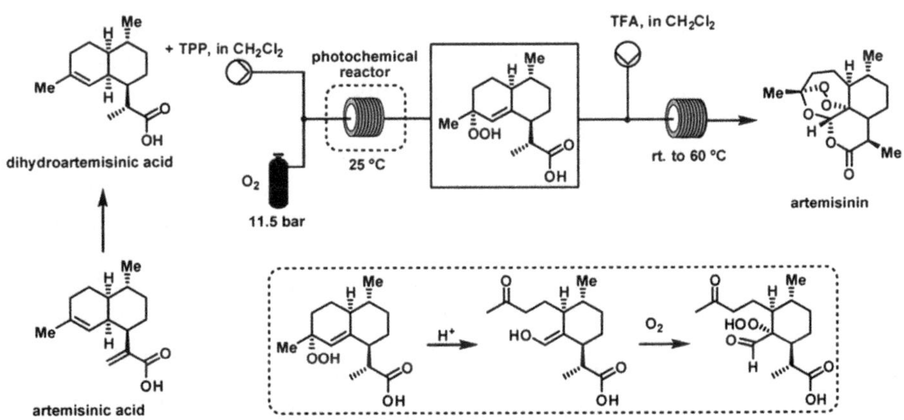


**Figure 6.6** Continuous photochemical process for conversion of DHAA into the antimalarial drug artemisinin.
Adapted from ref. 40.

combination therapy as a first-line defense against multi-drug resistant strains of the *Plasmodium* parasite. Historically, access to artemisinin has been tied to its natural source, the plant *Artemisia annua*, but variability in crop yields from season to season result in an unreliable supply of artemisinin and therefore a volatile market. DHAA can be accessed in one chemical step from artemisinic acid, which in turn can be isolated from fermentation of an engineered yeast strain developed by Keasling and coworkers.[21] Seeberger's continuous conversion of DHAA into artemisinin is initiated by the activation of molecular oxygen to the singlet state. Lévesque and Seeberger executed the reaction between DHAA and photoinduced singlet oxygen in a flow reactor with LED lighting. After optimization of reaction parameters, they were able to produce 1.5 mmol of the ene reaction product per minute, in 91% conversion and 75% yield. The output of this photochemical reaction was coupled downstream with the remaining reagents and reactor components necessary to convert the ene reaction product into artemisinin, producing an integrated continuous flow system for the transformation of DHAA into the final product, artemisinin. Throughput estimations suggest that a single version of this continuous flow system could produce 200 g of artemisinin per day in the laboratory unit.

### 6.3.4 Hazardous Reactions

Continuous flow techniques can be exploited to improve the safety profile of chemical processes, making flow chemistry an enabling technology for reactions that would otherwise be too hazardous to be performed on scale. The precise control over reaction temperature is a recurring theme in continuous operations. The improved heat transfer available with a flow reactor can facilitate the safe execution of highly exothermic reactions, reducing the possibility of a thermal runaway.[22] Flow chemistry can also make an impact for chemical transformations involving the evolution of gas, as concerns about pressure build-up in the head space of a batch reactor are alleviated in a flow system.[23] Perhaps the most compelling benefit of continuous chemistry from a safety perspective is that hazardous reagents and intermediates can be generated *in situ* in small amounts and reacted immediately. Because the overall quantity of these species is limited in a flow reactor as compared to a bulk solution preparation, the build-up of potentially dangerous or explosive compounds is greatly minimized. For example, flow technology empowers chemists to safely develop and scale up processes for diazomethane generation and use, azide formation, and reactions that may generate peroxides.[24–27]

Ozonolysis is a fundamental and versatile reaction in organic chemistry, yet is often overlooked in favor of less environmentally friendly processes due to safety concerns and operational constraints. Ozonide intermediates are potentially explosive and shock sensitive, making ozonolysis a good candidate for the development of a continuous process that would avoid the accumulation of ozonides or related peroxy compounds. Several industrial and academic groups have recognized this opportunity. In addition to

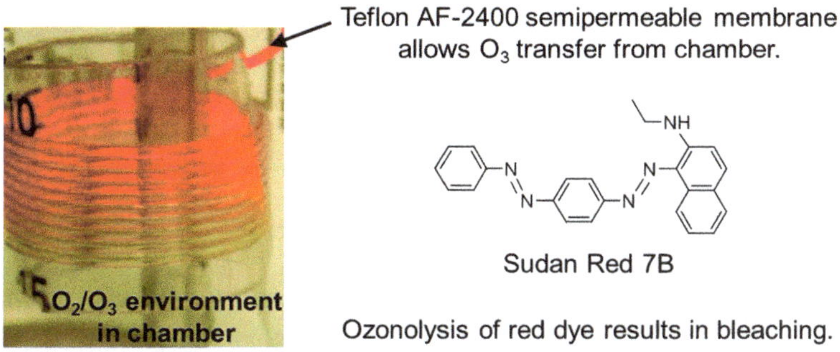


**Figure 6.7** Ozonolysis in flow using semipermeable tubing.
Adapted with permission from ref. 27. Copyright (2010) American Chemical Society.

commercially available flow reactors for ozonolysis, such as the ThalesNano IceCube™, there have been several recent reports describing the development of specialized equipment for continuous ozonolysis processes.[26] In one case, Ley and coworkers used semipermeable PTFE tubing to assemble a flow reactor that takes advantage of gas-to-liquid transfer rather than the mechanical mixing of the two phases.[27] The consistency and efficiency of the resulting system was demonstrated using ozone-mediated bleaching of a red dye (Figure 6.7). A solution of the dye was pumped through the gas-permeable tubing, which was coiled in a chamber connected to an ozone generator. An integrated quenching step could be incorporated at the end of the flow stream, ensuring that overall ozonide/peroxide levels remain low. Using a flow system such as this, chemists can make use of ozonolysis as a greener oxidation strategy than alternatives involving heavy metals such as osmium tetroxide.

A similar reactor design was recently used for the continuous production and consumption of diazomethane, an extremely hazardous but highly useful methylating reagent.[24] In addition to its shock sensitivity, diazomethane gas is also notably toxic, a problem that is exacerbated by its volatility. In batch reactions, diazomethane can be generated from Diazald® (*N*-methyl-*N*-nitroso-*p*-toluenesulfonamide) and aqueous base using designated reaction kits that have smoothed joints in place of ground-glass joints; however, this process is limited to small scales. Kappe and coworkers combine Diazald® and potassium hydroxide reagent streams in flow, generating diazomethane *in situ* within the same gas-permeable membrane tubing pioneered by the Ley laboratory. This gas-permeable tube resides within an exterior PTFE tube carrying the substrate solution; the reaction takes place after diazomethane diffuses across the membrane and into the reaction stream (Figure 6.8). Kappe and his team demonstrated the versatility of this process in methylation reactions, cycloaddition reactions, and cyclopropanation reactions. Moreover, the interior semipermeable tube is hydrophobic in nature, so water molecules do not pass through the membrane. Accordingly, even though diazomethane is generated under

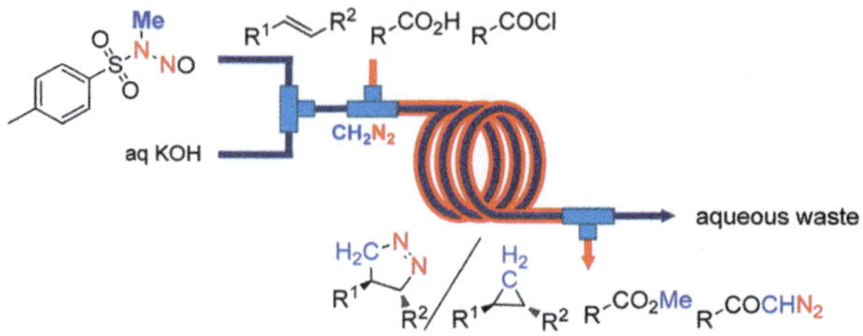


**Figure 6.8** Generation and utilization of diazomethane in flow using semipermeable tubing.
Adapted with permission from ref. 24. Copyright (2013) American Chemical Society.

aqueous conditions, it is supplied to the reaction flow as an anhydrous reagent, enabling processes such as the Arndt–Eistert reaction, which transforms water-sensitive acyl chlorides into α-chloroketones.

## 6.4 Equipment Requirements

The advantages of continuous process chemistry have spurred a growing interest among researchers in the pharmaceutical community. Recent advances in equipment design and a broader range of commercial options have provided the accessibility of flow chemistry to chemists who might otherwise shy away from alternatives to the traditional round bottom flask. The major components of a continuous unit are the individual reactors, the pumps that deliver the reactants to the reactor, and the mixing systems used to bring the reactants into intimate contact. In many cases, the mixing systems have been incorporated into the reactors.

### 6.4.1 Reactors

Various continuous reactor configurations have been developed in recent years that are specifically designed to leverage heat and mass transfer. These include microfluidic reactors, continuous oscillating baffle reactors (COBRs), spinning disc reactors (SDRs), static mixers, plug flow reactors (PFRs), and continuous stirred tank reactors (CSTRs). Figure 6.9 provides some general guidance for the application of these reactor systems to address specific process chemistry issues associated with heat and mass transfer.

For homogeneous reactions that do not require extreme heating and cooling requirements, PTFE tubing fitted with a static mixer has been used very successfully.[28] However, many chemical reactions are carried out in multi-phase systems, and reaction slurries can be particularly problematic. In these cases dynamic mixing (CSTR, COBR) is often required to facilitate mass transfer. An excellent example of a CSTR application is the Barbier/Grignard flow process developed by Eli Lilly for the production of

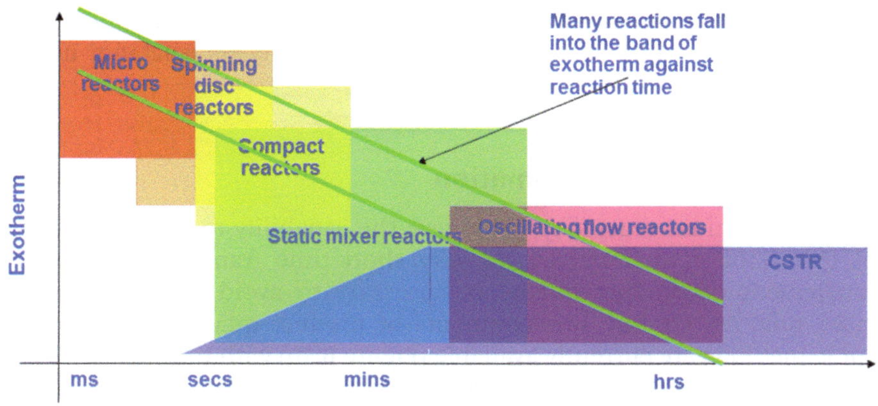


**Figure 6.9** Suitability of various flow reactor types for different process chemistry issues. Figure courtesy of Huw Thomas.

edivoxetine · HCl.[29] In this specific case, a 30% reduction in mass intensity and magnesium used over the traditional batch process was achieved.

## 6.4.2 Pumps

An essential and often problematic aspect of any continuous reactor system is the pumping component that delivers the reactants to the reaction zone. Piston pumps and syringe pumps are the most common choices for most liquid phase reactions and are capable of delivering liquids with a high level of accuracy. Piston pumps are also capable of delivering reactants against extremely high reactor pressure, which is particularly useful when attempting to operate above the atmospheric boiling point of the reaction solvent. Careful consideration should be given to the materials of construction for these pumps to ensure their compatibility with specific reactants. In many cases, jacketed systems can be obtained to pre-heat or cool reactants. However, piston and syringe pumps are typically not compatible with liquids containing significant quantities of solids.

In cases where solids are present in the reactants, peristaltic pumps can provide a viable option. However, peristaltic pumps lack the level of accuracy that is attainable with piston and syringe pumps. These pumps also have limited ability to pump against pressure, which restricts the operating range of the entire reactor system. Vapourtec recently introduced a peristaltic pump that is controlled by a rotor system that produces a relatively constant flow rate by adjusting its rotation rate. Ley and coworkers illustrated the utility of this system in accessing air-sensitive substances in flow by completing a telescoped continuous synthesis of (*E*/*Z*)-tamoxifen in which organolithium reagents and intermediates figured prominently.[30] The addition of functional peristaltic pump options to the flow chemist's toolbox improves the prospects for developing intermediate-scale flow processes that require reactants containing solids. Likewise, the judicious selection of reaction solvents and co-solvents may also provide a prospective approach to broadening pump options by ensuring the solubility of reactants in the reaction media.

### 6.4.3 Analytics and Automation

Because continuous reactor systems have the capability to produce significant quantities of material over a very short time frame, instantaneous measurement of product quality is necessary to avoid accumulation of impure product. On-line measurements of product quality by chromatographic methods are currently available (Figure 6.10) with response times of 3–5 min. Continuous systems can be augmented with sensor technologies with response times that approximate real time measurements. Typically, these sensor systems utilize one or more spectroscopic methods such as Raman, near-infrared, or ultraviolet to evaluate reaction profiles. The

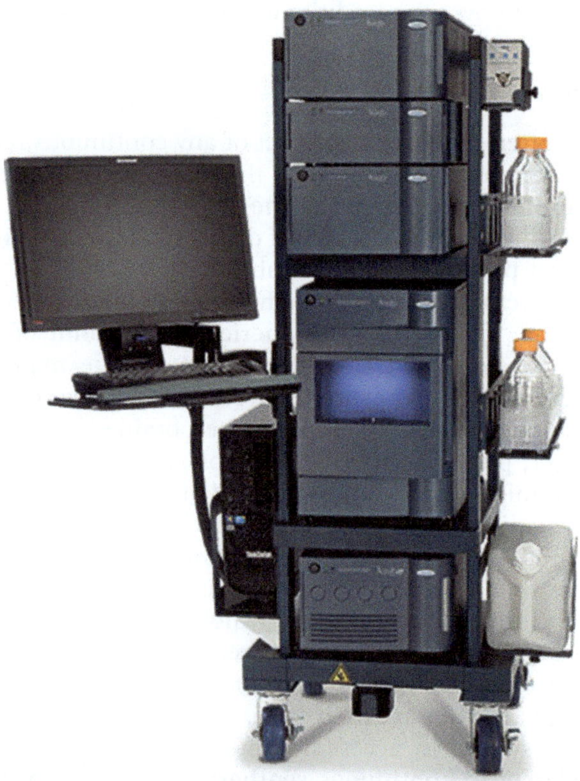

**Figure 6.10** The Waters PATROL UPLC Process Analysis System for real-time chromatography of continuous processes.
Figure courtesy of the Waters Corporation.

spectroscopic sensor of choice can be mounted onto in-line ball probes such as the NeSSI system developed by the Center of Process Analytical Chemistry (CPAC) at the University of Washington.[31] Used in tandem, these complementary methods can be a very powerful package to rapidly verify the structures of drug substances in flow or quickly optimize reaction conditions for an existing chemical process. If required, the probes can also be continuously calibrated against external standards.

### 6.4.4 Other Considerations

As previously stated, one of the more pressing challenges of continuous chemical processing is the successful management of solids in flow. Buildup of solids within reaction channels or at back pressure regulators can lead to obstruction of the flow path and clogging of the reactor. The state of the art has improved tremendously over the last few years and simple workarounds, such as sonication[32] or the introduction of a solubilizing solvent,[33] have been employed to prevent buildup of blockages and minimize adhesion of particles to the reactor walls. In many cases, increasing/optimizing the reactor channel diameter will not only reduce solids buildup but can also improve the process throughput.

Another approach involves the use of two immiscible solvents, one of which serves as a carrier phase for small droplets of the reaction solvent, effectively restricting solids to these microdroplets. To illustrate the effectiveness of this technique, McQuade and coworkers synthesized the solid dye indigo within aqueous reaction droplets carried by mineral oil.[34] Reactors designed specifically for processing of solids are also evolving. For example, the Coflore® agitating cell reactor (ACR) employs freely moving agitators within a reaction flow path, all contained within a reactor block that is supported on a shaking motor (Figure 6.11). Ley and coworkers have demonstrated the power of this technology by preparing meso-scale quantities (208 g) of *N*-iodomorpholinium hydroiodide salt in just 9 h, noting also that this ACR has the potential to offer a robust and general solution to the problem of handling solids in flow.[35]

With proper design, flow reactors can be used in tandem to generate multi-step, continuous synthesis routes.[6,36] This approach is inherently aligned with "green" principles – by eliminating the need to isolate intermediates, much of the energy input and waste associated with multi-step processes can be avoided. Early adopters of this strategy include the Lectka group, who employed gravity-driven columns in parallel and in sequence to demonstrate asymmetric catalysis[37] and the multi-step solid-phase synthesis of a pharmaceutically active drug candidate.[38] Success in this arena requires significant forethought to ensure that reaction byproducts are compatible with downstream steps. In-line cartridges can be used to quench unused reagents and/or scavenge byproducts that would otherwise be detrimental.[39] The Ley group has been prolific in this arena, exploiting scavenger resins

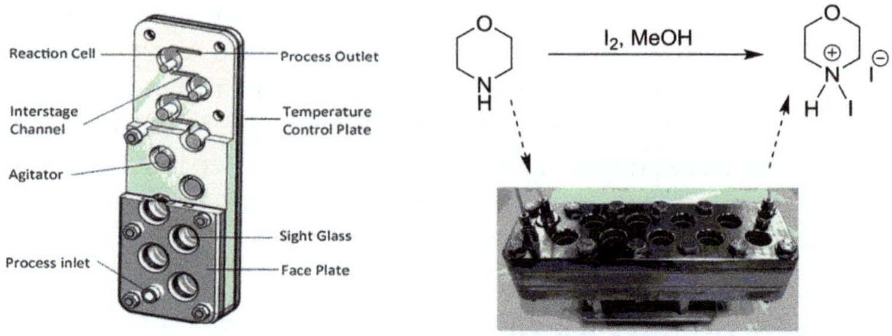


**Figure 6.11** Use of a Coflore® ACR in the flow synthesis of an ammonium salt. Adapted with permission from ref. 35. Copyright (2011) American Chemical Society.

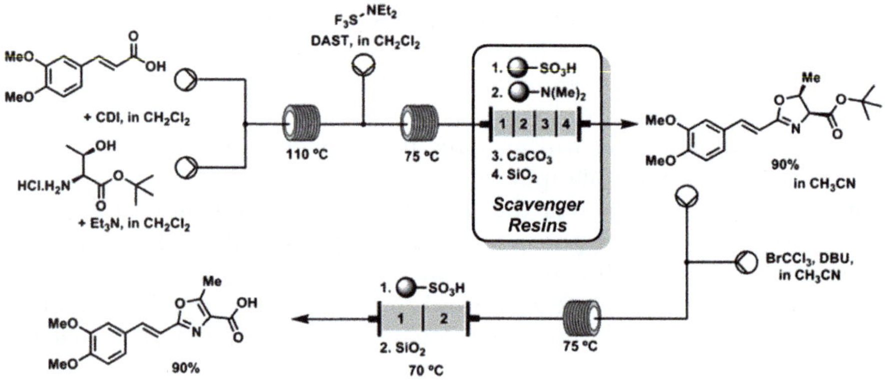


**Figure 6.12** Use of scavenger resins in a multistep flow synthesis. Adapted from ref. 40.

and reagent cartridges in the flow synthesis of several natural products and related compounds over the years (Figure 6.12).[40]

Despite the broad availability of solid-supported reagents and scavengers, not all reactions are amenable to this idealized approach for product isolation. Recent engineering advances have led to practical methods for continuous reaction work-up and product purification regimes, effectively addressing some of the limitations of telescoping multiple reaction steps in flow. Biphasic extraction techniques are a standard purification method in traditional synthetic organic chemistry. The removal of water-soluble reaction components is often a key step to isolating the desired intermediate for further transformations in batch processes. These principles have recently been applied to continuous processes as well. The use of hydrophobic membranes has been pioneered by Jensen and coworkers to develop a continuous extraction system for a flow microreactor.[41] This micro-separator was employed in a continuous synthesis of carbamates, removing water-soluble components *via* an aqueous waste stream. Continuous extraction techniques have also been demonstrated at pilot scale. May and coworkers at Eli Lilly introduced liquid–liquid extraction as one of the continuous unit operations downstream of a hydrogenation reaction.[11] Using a three-stage mixer–settler system, they were able to wash the reaction solution, isolating the product from water-soluble waste prior to implementing a continuous crystallization to purify the final product (Figure 6.3). Other continuous post-reaction workup techniques have been developed as well. For example, continuous distillations have been used to effect in-line solvent switches and to aid in product purification.[42,43]

Purification by continuous crystallization is becoming an attractive option in the pharmaceutical industry. Mixed-suspension, mixed-product removal (MSMPR) crystallizers such as the one used by Lilly (Figure 6.3),[29] have commercial applicability, and oscillatory baffled crystallizers (OBC) offer the opportunity to conduct continuous crystallization in a laboratory setting with excellent control over particle size and impurity profile. Continuous crystallization processing has been demonstrated for several APIs[11,44,45] and has also been incorporated as a unit operation in integrated continuous manufacturing efforts.[46]

## 6.5 Continuous Chemical Processing in Drug Discovery and Development

### 6.5.1 Drug Discovery

Medicinal chemists are charged with the task of generating relatively small amounts of a relatively large number of compounds, often within a very short timeframe. Accordingly, discovery routes are often unoptimized with respect to yield, waste generation, and environmental impact. Investing in a continuous synthesis strategy from the outset, where appropriate, has the potential to result in faster access to structural analogs based on a lead compound. For example, a team of researchers at Abbott Laboratories developed a segmented flow approach to library synthesis, illustrating the high-throughput potential of continuous synthesis technology.[47] In segmented flow, individual reaction "plugs" are separated by immiscible solvent spacers. By utilizing synchronized injection loops to introduce variable substrate solutions into a flow of the immiscible spacer solvent, sequential reaction droplets were generated and spatially separated along the length of the flow reactor. A distinct library member was constructed within each 200 μL droplet, and isolated yields were often sufficient to support activity studies (tens of milligrams). Twenty-member thiazole and pyrazole libraries were synthesized to demonstrate the efficiency of this strategy, and library generation was complete in half the time as compared with a conventional batch approach.

Similarly, Hanson and Organ have championed the application of microwave-assisted continuous-flow organic synthesis (MACOS) as a tool for medicinal chemistry applications, including diversity-oriented synthesis.[48] Using the MACOS platform, they have demonstrated the benefits of flow technology for parallel synthesis by optimizing the continuous preparation of substituted sultams and applying this MACOS protocol in parallel to rapidly generate a 50-member library of these pharmaceutically relevant cyclic sulfonamides. Ley and coworkers have also used flow chemistry to generate libraries of various heterocyclic compounds from simple acyclic precursors.[49] Their synthetic strategies are often multistep, taking advantage of immobilized reagents and scavenging resins to improve product isolation/purity. Eliminating or reducing traditional work-ups and intermediate isolation, even on discovery scale, improves the environmental footprint of a library synthesis. Moreover, Ley and coworkers noted the ease of transferring these processes to multi-gram scale, suggesting that the early adoption of flow techniques can pay dividends at initial process development stages as well.

Continuous flow technology has the potential to impact the drug discovery process beyond library development. Roth and coworkers have applied electrochemistry in flow to mimic the *in vivo* biological processing of APIs by cytochrome P450 (CYP450) enzymes in the liver.[50] By capitalizing on the inherent advantages of microfluidics devices with respect to electrosynthesis, such as a high electrode surface-to-solution volume ratio and improved resistivity profiles, researchers were able to synthesize preparative quantities of oxidized drug metabolites for characterization. The authors note that, in addition to expediting the identification of metabolized pharmaceuticals for ADME studies and providing an alternative technology from which to access new compound scaffolds, continuous-flow electrosynthesis can improve the drug discovery process by facilitating early evaluation of the CYP450 oxidative liabilities of lead compounds prior to *in vivo* testing.

Yet another benefit to continuous chemical synthesis in a drug discovery setting is its compatibility with automation technologies. Researchers at the Sanford-Burnham Medical Research Institute have recognized this potential, detailing an automated multi-step flow synthesis of indolylthiazoles using a microfluidics platform.[51] More recently, flow chemistry has been utilized to support an exciting new approach to automation in drug discovery. Selway and coworkers have integrated the chemical synthesis, purification, and bioassay of drug targets into a single continuous feedback loop; the information gained is then used to refine a bioactivity prediction model (Figure 6.13).[52] In this case, the use of continuous chemical processes enables what is essentially a completely automated SAR study. The authors applied their strategy to Abl kinase, a well-studied and clinically relevant target and were able to rapidly identify inhibitors with a novel binding motif, reinforcing the potential for continuous flow chemistry to become a key enabling technology in drug discovery.

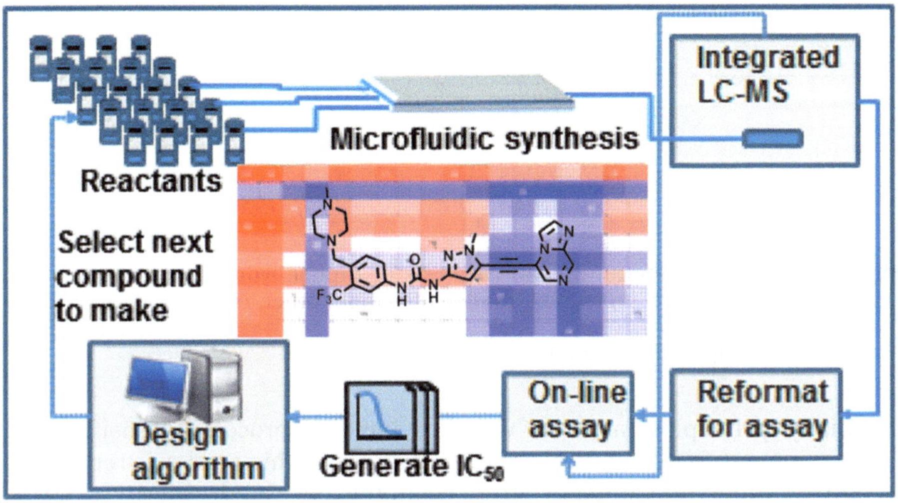


**Figure 6.13** Automated synthesis, purification, bioassay, and predictive modeling identifies a novel binding motif for Abl kinase inhibitors.
Reproduced with permission from ref. 52. Copyright (2013) American Chemical Society.

## 6.5.2 Process Development

In contrast to the mission of drug discovery, the objective of pharmaceutical process development is to generate chemical processes that are capable of producing significant quantities of drug substance in a safe and economically viable manner. Discovery routes are often inappropriate for scale-up due to safety concerns or operational steps that would be inefficient at the pilot or manufacturing plant scale. Process chemists work to modify or completely revamp target syntheses, with an aim to optimize yield, minimize waste, streamline unit operations, and improve safety profiles. As often has been said before, green chemistry is just good process chemistry!

The potential benefits of applying continuous flow chemistry during process optimization and pharmaceutical manufacturing have been documented in examples from both academic and industrial settings. Snead and Jamison recently improved upon established processes for making diphenhydramine hydrochloride (Figure 6.14).[53] Revisiting the manufacturing approach to this API from a flow chemistry perspective, the authors were able to run the key reaction neat, eliminating the hazards associated with the chlorobenzene solvent often used in manufacturing of diphenhydramine · HCl and minimizing the overall waste associated with the process. Furthermore, flow conditions were devised to minimize post-synthesis processing steps, first by applying a continuous in-line extraction method, but later by crystallizing the hydrochloride salt directly from the reaction stream. Flowing molten ammonium salts neat, a strategy that would not be practical in a large-scale batch setting, enabled the authors to highlight several green

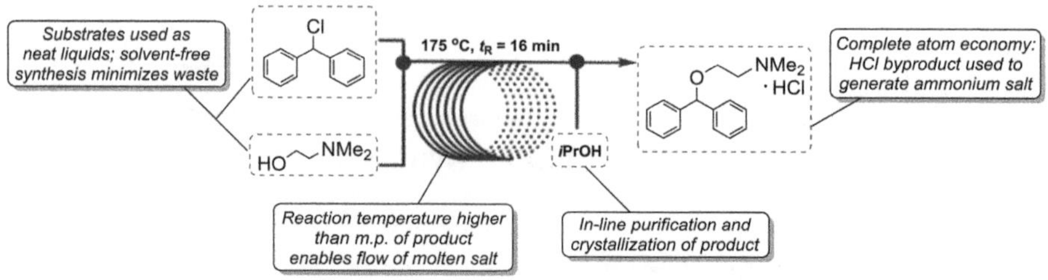


**Figure 6.14** Atom-economical flow synthesis of diphenhydramine hydrochloride, the active ingredient in Benadryl and other over-the-counter medications.
Adapted from ref. 53.

chemistry principles such as waste reduction, process intensification, and equipment footprint minimization. Moreover, this synthesis represents a rare example of complete atom economy, as the hydrogen chloride generated by displacement is used directly to formulate the HCl salt of the product.

Process chemists at Eli Lilly have also leveraged continuous processing to gain clear environmental and economic benefits. Charged with scale-up of a substituted imidazole intermediate *via* cyclization of a β-ketoamide precursor that required high temperature, they were able to exploit the efficient heat transfer afforded by a plug flow reactor (PFR), to develop a first-generation continuous process that scaled linearly from a research-scale PFR (1.75 mL) to a pilot-scale PFR (7.1 L).[54] Researchers were also able to switch to a lower-boiling solvent, which aided in product isolation. By implementing flow chemistry, a rapid cGMP delivery of 29 kg of the substituted imidazole was achieved using equipment with a laboratory hood (rather than a pilot plant) footprint. A second generation process developed by the same team highlights the benefits of automated reaction sampling and analysis for rapid optimization of reaction conditions. This more efficient synthetic route included an additional continuous deprotection step using supercritical conditions in flow to achieve a single homogeneous phase despite the evolution of gases during the reaction. Demonstration runs using the second generation process suggest that a laboratory hood footprint would be sufficient to produce manufacturing-scale quantities (1–2 MT per year) of the desired pathway intermediate.

Chemists and chemical engineers at MIT have recently illustrated the potential economic and environmental benefits of integrating a full set of continuous chemical reactions and post-synthesis processes. In reporting the first example of end-to-end continuous pharmaceutical manufacturing, Trout and coworkers created and operated a manufacturing plant for tablets of the antihypertensive drug aliskiren fumarate.[55,56] Beginning with a synthetic intermediate, this fully continuous process incorporates chemical transformations, purifications, formulations, and finally tableting. The number of unit operations was reduced from 21 to 14, mainly due to exploiting the advantages of continuous flow technologies during post-synthesis processing. In addition to a more compact operating footprint and a dramatically shorter processing time, this continuous manufacturing approach has the added benefit of throughput flexibility. The plant could produce anywhere from 20 to 100 g $h^{-1}$ of the final API by adjusting control set-point values, enabling a faster response time to changing demand. Finally, Trout and coworkers point out that, in general, stockpiled intermediates can be minimized by adopting a continuous manufacturing regime, which is especially important in cases with hazardous compounds or potent APIs.

## 6.6 Barriers to Implementation

There are several important reasons why continuous chemical processing has not been widely adopted by the pharmaceutical sector thus far. One important element of consideration is the limited time available for identification and process development of a new drug. Discovery chemists face intense pressure to construct a large number of new compounds very quickly – sometimes with variations on a single scaffold but sometimes sets with large structural diversity. Moreover, there is a very narrow window of time between when a molecule is identified as a drug candidate and when the process is fixed as component of a New Drug Application. Perhaps the most daunting yet scalable barrier to the widespread application of continuous chemistry in a drug discovery context is the comfort level and experience of the chemists themselves. Most medicinal chemists are trained in more conventional techniques and have limited, if any, exposure to flow chemistry during their formal education. Given these constraints, discovery chemists tend to focus on traditional batch processing platforms. The batch philosophy is often established early in laboratory round bottom flasks and is rapidly propagated into larger batch equipment through scale up. In actuality, continuous chemical process development has the potential to add value to chemical process optimization through rapid screening of process parameters. However, most process chemists also lack training in continuous methods and migrate to traditional batch glassware to meet their objective.

A second but equally important issue is the general lack of continuous manufacturing assets in pharmaceutical manufacturing operations. Furthermore, the existing capital investments in batch manufacturing represent a significant barrier to the implementation of continuous processing in pharmaceutical operations. However, many pharmaceutical companies are developing hybrid manufacturing strategies, which take advantage of the inherent reproducibility of continuous chemical processes with downstream unit operations such as crystallization, liquid/solid separation, and drying (Figure 6.3). This approach also provides the ability to retain discrete batch to batch integrity and traceability.

For the conversion of existing commercial batch processes into continuous, additional regulatory hurdles exist. In some cases, the general lack of batch to batch reproducibility in commercial operations necessitates this transition. Post-approval process changes such as these require supplemental regulatory approval, which can be onerous. Despite this and other barriers to implementation, increasing interest in flow chemistry across academia and industry suggests that the application of continuous processing will continue to gain momentum.

## 6.7 Concluding Remarks

Continuous processing is an evolving element of pharmaceutical synthesis that offers significant opportunities to streamline the drug discovery process. This approach provides a key component to the synthetic tool box by accessing important pharmacophores *via* "kinetic pathways" that are inaccessible through traditional batch methods. By incorporating this approach into the drug discovery process, synthetic methods can be more effectively translated into scalable processes that have the capability to embrace the elements of process intensification and sustainability.

## References

1. C. Jimenez-Gonzalez, P. Poechlauer, Q. B. Broxterman, B.-S. Yang, D. am Ende, J. Baird, C. Bertsch, R. E. Hannah, P. Dell'Orco, H. Noorman, S. Yee, R. Reintjens, A. Wells, V. Massonneau and J. Manley, *Org. Process Res. Dev.*, 2011, **15**, 900–911.
2. A. I. Stankiewicz and J. A. Moulijn, *Chem. Eng. Prog.*, 2000, 22–34.
3. R. A. Sheldon, *Green Chem.*, 2007, **9**, 1273–1283.
4. R. L. Hartman, J. P. McMullen and K. F. Jensen, *Angew. Chem., Int. Ed.*, 2011, **50**, 7502–7519.
5. J. Wegner, S. Ceylan and A. Kirschning, *Chem. Commun.*, 2011, **47**, 4583–4592.
6. D. T. McQuade and P. H. Seeberger, *J. Org. Chem.*, 2013, **78**, 6384–6389.
7. B. Egle, J. d. M. Munoz, N. Alonso, W. M. De Borggraeve, A. de la Hoz, A. Diaz-Ortiz and J. Alcazar, *J. Flow Chem.*, 2014, **4**, 22–25.
8. K. Mennecke and A. Kirschning, *Beilstein J. Org. Chem.*, 2009, **5**, 21.
9. G. Sipos, V. Gyollai, T. Sipocz, G. Dorman, L. Kocsis, R. V. Jones and F. Darvas, *J. Flow Chem.*, 2013, **3**, 51–58.
10. A. Bogdan and D. T. McQuade, *Beilstein J. Org. Chem.*, 2009, **5**, 17.
11. M. D. Johnson, S. A. May, J. R. Calvin, J. Remacle, J. R. Stout, W. D. Diseroad, N. Zaborenko, B. D. Haeberle, W.-M. Sun, M. T. Miller and J. Brennan, *Org. Process Res. Dev.*, 2012, **16**, 1017–1038.
12. D. Webb and T. F. Jamison, *Org. Lett.*, 2012, **14**, 568–571.
13. J. Wu, X. Yang, Z. He, X. Mao, T. A. Hatton and T. F. Jamison, *Angew. Chem., Int. Ed.*, 2014, **53**, 8416–8420.
14. G. Zadel and E. Breitmaier, *Angew. Chem., Int. Ed.*, 1992, **31**, 1035–1036.
15. J. Yoshida, A. Nagaki and T. Yamada, *Chem. – Eur. J.*, 2008, **14**, 7450–7459.

16. J. Yoshida, Y. Takahashi and A. Nagaki, *Chem. Commun.*, 2013, **49**, 9896–9904.
17. H. Usutani, Y. Tomida, A. Nagaki, H. Okamoto, T. Nokami and J. Yoshida, *J. Am. Chem. Soc.*, 2007, **129**, 3046–3047.
18. H. Kim, A. Nagaki and J. Yoshida, *Nat. Commun.*, 2011, **2**, 264.
19. T. Asai, A. Takata, Y. Ushiogi, Y. Iinuma, A. Nagaki and J.-i. Yoshida, *Chem. Lett.*, 2011, **40**, 393–395.
20. F. Lévesque and P. H. Seeberger, *Angew. Chem., Int. Ed.*, 2012, **51**, 1706–1709.
21. D. K. Ro, E. M. Paradise, M. Ouellet, K. J. Fisher, K. L. Newman, J. M. Ndungu, K. A. Ho, R. A. Eachus, T. S. Ham, J. Kirby, M. C. Chang, S. T. Withers, Y. Shiba, R. Sarpong and J. D. Keasling, *Nature*, 2006, **440**, 940–943.
22. C. E. Brocklehurst, H. Lehmann and L. La Vecchia, *Org. Process Res. Dev.*, 2011, **15**, 1447–1453.
23. A. S. Kleinke and T. F. Jamison, *Org. Lett.*, 2013, **15**, 710–713.
24. F. Mastronardi, B. Gutmann and C. O. Kappe, *Org. Lett.*, 2013, **15**, 5590–5593.
25. B. Gutmann, J. P. Roduit, D. Roberge and C. O. Kappe, *Angew. Chem., Int. Ed.*, 2010, **49**, 7101–7105.
26. A. D. Allian, S. M. Richter, J. M. Kallemeyn, T. A. Robbins and V. Kishore, *Org. Process Res. Dev.*, 2011, **15**, 91–97.
27. M. O'Brien, I. R. Baxendale and S. V. Ley, *Org. Lett.*, 2010, **12**, 1596–1598.
28. A. R. Bogdan, S. L. Poe, D. C. Kubis, S. J. Broadwater and D. T. McQuade, *Angew. Chem., Int. Ed.*, 2009, **48**, 8547–8550.
29. M. E. Kopach, D. J. Roberts, M. D. Johnson, J. McClary Groh, J. J. Adler, J. P. Schafer, M. E. Kobierski and W. G. Trankle, *Green Chem.*, 2012, **14**, 1524–1536.
30. P. R. D. Murray, D. L. Browne, J. C. Pastre, C. Butters, D. Guthrie and S. V. Ley, *Org. Process Res. Dev.*, 2013, **17**, 1192–1208.
31. CPAC, http://cpac.apl.washington.edu/, 2014. (Accessed February 23, 2015).
32. R. L. Hartman, J. R. Naber, N. Zaborenko, S. L. Buchwald and K. F. Jensen, *Org. Process Res. Dev.*, 2010, **14**, 1347–1357.
33. C. B. Kelly, C. X. Lee and N. E. Leadbeater, *Tetrahedron Lett.*, 2011, **52**, 263–265.
34. S. L. Poe, M. A. Cummings, M. P. Haaf and D. T. McQuade, *Angew. Chem., Int. Ed.*, 2006, **45**, 1544–1548.
35. D. L. Browne, B. J. Deadman, R. Ashe, I. R. Baxendale and S. V. Ley, *Org. Process Res. Dev.*, 2011, **15**, 693–697.
36. D. Webb and T. Jamison, *Chem. Sci.*, 2010, **1**, 675–680.
37. A. M. Hafez, A. E. Taggi and T. Lectka, *Chem. – Eur. J.*, 2002, **8**, 4114–4119.
38. S. France, D. Bernstein, A. Weatherwax and T. Lectka, *Org. Lett.*, 2005, **7**, 3009–3012.
39. A. R. Longstreet, S. M. Opalka, B. S. Campbell, B. F. Gupton and D. T. McQuade, *Beilstein J. Org. Chem.*, 2013, **9**, 2570–2578.

40. J. C. Pastre, D. L. Browne and S. V. Ley, *Chem. Soc. Rev.*, 2013, **42**, 8849–8869.
41. J. G. Kralj, H. R. Sahoo and K. F. Jensen, *Lab Chip*, 2007, 7, 256–263.
42. R. L. Hartman, J. R. Naber, S. L. Buchwald and K. F. Jensen, *Angew. Chem.*, 2010, **122**, 911–915.
43. M. D. Hopkin, I. R. Baxendale and S. V. Ley, *Chem. Commun.*, 2010, **46**, 2450–2452.
44. A. J. Alvarez, A. Singh and A. S. Myerson, *Cryst. Growth Des.*, 2011, **11**, 4392–4400.
45. S. Ferguson, F. Ortner, J. Quon, L. Peeva, A. Livingston, B. L. Trout and A. S. Myerson, *Cryst. Growth Des.*, 2014, **14**, 617–627.
46. H. Zhang, R. Lakerveld, P. L. Heider, M. Tao, M. Su, C. J. Testa, A. N. D'Antonio, P. I. Barton, R. D. Braatz, B. L. Trout, A. S. Myerson, K. F. Jensen and J. M. B. Evans, *Cryst. Growth Des.*, 2014, **14**, 2148–2157.
47. C. M. Thompson, J. L. Poole, J. L. Cross, I. Akritopoulou-Zanze and S. W. Djuric, *Molecules*, 2011, **16**, 9161–9177.
48. F. Ullah, Q. Zang, S. Javed, A. Zhou, C. A. Knudtson, D. Bi, P. R. Hanson and M. G. Organ, *J. Flow Chem.*, 2012, **2**, 118–123.
49. M. Baumann, I. R. Baxendale and S. V. Ley, *Mol. Diversity*, 2011, **15**, 613–630.
50. R. Stalder and G. P. Roth, *ACS Med. Chem. Lett.*, 2013, **4**, 1119–1123.
51. N. Pagano, M. L. Heil and N. D. Cosford, *Synthesis*, 2012, **44**, 2537–2546.
52. B. Desai, K. Dixon, E. Farrant, Q. Feng, K. R. Gibson, W. P. van Hoorn, J. Mills, T. Morgan, D. M. Parry, M. K. Ramjee, C. N. Selway, G. J. Tarver, G. Whitlock and A. G. Wright, *J. Med. Chem.*, 2013, **56**, 3033–3047.
53. D. R. Snead and T. F. Jamison, *Chem. Sci.*, 2013, **4**, 2822–2827.
54. S. A. May, M. D. Johnson, T. M. Braden, J. R. Calvin, B. D. Haeberle, A. R. Jines, R. D. Miller, E. F. Plocharczyk, G. A. Rener, R. N. Richey, C. R. Schmid, R. K. Vaid and H. Yu, *Org. Process Res. Dev.*, 2012, **16**, 982–1002.
55. S. Mascia, P. L. Heider, H. Zhang, R. Lakerveld, B. Benyahia, P. I. Barton, R. D. Braatz, C. L. Cooney, J. M. Evans, T. F. Jamison, K. F. Jensen, A. S. Myerson and B. L. Trout, *Angew. Chem., Int. Ed.*, 2013, **52**, 12359–12363.
56. P. L. Heider, S. C. Born, S. Basak, B. Benyahia, R. Lakerveld, H. Zhang, R. Hogan, L. Buchbinder, A. Wolfe, S. Mascia, J. M. B. Evans, T. F. Jamison and K. F. Jensen, *Org. Process Res. Dev.*, 2014, **18**, 402–409.

CHAPTER 7

# Applying Green Chemistry Principles in Biologics Drug Development

KRISTI L. BUDZINSKI

Genentech, Inc., Environmental, Health, and Safety, 1 DNA Way, MS 71, South San Francisco, CA 94080, USA
Email: Budzinski.kristi@gene.com

## 7.1 Introduction

Traditional pharmaceutical drugs are small molecule drugs manufactured through chemical synthesis which have a well-defined chemical structure. With the advent of molecular biology, a new class of compounds, termed biologics, has emerged for treating disease. Biologics are therapeutic biological compounds encompassing any protein, virus, vaccine, blood product, or gene transfer product. In contrast to small molecule drugs, biologics tend to be very large, complex molecules or mixtures of molecules that are difficult to characterize because they are manufactured by living systems such as mammalian or microbial cells, plant cells, or animal cells. As these compounds are produced *via* fermentation as opposed to chemical synthesis, they have been viewed as natural products and, therefore, inherently greener than small molecule pharmaceuticals. In fact, the industrial and specialty chemical industries are adopting biotechnology processes such as whole cell or enzyme biocatalysis to reduce their environmental footprint because biotechnology methods require less solvent and may provide a more efficient synthesis.[1] However, biologics still have a significant environmental

RSC Drug Discovery Series No. 46
Green Chemistry Strategies for Drug Discovery
Edited by Emily A. Peterson and Julie B. Manley

Published by the Royal Society of Chemistry, www.rsc.org

footprint particularly in areas of water consumption, solid waste generation, energy usage, and waste water toxicity. Applying the framework of green chemistry and engineering to biologics development and manufacturing can help greatly reduce the environmental footprint of the biopharmaceutical industry.

### 7.1.1 Principles for Green Biologics

In their original format, the twelve principles of green chemistry are not entirely applicable to the world of biologics.[2] However, a number of the 12 principles can be modified to provide a framework for Green Biologics, an environmentally sustainable approach to biologics research, development, and manufacturing. This new set of principles can be used to evaluate new technology and process developments across a range of scales to highlight more sustainable methods for developing biologics (Table 7.1).

With the incredible growth of the biologics market, US sales alone saw an increase of 18.2% over 2011 compared to a pharmaceutical sector growth of 2.5%, there is an important need to understand the full environmental impact of biologics manufacturing and find opportunities to reduce this impact.[3] As the market grows, so does the competition; therefore, biopharma companies need to reduce cost and increase efficiency to remain competitive in an increasingly crowded market. Using a framework to reduce environmental impact of new processes and systems can often go hand-in-hand with cost savings and increased efficiency. The pharmaceutical industry has experienced these benefits first hand with examples in convergent synthesis,[4] continuous processing,[5] and solvent recycling.[6] The growth of the biologics market provides the biopharmaceutical industry the opportunity to invest in innovative processes and facilities to improve their environmental footprint and gain a competitive edge.

**Table 7.1** Eight principles for Green Biologics.

| Principle | Description |
|---|---|
| 1 | Develop and monitor key parameters to ensure continuous process improvement to achieve the desired product quality attributes |
| 2 | Practice process intensification to achieve improvements in operational and process efficiency |
| 3 | Processes and operations should be designed to maximize reuse and recycle of resources such as water, raw materials, and consumables |
| 4 | Minimize overall waste generation and create the least number of waste streams possible |
| 5 | Design processes to use and generate less hazardous substances |
| 6 | Use raw materials that are reusable or renewable and recyclable rather than depleting |
| 7 | Design processes to minimize risk of accidents, exposures, or environmental releases |
| 8 | Processes and systems should be designed for and operated for overall energy efficiency |

## 7.1.2 Producing Biologics

Almost all biologics are manufactured by a living system, with the exception of peptides and nucleotides that are produced by chemical synthesis, usually solid-phase synthesis. Thus, the final biologic product is highly sensitive to the manufacturing process. Therapeutic proteins, plasmid DNA, and vaccines are typically produced by fermentation using microbial cells or mammalian cells. Monoclonal antibodies (mAb) are a general class of therapeutic proteins with a defined structure and molecular weights around 150 kDa, which represent the largest sector of the biologics industry, constituting almost 40% of the total biologics market in 2013. Generalized manufacturing schemes for therapeutic proteins are shown in Figure 7.1 for proteins produced by mammalian cell culture[7] (A) and microbial fermentation (B).[8] Here we focus on the production of mAbs in mammalian cell

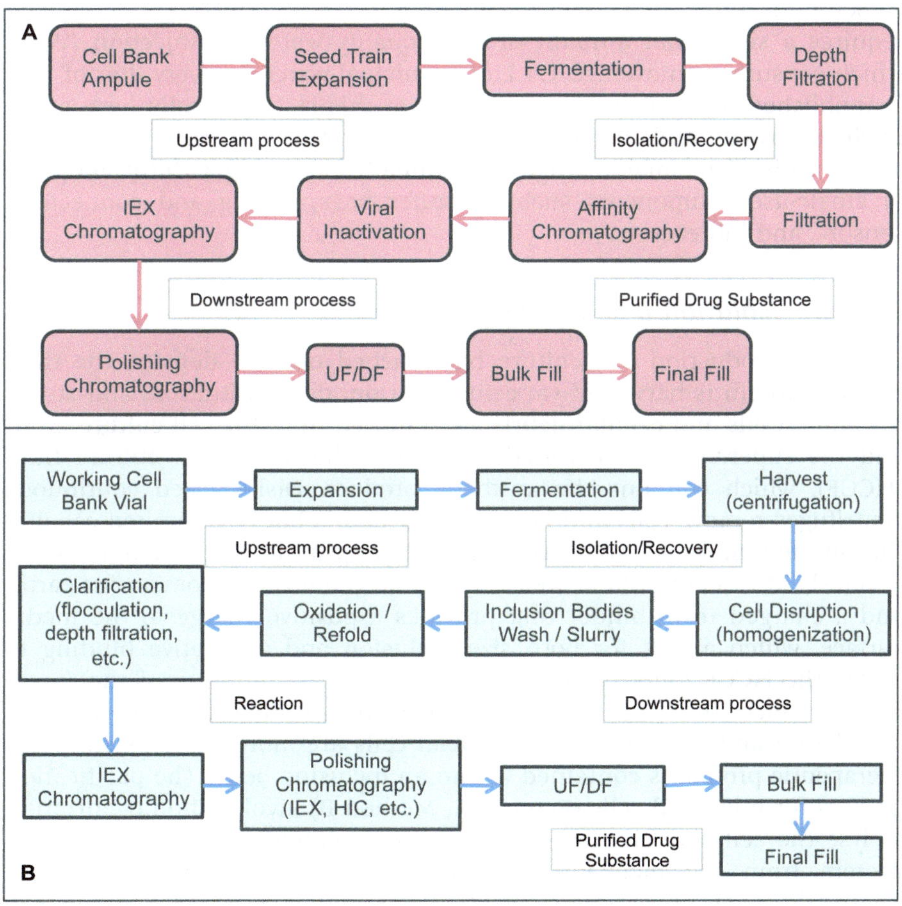


**Figure 7.1** Production scheme for (A) mammalian cell culture and (B) microbial cell fermentation.

culture to serve as the case study for biologics; 50–60% of all protein therapeutics are produced using this method.[9]

### 7.1.2.1 Upstream Process

Current industrial manufacture of therapeutic proteins using mammalian cell culture has centered on a platform process that uses suspension culture, typically Chinese Hamster Ovary (CHO) cells,[10] in a stirred tank bioreactor to generate the protein of interest. The platform production process is operated in fed-batch mode and occurs in three separate phases, seed train, inoculum train, and production phase. The seed train provides fresh cells from the cell bank ampule for the inoculum train, where the volume is expanded to generate a sufficient number of cells that are used in production phase. The production process requires 7–14 days with current cell lines producing 1–5 g $L^{-1}$ of mAb in production vessels ranging from 5000 to 25,000 L for a total of 15–100 kg protein per batch.[11] Upstream processing requires a significant amount of high quality water for injection (WFI), which consumes an average 1.4 L of municipal water for every liter of WFI (unpublished data). Raw materials used at this stage typically have a low intrinsic toxicity and include salts, vitamins, trace minerals, sugars, and amino acids. Consumable usage is continually increasing with the adoption of single-use components such as buffer bags, tubing and connectors, sensors, and bioreactors.

### 7.1.2.2 Isolation and Recovery

Once the production cell culture has reached optimal density, the therapeutic protein is harvested *via* centrifugation, depth filtration, and sterile filtration. Cells and cellular debris are removed from the cell culture broth and the supernatant is clarified to produce harvested cell culture fluid (HCCF), which contains the antibody product. Disk-stacked continuous centrifuges remove cells and large cellular debris but have limited capabilities in the removal of submicron particles. Depth filtration is employed for this task. Most depth filters are composed of cellulose, diatomaceous earth, and a charged resin binder that provides a positive charge to the media surface, which allows for both size exclusion and adsorptive binding to clarify the HCCF following centrifugation.[12] Sterile filtration of the supernatant reduces the bioburden for the downstream processing steps.

Isolation and recovery from microbial cells are more complex since the therapeutic protein is contained within an inclusion body. The purification process for microbial cells (Figure 7.1, Method B) involves homogenization to lyse the cells, differential centrifugation, solubilization to separate the protein from the inclusion bodies, and renaturation of the protein. These additional processing steps often require increased water usage, raw materials (*e.g.*, urea, guanidine hydrochloride) and consumables (filters, resins, *etc.*).[13]

### 7.1.2.3 Downstream Process

mAbs are typically purified by a three-step chromatography process that consists of protein A affinity chromatography, ion exchange chromatography (IEX), and virus filtration.[14] Each chromatography step uses between 1400 and 2000 kg water per kg protein, accounting for a significant portion of the water required to produce a purified biologic. Protein A chromatography exploits the affinity between staphylococcal protein A and the $F_c$ region of IgG-type antibodies.[15] It is used as the initial capture step due to its efficiency, stability, and reproducibility.[12] This step results in a highly pure, more stable product; however, impurities remain due to non-specific binding of contaminants to the resin, protein A leaching, and buffer and wash solution compositions. Thus, several chromatography polishing steps are typically employed after protein A affinity chromatography.

Ion exchange chromatography (IEX) is the most commonly utilized polishing step to remove high molecular weight aggregates, charge-variants, residual DNA, host cell proteins, protein A leachate, and virus particles.[16] IEX chromatography separates components based on charge and can be divided into anion exchange (AEX) chromatography, which employs a resin with positively charged ligands, and cation exchange (CEX) chromatography, which employs resins that have negatively charged ligands immobilized on their surface. IEX columns can be operated in flow-through mode where impurities bind to the resin and product flows through or in bind-and-elute mode where the product binds to the resin and impurities flow through the column.[12] The selection of resins is dependent on binding capacity, required resolution, and feasibility of use at manufacturing scale.

Hydrophobic interaction chromatography (HIC) typically follows IEX chromatography as it is a complementary technique that is especially useful for removing aggregates.[16] HIC separates molecules based on the interaction of hydrophobic ligands (phenol or butyl) and hydrophobic patches on the surface of proteins. To expose the hydrophobic regions, the sample is loaded under aqueous conditions in the presence of a kosmotrophic salt (ammonium sulfate, sodium citrate, potassium phosphate), which decreases the solvation of protein molecules, causing them to bind the resin.[17] The more hydrophobic the molecule, the less salt required for binding and, thus, a gradient of decreasing salt concentration is used to elute the product.[12] The main drawback to this technique is the use of kosmotrophic salts, which can corrode stainless steel tanks and create disposal concerns in municipal waste water treatment facilities.[18] The presence of these salts in the elution pool may require significant dilution or an ultrafiltration/diafiltration unit operation between processing steps.[19] Another choice for the polishing step is multimodal chromatography which combines multiple interaction types (ionic interaction, hydrogen bonding, and hydrophobic interaction) into a single resin, which makes these resins attractive for solving purification problems with high or low conductivity or pH.[12] The application of multimodal chromatography after protein A chromatography can help reduce the number of chromatographic steps; however, the complex interaction chemistry requires a more comprehensive experimental approach.[20]

Viral clearance is an extremely important part of the production purification process. Mammalian cells produce endogenous retroviruses and may become infected with adventitious viruses during processing, producing anywhere from $10^3$ to $10^9$ virus particles per mL of cell culture fluid.[14] Industry standard is less than one virus particle per million doses translating to a 12–18 $\log_{10}$ clearance of endogenous retroviruses and 6 $\log_{10}$ clearance of adventitious viruses.[12] Thus orthogonal virus clearance steps with complementary mechanisms are used during the purification to achieve this goal. Virus filters use a size-based clearance mechanism to retain the virus particle (18–26 nm) and pass the mAb (hydrodynamic diameter 8–12 nm). Pre-filtration of the feed solution is required to remove any impurities such as aggregates, DNA, or denatured product to prevent fouling and increased virus retention. Because of the high degree of performance required of viral filters, they are often subjected to integrity testing before and after use by the end-user and filter manufacturer.[21]

#### 7.1.2.4 *Purified Drug Substance*

The final purification operation is ultrafiltration/diafiltration (UF/DF), which concentrates the product and exchanges it into the final formulation buffer. The membranes used for UF/DF are made from various synthetic polymers such as polysulfone, poly(vinylidene fluoride), and regenerated cellulose, which are resistant to abrasive cleaning solutions allowing them to be reused without deterioration or cross contamination. The product is filtered through a final bulk filtration step that ensures the sterility of the product pool. The pool is then frozen for transfer to the appropriate fill and finish site.

### 7.1.3 Microbial Fermentation

The major differences between mammalian cell culture and microbial fermentation are the length of time required for the fermentation, secreted protein product *versus* internalized protein product, and the use of affinity chromatography in the downstream purification process. Protein production in microbial cells usually occurs within 2 days as opposed to the 14 days required for production by mammalian cell culture. In microbial cells, the therapeutic protein is contained within inclusion bodies, requiring an additional cell lysis step and renaturation step to recover the product in its correct form. The wide range of therapeutic proteins produced through microbial fermentation increases the complexity of the downstream purification process particularly because there is a lack of applicable high affinity chromatography similar to the protein A affinity for IgG antibodies.

## 7.2 Environmental Assessment of Biologics Development

The general scheme of therapeutic protein production outlined earlier highlights the major differences between biologics and small molecule drugs. Biologics require significantly more process water due to cell culture production and downstream purification. Raw material requirements are generally non-hazardous and consist mainly of sugars, salts, trace minerals, and supplements. Little organic solvent is used in the production process, mainly for equipment cleaning and sterilization, compared to small molecule drug production. Consumable usage also tends to be higher in biologics production due to sterility requirements. The development and integration of single-use technology such as buffer bags, disposable bioreactors, single-use filters, and even single-use process sensors is increasing the solid waste footprint of biologics production while reducing the water consumption.

An initial environmental assessment of therapeutic protein production by Ho *et al.* in 2011 identified that water consumption is the largest contributor to the environmental footprint of biologics. For the typical mAb production process, water consumption can range from 3000 to 7000 kg per kg mAb produced. Cell culture water usage accounts for 20–25% of the total water consumed while downstream chromatography accounts for ~50% of the total. Consumables ranges from 2 to 20 kg per kg of therapeutic protein produced while ancillary substances (salts, buffers, *etc.*) can range from 100 to 400 kg per kg of protein.[22] Using this information, the environmental impact of the biopharmaceutical industry can be compared with other industrial and specialty chemical industries. Roger Sheldon defined the *E*(nvironmental)-factor (eqn (7.1)) to assess the environmental impact of manufacturing processes in the late 1980s and it has been used by numerous industries to quantify their impact:

$$E\text{-factor} = \frac{\text{kg process waste}}{\text{kg product}} \tag{7.1}$$

When accounting for process water (without recycling), it has been estimated that the biopharmaceutical industry has an *E*-factor (kg process waste per kg product) of greater than 10 000 (Table 7.2).[1] This is roughly two orders of magnitude higher than the pharmaceutical industry, resulting from the

**Table 7.2** Comparison of environmental impact of industrial and specialty chemical industries.

| | Industrial chemical | Specialty chemical | Pharmaceutical chemical | Biopharmaceutical chemical |
|---|---|---|---|---|
| *E*-factor (kg waste per kg product) | <1–5 | 50–50+ | 25–100+ | >10 000 |
| Product tonnage | $10^4$–$10^6$ | $10^2$–$10^4$ | 10–$10^3$ | $10^{-3}$–10 |

large volumes of water used. Biopharmaceutical processes tend to be at least 80% defined by the time the therapeutic reaches clinical trial stage 2b; therefore, it is critical to encourage the use of green biologic principles in research and development stages.

## 7.3 Implementing Green Biologics Principles

### 7.3.1 Research and Development

With such a large *E*-factor, there are numerous opportunities to improve the environmental footprint during development of a biologic. In 2010, the United Nations (UN) passed a resolution recognizing the right of every human to have access to clean water and sanitation.[23] Biopharmaceutical companies have recognized the impact their processes have on local water quality and quantity, with many committed to reducing their water consumption.[24] Previous environmental assessments have suggested that during fermentation alone approximately 100 kg of water is required per kilogram of unprocessed protein, accounting for 20–25% of the total water consumed.[22] The purification process consumes even more water, with column chromatography and ultrafiltration/diafiltration consuming 100–1000 kg of water per kilogram of purified protein, accounting for >50% of the total water.[13] The additional water consumption is due to the need for purified water and/or water-for-injection (WFI). These systems generate significant water discharge during the reverse osmosis purification cycle. Some facilities have been able to reduce water discharge due to purification systems by reusing the reject water for other operations such as cooling towers.[24d]

Fermentation is inherently water intensive as the cells are grown suspended in an aqueous medium. Conventional seed train starts with a cell bank ampoule thawed into a 20 L seed train bioreactor (STB), which is operated for the duration of the production campaign (weeks–months). Cells harvested from the STB are used to inoculate the first stage of the inoculum train which is then expanded until it is ready to be transferred to the production vessel (2000 L). Cells in the STB undergo batch refeed or solera where a pre-calculated volume of cells is removed and replaced with fresh media every 3–4 days. Over the course of the production campaign, the solera process can consume 400–800 L of water. An example of principle 2 (Table 7.1), process intensification, in action is the alternative approach to seed train which utilizes frozen bags of high density cell culture to inoculate directly into an 80 L bioreactor (the first stage of the inoculum train). FASTEC (frozen accelerated seed train for execution of a campaign) applies perfusion operations to generate high cell density cultures ($>70+10^6$ viable cells per mL) with final viability of >80% that can be frozen in multiple 150 mL cryopreservation bags termed FROSTIs (frozen seed train intermediates) which are used for inoculation (Figure 7.2).[25] The high density and high viability keep the number of FROSTIs that must be thawed for inoculation to a minimum while delivering equivalent product titer and quality.

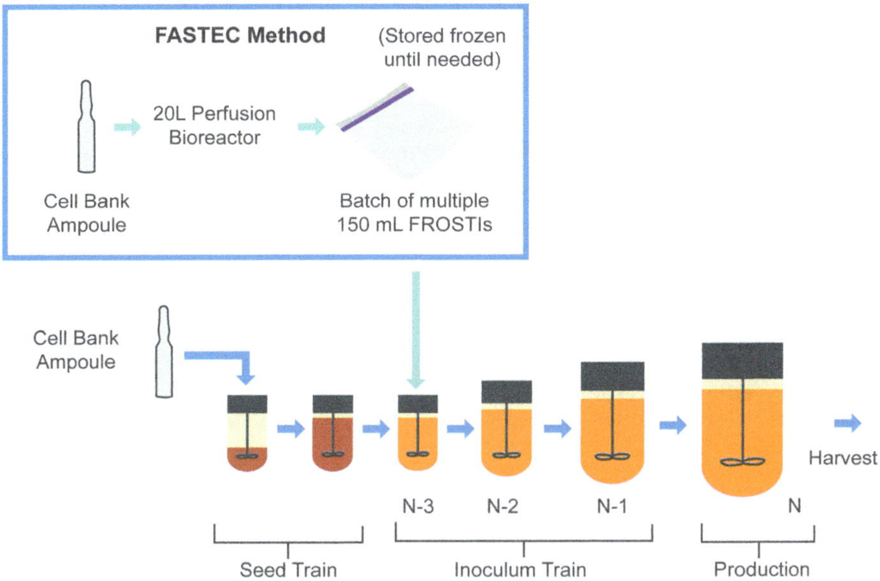


**Figure 7.2** Process flow for conventional seed train method compared to using the FASTEC method for cell culture initiation.

While pharmaceutical companies have successfully implemented continuous flow methods for small molecule synthesis, this approach has been slow to penetrate the biotechnology industry.[26] The most prevalent application of continuous processing is the perfusion bioreactor, in which a suspension cell culture is continuously fed fresh media while the spent culture media is continuously harvested. Cells are filtered from the harvest stream and returned to the bioreactor to maintain a constant culture volume, producing very high cell densities that can produce more product over longer periods of time. An initial environmental analysis of perfusion bioreactors, however, showed that perfusion mode can consume >35% more water and 50% more consumables per kilogram of mAb than traditional fed-batch processes.[22]

Developments in continuous chromatography allow for the integration of perfusion bioreactors with continuous downstream processing, potentially reducing the overall environmental footprint while increasing flexibility. Genzyme demonstrated the integration of a perfusion bioreactor with a four column periodic counter-current (PCC) chromatography system for continuous mAb production and recovery. The system was operated over a 70 day period with a maximum productivity of 1 g $L^{-1}$ reached between days 30 to 40 and maintained roughly 80% recovery. This process was able to increase chromatography capacity utilization by 20%, reduce buffer usage by 25%, and decrease column size (and therefore water consumption) by 75-fold compared to batch mode.[27]

**Table 7.3** Typical components used for therapeutic protein manufacturing.

Upstream fermentation
- Salts (sodium chloride, potassium/sodium phosphate, inorganic & organic salts, amino acids)
- sugars
- antibiotics
- complex protein source (yeast extract, serum)
- trace minerals
- antifoam

Downstream: purification/renaturation
- Salts (organic, inorganic, amino acids, ammonium sulfate)
- acids (organic, inorganic)
- chelators (EDTA)
- solvents (ethanol, acetonitrile)
- denaturants (urea, detergents)

Cleaning bioprocess equipment
- Acids (phosphoric, sulfuric, acidified sugar solutions)
- bases (sodium/potassium hydroxide, ammonium hydroxide)
- disinfectants (bleach, quaternary ammonium)

Cell culture is typically considered a benign production method since the feed components must be life-supporting or life-promoting (see Table 7.3 for a list of components used in the manufacture of therapeutic proteins). Despite this, when the process is scaled for manufacturing, the concentration of some components can be toxic to aquatic life or interfere with wastewater treatment facilities. Trace metals, for example, are often a component of fermentation growth media, with higher levels utilized for *Escherichia coli* and *Saccharomyces* systems than for CHO systems, which can reach concentrations above $LC_{50}$ or the no observed effect concentration (NOEC) causing discharge concerns.[28] Additionally, cell growth produces by-products such as ammonia that accumulate during fermentation, which create discharge issues on an industrial scale. During the cell culture development phase, the use of stoichiometric mathematical models to design media that is utilized more efficiently by the cells can decrease the formation and accumulation of ammonia, which presents concerns for waste water discharge depending on the local municipal limits for ammonia.[29] Media blending executed on a high-throughput cell culture platform can rapidly identify optimal media composition for increased antibody titer and cell growth.[30] Using inline analytical monitoring during process development to detect changes in media components can pinpoint process-induced changes and can enable detection of the fermentation endpoint, ensuring that the precise amount of media feed is used to generate the desired amount of therapeutic protein.[31]

Downstream process operations utilize a wider range of components for processing and formulation buffers compared to cell culture media (Table 7.3) and should be examined for environmental impact. Roche recently conducted an environmental risk assessment for five components that are common or becoming more common in production processes: morpholinoethanesulfonic acid (MES), morpholinopropane sulfonic acid (MOPS), 1,4-piperazine-diethanesulfonic acid (PIPES), ethylenediaminetetraacetic acid (EDTA), and gentamycin. Through a literature search and in-house testing, the biodegradability, ecotoxicity, and predicted no effect concentration (PNEC) values were derived for each compound. Comparing the predicted environmental concentration (PEC) for production sites in Basel, Switzerland and Penzburg, Germany against the calculated PNEC revealed risk characterization ratios of less than 1, suggesting that there is no significant environmental risk to waste water treatment plants (WWTP) or receiving waters from these compounds.[32] The literature yielded little information, except concerning EDTA, which highlights a need for further environmental testing on commonly used substances to determine any environmental risk before the process is locked to ensure that discharge water can be disposed without risk to WWTP or receiving waters.

As the chain of chromatography steps used for purification accounts for over half the water usage, implementing process enhancement strategies can have significant environmental benefits. Replacing the current three-column chromatography platform with a two-column chromatography conventional unit operation platform utilizing high resin/membrane loadings can reduce water consumption and the total amount of consumables (filters, resins, pre-filters) used.[22] The two-column process starts with protein A capture using a high product load and a single polishing step using an AEX resin operated in an isocratic mode called weak partitioning chromatography (WPC). WPC mode increases the amount of product bound to the resin and the amount of impurities removed.[33] Product yield is maximized by a short isocratic wash – average yields of 90% were obtained for clinical production.[12]

Non-affinity purification processes have been developed that eliminate the protein A chromatography step, reducing material costs but also giving a smaller plant footprint. In one two-column process, a CEX resin is used as the initial capture step followed by an AEX resin for polishing followed by high-performance tangential flow filtration (HPTFF).[12] The use of HPTFF allows for the concentration, formulation, and purification of the product in a single step, and provides the ability to remove host cell proteins and DNA impurities at levels that match affinity chromatography purification processes and process yields of roughly 80%.[14] To meet the viral safety requirements, a heat inactivation step was employed as a replacement for the low pH elution from the protein A resin, which inactivates certain viruses. While heat inactivation may require more energy than the low pH elution step, it reduces the use of acidic buffer components (or basic buffer components for alkali-stabile protein A resins) thereby reducing water consumption and risk to municipal water discharge limits.

Principles 4 and 5 encourage the reduction in concentration or hazardous nature of ancillary substances to reduce the environmental impact of biopharmaceutical manufacturing. One example of this approach is the use of hydrophobic interaction chromatography (HIC) with no kosmotrophic salt in the mobile phase. Kosmotrophic salts are used to promote product binding to the column although they can corrode stainless steel and present disposal issues. The presence of these salts in the elution pool may require significant dilution or an ultrafiltration/diafiltration unit operation between processing steps.[19] To mitigate these issues, a very hydrophobic resin was chosen for the stationary phase, while the pH of the mobile phase was modulated such that the mAb of interest flowed through while the impurities bound to the column. This method achieved step yields of ~80% and similar product quality to the conventional HIC purified product pool.[17] The integration of novel methodologies with conventional techniques can decrease the environmental impact of bioprocessing without significant investment in new processing equipment or vastly different unit operations.

Another example of principle 3 (Table 7.1), novel methodologies applied to conventional techniques, is the reuse of chromatography resins for multiple products during downstream process development. Protein A resin has an estimated lifetime of at least 200 cycles in a manufacturing setting.[12] Despite its lifetime, protein A resin still accounts for roughly 35% of raw material costs for downstream purification at manufacturing scales.[34] Additionally, at pilot plant and clinical scales it is estimated that only 10% of the potential lifetime is utilized, increasing the cost burden for development. Genentech implemented a resin cleaning protocol that successfully showed that protein A resin can be reused for multiple products at pilot plant and clinical scales.[35]

This new resin cleaning protocol takes 4.5 h at 20 column volume (CV) per hour flow rate and includes pH cycling and static holds to effectively wash the column. Numerous assays were performed to determine the amount of product carryover, the presence of impurities such as host cell proteins, viruses, DNA, and leached protein A. These tests revealed that the cleaning protocol sufficiently reduced protein carryover to less than 1 ppm and provided mAbs in greater than 90% yield.[35] Thus, the full capacity (~200 cycles) of protein A columns can be utilized resulting in reduced resin costs, storage, and disposal.

### 7.3.2 Production Considerations

In addition to the methodologies discussed above, there are emerging production technologies that have the potential to significantly impact the environmental footprint of the therapeutic protein industry. Single-use technologies have increased in adoption with some new facilities being constructed solely of disposable equipment. The interest in single-use technology stems from its flexibility, reduced cost compared to stainless steel, and, interestingly, their environmental footprint.[36] While the single use nature of these materials increases the solid waste output of a facility, they dramatically reduce the need for cleaning and sanitization, reducing the amount of water, chemicals, and energy utilized by the plant. Single-use systems have been estimated to reduce new facility capital costs by up to 40%[37]

while operational plants benefit from faster campaign turnaround times and increased operational flexibility.[38] Drawbacks to single-use technology arise from the material of construction itself – plastics – and include leachables and extractables, which can affect cell growth and productivity, reduced capabilities for heat transfer, mass transfer, operating pressure, and scale up challenges.

Numerous studies have been carried out to determine the environmental impact of single-use technology compared to stainless steel facilities.[39] These studies are highly dependent on their boundary conditions, typically comparing point-of-use within the facility where single-use technology has a significant advantage. One study that considered the energy required for sterilization, cleaning, and raw material production revealed that single use systems consume half as much energy as traditional stainless steel systems.[39b] The environmental impact of single-use technology has also been examined using life cycle assessment (LCA). This more comprehensive approach was applied to the full process train at 100, 500, and 2000 L scales based on a ten-batch campaign assuming 6 g $L^{-1}$ titers. The scope of the LCA included HVAC requirements for each facility (scaled accordingly), transportation, packaging, end-of-life, and sterilization of equipment (clean/steam in place [CIP/SIP] for stainless steel and gamma irradiation for single-use). The results focused on global warming potential (GWP), cumulative energy demand, and water usage. The majority of the environmental impact occurs during the use phase; single-use systems had a 38% lower GWP due to the lack of CIP/SIP requirements. Water usage was significantly reduced for single-use systems for the same reason. Cumulative energy demands followed the same pattern as GWP because almost all of the GWP is related to energy production and usage.[39a] A follow up LCA analysis was conducted to determine the environmental impacts on ecosystems, human health, and resources.[40] In all three categories, single-use systems exhibited lower impact than traditional systems and, again, the largest impact was during use phase. Interestingly, supply chain impacts contribute 4–8% of the total lifecycle impacts and end-of-life impacts contribute less than 1% for both single-use and traditional processes.[40]

Lifecycle assessments are useful as they highlight the area of highest impact, which, for mAb production, is the CIP/SIP requirements. However, single-use technology is still resource depleting as the feedstock is fossil fuel derived. Additionally, many single-use components are manufactured using various plastics that cannot be easily recycled.[36] Other disposal options have been evaluated including incineration, incineration with cogeneration, pyrolysis for fuel, and landfill, although these methods are highly dependent on availability of facilities and logistics for transport. Some producers are developing take-back programs (EMD Millipore) or novel recycling programs (Sharps Compliance, Inc.) that create new materials for use in other applications.[41] Collaborations between producers and consumers, such as these, will only serve to further decrease the environmental impact of single-use technology.

Process analytical technology (PAT) is an initiative launched by the FDA in 2004 to encourage biopharmaceutical companies to adopt on-line or in-line monitoring tools to ensure product quality at the end of the process. PAT concepts can be applied during upstream unit operations to ensure drug substance quality and manufacturing costs and downstream unit operations to improve operational efficiency.[42] The upstream operations are arguably the most important as this is the stage at which the product and product impurities are produced. PAT applications have been used to quantify the mAb titer to indicate harvest time,[43] monitor metabolic flux of cells,[44] and monitor the concentration of live cells in culture.[45] Downstream applications include monitoring the protein A resin performance to predict real-time column yield decay or column integrity failures[46] and for determining which chromatography fractions to pool to achieve desired product quality attributes.[47] The application of PAT to downstream processes has yet to be fully exploited partially due to the short process times of these unit operations; however, the development of PAT applications could significantly enhance the efficiency of downstream process in terms of cleaning and material use.[42]

### 7.3.3 Facilities and Technology Support

Applying Green Biologics principles in the research and development phase along with innovative production monitoring techniques will significantly contribute to a reduction in the environmental footprint of the biologics industry. However, these initiatives must be joined with sustainable facilities that support energy and water conservation, waste separation and reduction, and automation strategies. Biologics production relies heavily on vendor partnerships for specialized equipment such as filters, resins, and now single-use components. For a truly sustainable industry, vendors must be incorporating the principles of green chemistry and engineering into their methods and using LCA to deliver a greener product.

Laboratory facilities consume significantly more energy than office areas on a square foot basis; it is estimated that laboratories consume 3–4-times the amount of energy as office spaces due to increased requirements for heating, ventilation, and air conditioning (HVAC), and supporting infrastructure such as freezers and autoclaves. In an industrial setting where research and production occur at the same site, these activities can account for $\sim$90% of the site's energy consumption (Figure 7.3A).[48] Energy audits conducted by S-Labs, (Safe, Successful and Sustainable Laboratories) an initiative of HEEPI (Higher Education for Environmental Performance Improvement) revealed how laboratory energy consumption is split and that the split varies depending on the laboratory's field of study. Compared to chemistry laboratories, life sciences laboratories use a small proportion of energy for HVAC (due to lower fume hood usage) but have increased power consumption from equipment such as freezers, incubators, and growth chambers (Figure 7.3B).

A

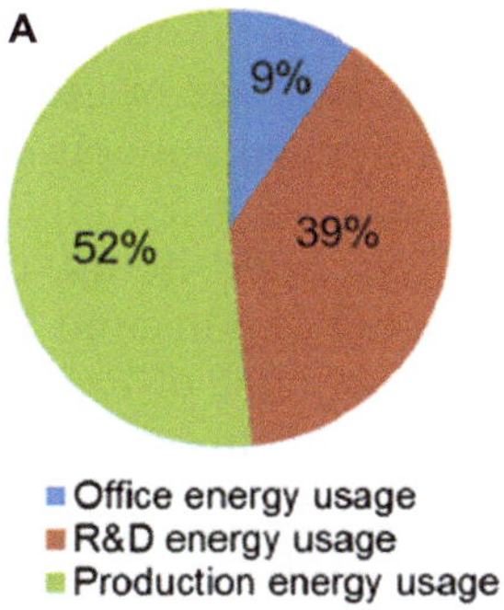


B

| Category | Life Sciences | Chemistry | Offices |
|---|---|---|---|
| HVAC-related | 45% | 60% | 3% |
| Space/ water heating | 20% | 20% | 74% |
| Equipment/ small power | 25% | 15% | 12% |
| Lighting | 10% | 5% | 11% |

**Figure 7.3** Energy use breakdown: (A) Site energy use breakdown for US biopharmaceutical company Genentech, Inc., a member of the Roche group. (B) Comparison of energy use by laboratory type compared to office energy use. HVAC = heating, ventilation, and air conditioning. Compiled by S-Labs in the UK.

Many life science research institutions and universities have undertaken programs to reduce energy usage from equipment, particularly from ultra-low temperature freezers. UC Davis began a campaign entitled "Store Smart" which focused on using laboratory cold storage as efficiently as possible.[49] This program turned into a national freezer challenge between universities, culminating in a program used by National Institutes of Health (NIH), Centers for Disease Control (CDC), and private life science companies to reduce cold storage energy usage. The challenge has four main components:

1. Good freezer management practices such as sample inventories and routine clean outs;
2. Freezer temperature tuning: raising the temperature from −80 to −70 °C or higher;
3. Retiring old freezers either by consolidating samples into fewer freezers or upgrading old freezers to new, energy efficient freezers;
4. Innovative storage techniques such as room temperature DNA/RNA storage, sample management software systems, or setting up large-scale sample repositories.

By participating in the 2012 Freezer Challenge, the CDC estimated that it would save more than $127 000 per year in operating cost and reduce energy consumption by 320 000 kWh.[50] Genentech is sponsoring an internal 2014 Freezer Challenge in support of the Roche corporate goal to reduce total energy consumption per employee by 10% by 2014 over 2009 levels.[51] A key aspect of the 2014 Freezer Challenge is the introduction of Stirling Ultracold freezers, which use an electrically driven free-piston Stirling engine instead of a cascade compressor system, reducing energy usage by ∼50% and generating half as much heat. Through the Challenge, Genentech hopes to replace 10% of its freezer fleet with Stirling Ultracold freezers to reduce site energy consumption.

Reducing energy consumption in laboratories requires education of laboratory staff through competitions such as the Freezer Challenge or green lab assessments that highlight equipment energy usage.[48] Then, laboratory managers can take concrete steps to reduce energy usage such as implementing programmable timers to turn equipment off after hours or purchasing more efficient equipment. The Labs21 program, sponsored by the US Environmental Protection Agency (EPA) and Department of Energy (DOE), provides several tools that can be used to assess and improve the environmental performance of laboratory spaces.[52]

Research is a dynamic environment with constantly shifting resource needs and new technology developments, ultimately generating a complicated waste profile that may include hazardous, biohazardous, and medical waste in addition to non-hazardous waste. Prudent Practices in the Laboratory defines four tiers for laboratory waste management: source reduction; reuse of surplus materials; reclamation, or recycling of materials; and, lastly, disposal.[53] These tiers are accounted for in Green Biologics Principles 3 and 4 (Table 7.1), which ask the researcher to minimize overall waste generation and reuse and/or recycle resources where possible. The laboratory environment must support these activities through appropriate waste collection, incorporation of new technology, and supplier partnerships.

Facilities can support source reduction through on-site reuse programs that promote a sharing culture. In the industry setting, this is an especially useful way to reduce waste from GMP (Good Manufacturing Practices), where material may be internally expired without being past the manufacturer's expiration date. These "GMP-expired" materials can be used in a research or development setting reducing new material costs and waste. Supplier innovation also is an excellent avenue for source reduction, but research institutions and industry must be willing to try reformulated products or set up systems that allow for bulk purchasing or reusable shipping containers. An excellent example of supplier innovation in source reduction is Life Technologies TaqMan Genomic Assays, which now ship at ambient temperature instead of cold shipping, reducing the use of dry ice and expanded polystyrene containers.

Recycling lab waste is an effective means to reduce non-hazardous waste in biopharmaceutical laboratories due to the use of benign raw materials. As laboratory waste recycling is not yet common, facilities must work closely with local recycling haulers to determine appropriate materials for the recycling stream and to ensure the health and safety of the haulers. Laboratory personnel require explicit training and comprehensive signage to effectively participate in a recycling program. Some suppliers are establishing take-back programs that provide a way to recycle materials that would otherwise be unrecyclable. In 2012 EMD Millipore began a Bioprocessing Plastics Recycling Program that accepts some single-use bioprocessing material, such as filters, media bags, and tubing for recycling or as an alternative energy source for cement kilns. During the pilot Millipore diverted ~11 000 lb per month for one customer, equivalent to a 5–7% increase in recycling and a 14% reduction in global warming potential from disposal compared to incineration.[54] Millipore's program to recycle ALL eligible bioprocessing materials, regardless of manufacturer, highlights the need for suppliers to work with customers and recycling vendors to establish greener methods for disposing of laboratory materials.

Reducing the use of hazardous materials can reduce the environmental impact of research while also improving lab safety and reducing disposal costs. In biology labs, ethidium bromide (EtBr) is used to stain DNA fragments separated during agarose gel electrophoresis. EtBr is a carcinogen, mutagen, and teratogen and must be disposed of as hazardous waste; however, numerous non-hazardous alternatives are available for researchers.[55] Published solvent selection guides help researchers choose greener organic solvents[56] when possible and environmental assessments of common biological buffers provide selection criteria for choosing greener buffers.[32]

The principles of green chemistry and green biologics can be applied to experimental design and execution, along with metrics such as *E*-factor and process mass intensity (PMI) to inform the scientist about the environmental impact of their research. Additionally, researchers must be supported by facilities that ensure a safe, healthy, and environmentally conscious research environment. Raising awareness about energy consumption, installing recycling bins, and promoting greener products are some initial steps that can be taken to increase awareness about the impact of scientific research on the environment, resulting in wider adoption of green science principles.

## 7.4 Metrics and Modeling Needs

As shown in the above examples, numerous methods and technologies are emerging that reflect one or more of the Green Biologic Principles as presented in this chapter. The question taking shape now is how to compare the relative "greenness" of process changes and new technology or methodology? Continuing down the sustainable path requires metrics, process modeling methodologies, and systems thinking.

Pharmaceutical companies realized the need for metrics to encourage adoption of green chemistry principles and practices and set out to define industry standards. Roger Sheldon introduced the *E*-factor in 1992 where the ideal *E*-factor would be zero, highlighting the ideal state of zero waste. The American Chemical Society Green Chemistry Institute's Pharmaceutical Roundtable (ACS GCIPR) adopted process mass intensity (PMI) as the industry-standard mass metric. PMI is defined as the total mass of materials used to produce a specified mass of product (eqn (7.2)) and was chosen because it emphasizes the benefit of green chemistry in adding value and increasing productivity over reducing waste:[57]

$$\mathrm{PMI} = \frac{\text{total mass in a process or process step (kg)}}{\text{mass of product (kg)}} \tag{7.2}$$

Using PMI, the ACS GCIPR conducted a benchmarking exercise that revealed that solvents contribute over 50% to the PMI for small molecule processes, which lead the Roundtable to focus on finding suitable solvent replacements.[58] This initial tool evolved into a holistic life cycle assessment of pharmaceutical processes to provide a comprehensive view of the environmental impact associated with the product manufacture.[59]

The time has come for the biopharmaceutical industry to embrace similar metrics. Initial environmental assessments of biopharmaceuticals suggest that approximately 1000 to >10 000 kg of water is used per kg of biologic produced,[13] 30–90 disposable containers may be used in a typical 2000 L mAb process,[36] and up to 500 g of trace metals in the wastewater discharged after harvest.[1] Given these data, a process mass intensity metric for biologics should include the total amount of water, the total mass of raw materials, and the total mass of consumables used in the process. Including cleaning water, raw materials, and consumables in the biologics PMI would allow companies to compare environmental impact as well as cost when developing new procedures such as the reuse of protein A resin *versus* wasting underutilized resin. Using such a PMI metric could provide important insight for companies comparing single-use processes against traditional processes and the associated cleaning required, a particularly important consideration for facilities located in arid climates. The business case for the incorporation of new technologies or methodologies may be strengthened by a PMI assessment, which would help highlight increased process efficiencies, reduced environmental risk, as well as waste streams.

A biologics PMI would provide a first-pass at understanding the environmental "hot spots" in upstream and downstream processes and allow for comparison between new or alternative processes. However, PMI still lacks the comprehensive analysis of the cumulative environmental impacts that would be provided by a life cycle assessment. Life cycle assessments have been used to assess the impact of single-use technologies[39a,40] although most of them have focused on a single unit operation.[39c,60] Conducting an LCA analysis during development would allow process engineers to (i) assess the environmental impact of different production technologies (single-use *versus* traditional) and methods (batch *versus* continuous); (ii) assess the impact of different disposal options such as landfill, recycling, or incineration; and (iii) identify stages or unit operations with the highest environmental burden.[61] The use of LCA would also allow process engineers to account for the environmental impact of raw materials used in fermentation and purification allowing for a more in-depth analysis of cell lines, purification methods, and ancillary substances.

## 7.5 Conclusion

Biologics continues to be a growing market within the pharmaceutical sector. The launch of biosimilars[62] and biobetters[63] will further expand the production of biologics and the environmental impact of biologics.

Additionally, other industries are adopting biotechnology techniques, termed white biotechnology, to reduce the use of hazardous raw materials and energy.[1] While the resulting environmental footprint of these adaptations is generally smaller than the original industrial technique, there is still a need to understand and address the environmental impact from biotechnology and biologics production, particularly around water consumption. The incorporation of single-use technology further emphasizes the needs for environmental impact assessment particularly around waste disposal options. Biologics manufacturers and suppliers can work collaboratively to bring innovative technologies to the market with an environmentally friendly end-of-life strategy in place. The development and industry acceptance of a mass-based metric and/or life cycle assessment analysis will inform these decisions and show the industry's commitment to a healthy planet for healthy people.[64]

Research laboratories can benefit from increased awareness of their environmental impact due to energy consumption of equipment, hazardous waste generation, and end-of-life options for consumables. Applying green biologics principles can also inform experimental design and process development, which will ultimately reduce the environmental impact of manufacturing platforms. With biologic drug development spanning 10–15 years and process lock occurring during late-stage clinical trials, it is important to determine the environmental impact early in the biologic lifecycle so that companies can reach sustainability goals, reduce cost, and increase efficiency. Incorporating innovative technology such as new cell lines (for an overview see "Towards greener therapeutic proteins"[22]), two-column chromatography processes for purification, reusing resins, filters, or other "single-use" technology when feasible, and assessing raw materials for environmental impact before committing them to the process will not only reduce the impact of biologics research, development, and production but increase employee engagement and patient well-being.

## Acknowledgements

I would like to thank Sa V. Ho, Sr. Research Fellow, Pfizer; Barbra Owen, Director, Biosafety & Corporate Biosafety Officer, Merck Sharp & Dohme; and Debbie O'Connor, Head of Purification and Drug Product Pilot Plant, Genentech for sharing their expertise and providing feedback on the chapter.

## References

1. B. Junker, *BioProcess Int.*, 2010, **8**, 62.
2. P. Anastas and N. Eghbali, *Chem. Soc. Rev.*, 2010, **39**, 301.
3. R. S. Aggarwal, *Nat. Biotechnol.*, 2014, **32**, 32.
4. P. J. Dunn, S. Galvin and K. Hettenbach, *Green Chem.*, 2004, **6**, 43.

5. P. Poechlauer, J. Colberg, E. Fisher, M. Jansen, M. D. Johnson, S. G. Koenig, M. Lawler, T. Laporte, J. Manley, B. Martin and A. O'Kearney-McMullan, *Org. Process Res. Dev.*, 2013, **17**, 1472.
6. C. S. Slater, M. J. Savelski, T. M. Moroz and M. J. Raymond, *Green Chem. Lett. Rev.*, 2011, **5**, 55.
7. J. R. Birch and A. J. Racher, *Adv. Drug Delivery Rev.*, 2006, **58**, 671.
8. A. L. Demain, *Trends Biotechnol.*, 2000, **18**, 26.
9. (a) F. M. Wurm, *Nat. Biotechnol.*, 2004, **22**, 1393; (b) G. Walsh, *Nat. Biotechnol.*, 2010, **28**, 917.
10. F. M. Wurm, *Nat. Biotechnol.*, 2004, **22**, 1393.
11. B. Kelley, *mAbs*, 2009, **1**, 443.
12. H. F. Liu, J. Ma, C. Winter and R. Bayer, *mAbs*, 2010, **2**, 480.
13. S. V. Ho, J. M. McLaughlin, B. W. Cue and P. J. Dunn, *Green Chem.*, 2010, **12**, 755.
14. A. Mehta, M. L. Tse, J. Fogle, A. Len, R. Shrestha, N. Fontes, B. Lebreton, B. Wolk and R. van Reis, *Chem. Eng. Prog.*, 2008, **104**, S14.
15. S. Hober, K. Nord and M. Linhult, *J. Chromatogr. B*, 2007, **848**, 40.
16. I. Arora, *Int. J. Emer. Tech. Adv. Eng. (IJETAE)*, 2013, **3**, 475.
17. S. Ghose, Y. Tao, L. Conley and D. Cecchini, *mAbs*, 2013, **5**, 795.
18. P. Gagnon, Polishing methods for monoclonal IgG purification, in *Process Scale Bioseparations for the Biopharmaceutical Industry*, ed. A. A. Shukla, M. R. Etzel and S. Gadam, Taylor & Francis, New York, 2006, pp. 491–505.
19. J. Chen, J. Tetrault and A. Ley, *J. Chromatogr. A*, 2008, **1177**, 272.
20. K. Eriksson, A. Ljunglöf, G. Rodrigo and E. Brekkan, *BioProcess Int.*, 2009, 52.
21. G. Kern and M. Krishnan, *BioPharm Int.*, 2006, **19**, 32–41.
22. S. V. Ho, J. M. McLaughlin, J. Pollock and S. S. Farid, Toward Greener Therapeutic Proteins, in *Biocatalysis for Green Chemistry and Chemical Process Development*, John Wiley & Sons, Inc., Hoboken, 2011, pp. 197–219.
23. The Human Right to Water and Sanitation, 2010, http://www.un.org/waterforlifedecade/human_right_to_water.shtml, (accessed 12/19/2014).
24. (a) Roche Sustainability http://www.roche.com/sustainability/for_communities_and_environment/environment.htm; (b) Pfizer Responsibility http://www.pfizer.com/responsibility/protecting_environment/protecting_the_environment; (c) GlaxoSmithKline Responsibility http://www.gsk.com/en-gb/responsibility/our-planet/; (d) Genentech for Good http://www.gene.com/good/sustainability (all accessed June 1, 2014).
25. G. Seth, R. W. Hamilton, T. R. Stapp, L. Zheng, A. Meier, K. Petty, S. Leung and S. Chary, *Biotechnol. Bioeng.*, 2013, **110**, 1376.
26. L. Malet-Sanz and F. Susanne, *J. Med. Chem.*, 2012, **55**, 4062.
27. V. Warikoo, R. Godawat, K. Brower, S. Jain, D. Cummings, E. Simons, T. Johnson, J. Walther, M. Yu, B. Wright, J. McLarty, K. P. Karey, C. Hwang, W. Zhou, F. Riske and K. Konstantinov, *Biotechnol. Bioeng.*, 2012, **109**, 3018.

28. B. Junker, *BioProcess Int.*, 2010, **9**, 36.
29. E. Jain and A. Kumar, *Biotechnol. Adv.*, 2008, **26**, 46.
30. M. Jordan, D. Voisard, A. Berthoud, L. Tercier, B. Kleuser, G. Baer and H. Broly, *Cytotechnology*, 2013, **65**, 31.
31. (a) K. Konstantinov, S. Chuppa, E. Sajan, Y. Tsai, S. Yoon and F. Golini, *Trends Biotechnol.*, 1994, **12**, 324; (b) G. Mazarevica, J. Diewok, J. R. Baena, E. Rosenberg and B. Lendl, *Appl. Spectrosc.*, 2004, **58**, 804.
32. J. O. Straub, D. Gysel, U. Kastl, J. Klemmer, M. Sonderegger and M. Studer, *Environ. Toxicol. Chem.*, 2012, **31**, 681.
33. B. Kelley, *Biotechnol. Prog.*, 2007, **23**, 995.
34. D. K. Follman and R. L. Fahrner, *J. Chromatogr. A*, 2004, **1024**, 79.
35. M. Ekta, W. Jay, K. Kapil and L. Tina, One Resin, Multiple Products: A Green Approach to Purification, in *Developments in Biotechnology and Bioprocessing*, American Chemical Society, 2013, vol. 1125, pp. 87–111.
36. B. Rawlings and H. Pora, *BioProcess Int.*, 2009, 7, 18.
37. J. M. País-Chanfrau, K. Zorrilla and E. Chico, *BioPharm Int.*, 2009, **22**, 62.
38. A. A. Shukla and U. Gottschalk, *Trends Biotechnol.*, 2013, **31**, 147.
39. (a) M. Pietrzykowski, W. Flanagan, V. Pizzi, A. Brown, A. Sinclair and M. Monge, *BioPharm Int.*, 2011, **24**, 30; (b) M. Mauter, *Bioprocess Int.*, 2009, **8**, 18.
40. W. Flanagan, M. Pietrzykowski, V. Pizzi, A. Brown, A. Sinclair and M. Monge, *BioPharm Int.*, 2014, **27**, 40.
41. (a) P. Thomas, EMD Millipore: Giving and Receiving Single-Use Systems, *PharmaManufacturing.com*, May 16, 2012, http://www.pharmamanufacturing.com/articles/2012/077/; (b) C. Scott, *BioProcess Int.*, 2010, **8**, 40.
42. A. S. Rathore, R. Bhambure and V. Ghare, *Anal. Bioanal. Chem.*, 2010, **398**, 137.
43. J. Zhang, H. Zhou, Z. Ji and F. Regnier, *J. Chromatogr. B*, 1998, **707**, 257.
44. O. Henry, A. Kamen and M. Perrier, *J. Process Control*, 2007, **17**, 241.
45. J. P. Carvell and J. E. Dowd, *Cytotechnology*, 2006, **50**, 35.
46. Y. Hou, C. Jiang, A. A. Shukla and S. M. Cramer, *Biotechnol. Bioeng.*, 2011, **108**, 59.
47. J. Gomes, V. R. Chopda and A. S. Rathore, *J. Chem. Technol. Biotechnol.*, 2015, **90**, 583.
48. L. Hopkinson, N. Lenegan, T. McGrath and M. Tait, *Energy Consumption of University Laboratories: Detailed Results from S-Lab Audits*, S-Lab (Safe, Successful and Sustainable Laboratories) initiative of HEEPI (Higher Education for Environmental Performance Improvement): www.goodcampus.org, 2011.
49. University of California, D. Take Action: Store Smart. http://sustainability.ucdavis.edu/action/conserve_energy/store_smart.html (accessed 9/25/2014).
50. L. York, In *2012 Centers for Disease Control and Prevention Lab Freezer Challenge*, 2013 I2SL Annual Conference, Minneapolis, MN, International Institute for Sustainable Laboratories, Minneapolis, MN, 2013.

51. Roche http://www.roche.com/sustainability/for_communities_and_environment/environment/our_she_goals_and_performance.htm (accessed 09/25/2014).
52. Labs21 Environmental Performance Criteria 3.0 http://www.i2sl.org/resources/toolkit/epc.html, accessed 2014.
53. Prudent Practices in the Laboratory: Handling and Management of Chemical Hazards: Updated Version, 2011.
54. K. B. Ignacio, Greening the Supply Chain. In *NAEM 2013 EHS Management Forum*, Montreal, Canada, 2013.
55. Program, M. E. O. G. C. P. P. Ethidium Bromide Alternatives Assessment. http://web.mit.edu/workinggreen/buy/lab.html (accessed 9/28/2014).
56. R. K. Henderson, C. Jimenez-Gonzalez, D. J. C. Constable, S. R. Alston, G. G. A. Inglis, G. Fisher, J. Sherwood, S. P. Binks and A. D. Curzons, *Green Chem.*, 2011, **13**, 854.
57. C. Jimenez-Gonzalez, C. S. Ponder, Q. B. Broxterman and J. B. Manley, *Org. Process Res. Dev.*, 2011, **15**, 912.
58. R. K. Henderson, J. Kindervater and J. B. Manley, In *Lessons learned through measuring green chemistry performance-The pharmaceutical experience*, 11th ACS green chemistry and engineering conference, Washington, DC, 2007.
59. C. Jiménez-González, C. Ollech, W. Pyrz, D. Hughes, Q. B. Broxterman and N. Bhathela, *Org. Process Res. Dev.*, 2013, **17**, 239.
60. (a) T. M. Mata, A. A. Martins, B. Neto, M. L. Martins, R. L. Salcedo and C. A. Costa, *Chem. Eng. Trans.*, 2012, **26**, 261; (b) J. C. Jobin and M. Krishnan, *BioProcess Int.*, 2012, **10**(Suppl. 5), 66.
61. S. V. Ramasamy, N. J. Titchener-Hooker and P. Lettieri, *Food and Bioproducts Processing* [Online], 2014, DOI: 10.1016/j.fbp.2014.03.009 (accessed June 1, 2014).
62. P. Declerck, *Generics Biosimilars Init. J.*, 2012, **1**, 13.
63. N. Dinwoodie, *BioPharm Int.*, 2011, **24**, 31.
64. D. C. Pencheon, *Trials*, 2011, **12**, 80.

CHAPTER 8

# *Recycling and Reuse in the Laboratory*

DANIEL T. RICHTER

Pfizer Global Research and Development, 10714 Science Center Drive, San Diego, CA 92121, USA
Email: daniel.richter@pfizer.com

## 8.1 Introduction: Recycling and Reuse in the Laboratory

In general, recycling is commonplace throughout the world; bottles, cans, paper, and other goods are recycled in large quantities daily. However, when one thinks about the discovery chemistry laboratory there is a much lower level of emphasis on waste generation and corresponding minimization techniques. This is borne out of the primary focus of a drug discovery laboratory, which is to verify that chemistry is successful and that the desired products can be prepared quickly for testing against a specified target. With the focus, for the most part, on the desired outcome of speed to target and not necessarily the best path to get there, yields and waste generation are often examined secondary to synthesizing the product. In the case of waste minimization, it is commonly more of an afterthought than a factor considered in the development of the synthesis. Often, active compounds are required in a greater quantity for additional biological testing and the expectation is that somewhere further along the pipeline someone will examine opportunities for optimizing the procedure both to improve the yield/isolation and to reduce the waste generated. While this is the

RSC Drug Discovery Series No. 46
Green Chemistry Strategies for Drug Discovery
Edited by Emily A. Peterson and Julie B. Manley

Published by the Royal Society of Chemistry, www.rsc.org

primary purview for a process chemist (to improve synthetic routes for commercial production), in a discovery lab situation there are many opportunities to make small changes to the prosecution of reactions and work-ups that can have a significant impact on the amount of waste generated. For example, consider the use of a test tube, the individual scientist may use only 1–200 disposable glass test tubes in a week for fraction collection during chromatography. While this is not necessarily a large amount of glass to be thrown away, when you consider this amount for every scientist working in a small laboratory (20 people) every week (5 days) of the year (48 weeks, due to holidays), that small amount becomes very concerning on an annual basis ($200\times20\times5\times48=960\,000$ test tubes). By considering how many research laboratories there are globally, it is easy to imagine how small changes on a discovery scale can lead to huge impacts on the environment.

As such, in a discovery lab setting there are a multitude of opportunities to reduce, reuse, or recycle many of the items that are used every day. In terms of minimization, previous chapters highlighted synthetic planning and discussed atom economy, reaction concentrations/stoichiometry, and processes such as one-pot reactions, reaction telescoping, crystallization *etc.* as opportunities to minimize waste by reducing or eliminating steps, reagents, equivalents, isolation operations, or chromatography aligned with the 12 principles of green chemistry.[1] Even with these amendments to the prosecution of chemistry there are still waste products generated from any synthesis that need to be addressed.

In this chapter, the focus will be on several areas of high impact that are common to all discovery laboratories where waste can be reduced through reuse or recycling of materials: solvents, precious metal catalysts, glass, chromatography columns, and cardboard and plastic. These areas of frequent use can yield the greatest returns in the goal of minimizing waste and environmental footprint in a drug discovery setting. Furthermore, an explanation of the practical limitations and issues with these options will be included as various factors may impact feasibility due to scale, cost, or safety and regulatory concerns.

## 8.2 Solvents

The largest amount of waste generated from chemical synthesis is from solvent used during synthesis work up and purification of compounds. In fact, solvents account for approximately 50% of waste generated from chemical synthesis of active pharmaceutical ingredients (APIs) across the pharmaceutical industry (Figure 8.1).[2]

In manufacturing processes the ability to recycle solvents is often made easier by the fact that large quantities of solvent are able to be isolated with a minimal amount of other contaminants. While these recovered solvents may not be used for cGMP processes without strict analysis to recertify the solvent batch, they may be reused to clean reactors or sent for recycling.[3] In a

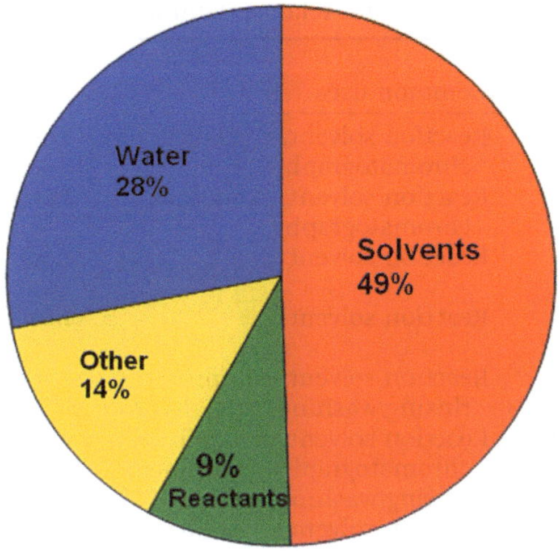


**Figure 8.1** Composition of waste generated from active pharmaceutical production as measured by PMI (process mass intensity).

discovery laboratory, reactions are carried out on a much smaller scale with a variety of solvents. In addition, purification by extraction or chromatography can further diversify the solvents leading to a complex waste mixture. The complex mixtures of solvents produced in a drug discovery laboratory and the likely contamination of this mixture with reagents, catalysts, and reaction byproducts increases the challenge of isolating a single solvent for any sort of recycling process. Mixtures can contain other lower boiling solvents, organic bases (such as pyridine, triethylamine), and water that create mixtures with solvent boiling points ranging from 35 °C for diethyl ether up to 189 °C for dimethyl sulfoxide (DMSO). Typically, these mixtures are disposed of as waste or burned as fuels, where costs for this process are associated with measured purity values.[4] More flammable mixtures require less energy to incinerate than mixtures with higher contents of low flammability solvents (Table 8.1) and as a result are of higher value. If solvents can be segregated and/or low flammability solvents reduced, the opportunity for reuse in alternative purposes increases accordingly.

In addition to carrying out reactions at higher concentrations as mentioned in previous chapters, to reduce solvent consumption by reuse or recycling, segregation is the most fundamental task to facilitate solvent recovery efforts. If solvents can be collected as primarily one component, it will simplify any recycling distillation process significantly. Alternatively, binary or in some cases ternary mixtures can be effectively separated by distillation as long as there is a reasonable difference between boiling points and no azeotropes exist. An azeotrope is a mixture of solvents that, when heated to boiling, distill as a mixture of those solvents; this is also known as a constant boiling mixture. An example of this is ethanol (EtOH)/water. EtOH boils at 78.3 °C and water boils at 100 °C, but when a mixture of these

**Table 8.1** Common solvent uses and relative flammability.

| Solvent | Common uses | Relative flammability | Boiling point (°C) |
|---|---|---|---|
| Acetonitrile | Reaction solvent, crystallization, chromatography | Low | 82 |
| Dichloromethane (DCM) | Reaction solvent, extraction, chromatography | Low | 40 |
| Dimethyl sulfoxide (DMSO) | Reaction solvent | Low | 189 |
| Dimethylformamide (DMF) | Reaction solvent | Low | 152 |
| Acetone | Reaction solvent, equipment rinsing/washing, crystallization | High | 56 |
| Methanol | Reaction solvent, crystallization, chromatography, equipment rinsing/washing | High | 65 |
| Ethanol | Reaction solvent, crystallization, chromatography | High | 78 |
| Isopropyl alcohol | Reaction solvent, crystallization chromatography | High | 82 |
| Hexane | Crystallization, chromatography | High | 69 |
| Heptane | Crystallization, chromatography | High | 98 |
| Ethyl acetate | Reaction solvent, extraction, chromatography, crystallization | High | 77 |
| Toluene | Reaction solvent, crystallization | High | 111 |
| Dioxane | Reaction solvent | High | 101 |
| Diethyl ether | Reaction solvent, crystallization, extraction | High | 35 |
| Tetrahydrofuran (THF) | Reaction solvent, crystallization | High | 66 |
| 2-Methyltetrahydrofuran | Reaction solvent, extraction | High | 80 |

solvents is heated, it boils at 78.2 °C, and the distillate is a mixture of 95% EtOH and 5% water. There are many azeotropic mixtures that can exist between the variety of solvents listed above, leading to a high level of complexity for isolation of a pure single solvent stream.[5]

The requirements for distillation are fairly simple: a solvent reservoir, distillation column, reception reservoir, and heating source. Perhaps the most simple and widely used apparatus is the rotary evaporator, a very commonly used system for solvent recovery and removal of solvents from reactions, extractions, or chromatography. Solvents can be recovered quite easily in this manner. If mixtures are present, this method would not have enough theoretical plates to allow for isolation of pure solvents and a more technical setup would be required. There are many variations of distillation setups available commercially that will distill solvents to a high grade of purity, certainly enough for use in exploratory chemistry laboratories. While use of recovered solvents in reactions would not be ideal without some characterization of purity, the solvents could be used for other common tasks such as crystallizations, chromatography, extraction, or cleaning of glassware.

Distillation appears to be an easy enough process to recover solvents for their reuse, but there are some factors to consider. First, it must make sense in terms of the ease of implementation such that it does not cost significantly more than purchasing a new bottle of solvent. Time spent also needs to be factored in. If it takes a whole day to recover 4 L of solvent by distillation, the cost will outweigh the gain. While some processes can be quite efficient, the amount of solvent volume needs to be tracked to make sure a process is going to pay off over time. Infrastructure and safety are also concerns that must be addressed in any solvent recycling program. To ensure this process is done safely, a designated area will need to be assigned for solvent collection and the proper equipment must be set up and installed. Due to the flammability of most solvents, to carry out this type of process it would need to be confirmed as safe by the environmental health and safety (EHS) group. In addition, the setup would need to be done in accordance with all fire codes and local, state, and federal safety and environmental regulations. This means that additional permits and inspections may need to be carried out before a process can be implemented. It is worth mentioning that in some locales this would be considered waste treatment, which adds a whole new set of regulations and permits.[6] These are not insurmountable plights to overcome, yet they tend to be cost prohibitive in relation to the value gained, and most research facilities would prefer to use scientist's time more productively on research. As a result, the best situation currently would be to separate major solvent streams and seek out a service that could recycle or at least reuse the solvents as fuel in another process. Ideally, a vendor that is geographically convenient would minimize costs associated with transportation.

## 8.3 Metals

Outside of the pharmaceutical industry metal recycling has been going on for a very long time from a variety of uses such as aluminum cans, catalytic converters from cars, electronic devices, scrap metal, *etc.* In the pharmaceutical industry, precious metals are used as catalysts for many chemical transformations such as hydrogenation, Heck, Stille, and Suzuki couplings, olefin metathesis, Buchwald–Hartwig aminations, and so on.[7] The advent of these catalytic reactions over the last 30 years has lead to an increased usage of precious and semi-precious rare earth metals, most commonly platinum (Pt), palladium (Pd), and rhodium (Rh), all with a relative abundance less than 20 parts per billion (ppb) in the earth's crust.[8] Due to their increasing scarcity and extremely resource intensive mining processes for extraction, recycling and reuse of these metals is of high value and importance. This is an area where interest is growing and investigations are being made into using more abundant metals such as iron (Fe) or nickel

(Ni) as alternatives. In fact the American Chemical Society Green Chemistry Institute's Pharmaceutical Roundtable has recently funded a research project with Professor Neil Garg at UCLA to investigate the use of nickel as a catalyst for Suzuki couplings in greener solvents.[9] In this instance a wide variety of substrates were found to undergo the coupling reactions in high yields using a readily available catalyst at lower catalyst loadings and mild conditions.

In many publications researchers show the effectiveness of a catalytic reaction, reporting high yields and mentioning that the catalyst can be recycled or reused many times without any significant impact on performance for future reactions. This provides an opportunity to reduce catalyst consumption and recycle spent catalysts. In most cases the catalyst is used in very small amounts, often 0.1 equivalents or less, such that reaction scale is a factor in this effort. Since the isolation of the catalytic materials can be a resource consuming process itself, the recovery of the catalyst depends on the type of catalyst, heterogeneous or homogeneous, to be recovered. Many references and reviews have been written on this subject such that in-depth detail addressing concepts such as TON (turnover numbers)/TOF (turnover frequency) will not be covered here.[10–12] What follows is a highlight of the practical application of available options for reuse and final recapture or recycling of catalyst containing materials in the drug discovery setting.

### 8.3.1 Heterogeneous Catalysts

Heterogeneous catalysts such as palladium on carbon, silica bound catalysts, or polymer supported catalysts can easily be filtered from a reaction and readily isolated for recycling or reuse. In the case of flow chemistry such as reactions performed with an H-cube™, catalysts are pre-packed into flow cartridges that can easily be removed from the system and reused multiple times before recycling.[13] Since flow chemistry in general is still in its infancy (see Chapter 6 on flow chemistry), this is an area that could impact catalyst recovery and recycling significantly in the future. For other supported catalysts, while there can be some metal leaching from the silica or polymer bound varieties, the recoveries are very high and catalysts perform well for repeated use in various transformations.[10] Although, over time, there is a gradual loss of performance that increases with each use and detailed monitoring of reactions would be required to track long-term activity to avoid low yields and failed reactions. However, even a single reuse of a catalyst would provide cost savings and yield significant overall impact to catalyst consumption.

One additional hurdle is the lack of broad availability of silica bound or polymer supported catalysts for the vast variety of catalytic reactions. Strem Chemicals, Inc. has produced an information booklet for available heterogeneous catalysts that includes technical notes for their isolation from reaction media.[14] This is not a comprehensive list but it does expose the vast difference between available heterogeneous catalysts *versus* available homogeneous catalysts. Furthermore, as many new homogeneous catalysts are developed that have enhanced properties or activities for certain substrates or types of reactions, this gap grows even larger.

### 8.3.2 Homogeneous Catalysts

Where homogeneous catalysts (typically those with bound ligands for specific coordination properties that also increase solubility of the species in organic solvents) are used, the isolation of these materials can be much more cumbersome since they cannot be simply filtered from the reaction and are often non-polar such that they are not captured easily on silica during chromatography. There are scavenger resins that can be used to purge Pd and other metals for recycling, but this prevents them from being reused.[15]

Current options are limited in terms of practical application for reuse of homogeneous catalysts in a discovery situation. However, in the absence of available non-precious catalysts the opportunity exists to recycle used homogeneous catalyst waste-contaminated materials. Concentrations of metals in laboratory use items (filter papers, gloves, Celite®, silica, catalysts trapped on a scavenger resin, *etc.*) are usually much higher by volume than those seen in the original mining process used to isolate the metal, rendering them a valuable commodity for the recovery of these metals. The amount of material that must be accumulated before this becomes feasible is on the order of hundreds of pounds to warrant the process to isolate the precious metal. A recent collection process at GlaxoSmithKline (GSK) yielded ~400 lb of catalyst containing materials that was recycled to recover Pd and Pt. The reclaimed metal covered the costs of shipping the material for processing. As an added benefit, the costs of treatment and disposal of the material as hazardous waste were eliminated.[16] Essentially this material needs to be a large volume with a high enough concentration of precious metal to compare with the original mining extraction process to produce that material. Scale is the primary issue; because catalysts are used in such small amounts, it is hard to isolate and accumulate enough material to warrant recycling from a small lab group. One solution would be to find other laboratories in your geographical area to network with and to arrange a pick up process that is shared to minimize costs. Additionally, while catalysts can be used repeatedly in many cases, the process to re-isolate them is often sufficiently tedious that it is not commonly practiced. Identifying more reasonable and efficient approaches to recycling and/or recapturing precious metal catalysts persists as an avenue in need of much exploration.

## 8.4 Glass

Glass recycling is a universally accepted process employed for items from soda bottles to dishes and bake ware. However, most of the glass used in a chemistry lab is a variety of borosilicate blends, such as Pyrex®, Kimax®, *etc.*

Even test tubes and Pasteur pipettes are made of borosilicate glass. Most other glass present in the laboratory from solvent or reagent bottles is either clear glass or brown glass that is readily recyclable. Since common recycling processes exist for these types of glass, only the options for handling borosilicate glass (test tubes, NMR tubes, pipettes) that are primarily limited to reuse will be discussed here.

Test tubes are used for a wide variety of purposes in many lab settings. In a chemistry laboratory, test tubes are primarily viewed as a single use item for fraction collecting during chromatography purification and small-scale experiments. As a result they need to be somewhat sturdy, resistant to fracture, and durable to temperature changes from as low as −78 °C to as much as 200 °C, as well as inert to solvents and other reagents. To achieve this robust versatility, most test tubes available are made from borosilicate glass. The addition of varying levels of boric oxide is what gives the test tube the enhanced sturdiness and thermal flexibility.[17] The attributes that make borosilicate glass useful for laboratory work also make recycling this glass a challenge as a result of the high temperatures required for the process. As such, recycling of borosilicate glass is not widely carried out at the time of writing.

The lifecycle of a test tube is fairly straightforward. Silica sand, soda, and ground lime are heated to the necessary temperature for processing and any additives (boric oxide) are added. Then, test tubes are molded or blown, cooled, checked for any flaws, packed, and then shipped. After use in the laboratory they are then shipped out as waste where they are incinerated to remove any chemicals or residues followed by disposal in a landfill as slag. This process highlights the problem with a test tube; it is a linear path from production to disposal that is very resource intensive given the short useful life of a test tube. While the glass itself is, for the most part, inert, the raw materials and waste generated from their production and disposal can be significant. Therefore the best way to reduce the impact of this process is to re-use them.

Test tubes can be easily cleaned and reused indefinitely for most chemical uses. This eliminates the production and disposal issues encountered with single use. If a tube is reused one time, a 50% reduction in the production and disposal waste will result since a new tube will not need to be made. Each reuse will continue to reduce these impacts significantly since they can be reused indefinitely with the main cause for replacement being breakage. This simple step can reduce solid waste generated from test tubes by >90%.

Although rinsing the tubes and reusing them will create more solvent waste, solvent waste can be recycled or, as mentioned before, incinerated as fuel for other processes,[4] and reusing the test tube significantly reduces the need for production and disposal and dramatically reduces solid waste generation. At Pfizer's La Jolla site, a program was adopted where tubes were rinsed and included in the normal glass washing process. The baskets of tubes that were washed required no additional water usage as they filled the gaps between the other types of glassware. As a result of this process, over 115 000 tubes were reused in the first year, eliminating 25 cubic yards (>3 tons) of solid waste per year while reducing associated costs of purchase and disposal by greater than $40 000 annually.[18] Despite this analysis, the rinsing and reuse of fraction tubes has not been widely implemented in pharmaceutical research, likely due to the perception that increased solvent usage and chemist time invested in rinsing the tubes is a higher cost than the increased waste from disposal of tubes. As an alternative to rinsing and reusing *all* tubes from a chromatography run, a convenient compromise is to simply reuse the tubes that contained the desired material, which were likely rinsed clean in collecting the desired compound. While this is not a complete mitigation of tube waste, it could reduce tube use if combined with glass wash procedures as described above. This is especially true when considering the advanced chromatography collection technology that allows chemists to collect only the fractions containing the desired product.

## 8.5 Silica Gel Chromatography Waste

Normal phase chromatography is the most popular and widely used purification method for the isolation of many organic compounds in drug discovery and academia. However, it is a highly waste intensive process due to the large amounts of solvents and stationary phase that are required compared to the amount of compound isolated. Since solvent recycling was discussed above, this section will consider the stationary phase, in most cases silica gel.

In general the amount of silica that is required to purify one gram of material is greater than ten times that amount. This means that the purification will generate significantly more solid waste than the reaction itself to produce the desired product. In the past, silica gel columns were packed by hand in glass columns, which made reuse of the material very difficult and unsafe given the small particle size when handling silica gel. In the last 20 years, however, pre-packed silica gel columns have been made available and are now widely used for chromatography. With the advances in available chromatography equipment, the need to self-pack columns has been eliminated and pre-packed columns are now standard. Furthermore, high performance pre-packed columns allow for higher flow rates and better separation while reducing the amount of silica and solvent required. While high performance columns are becoming more widespread, it is common for users to view these as a "single-use" item to be disposed of after one use. Given the high cost of these columns and the high cost of their disposal as chemical waste, any decrease in the amount used can yield significant cost savings and waste reduction. The advent of their design and usefulness has also afforded the option for their reuse. In much the same way that HPLC columns are reused for thousands of injections, these new cartridges can, in many cases, be flushed and reused multiple times without any significant loss in quality of separation, significantly reducing the waste generated from this process. The quality of performance is usually similar to the initial use, but can vary based on various factors discussed below.

To reuse a column effectively there are some simple steps that can be taken to ensure optimal results and alleviate any concerns about final purity. Most importantly is the use of thin-layer chromatography (TLC), which allows the user to determine the proper gradient and flush solvents to confirm that the majority of material is eluted from the column. Often this is simply an increase in the percentage of the polar eluent for one to two column volumes and then re-equilibration with the desired gradient for one column volume. This procedure is similar to that employed for preparative HPLC, an application where columns are reused for a multitude of separation runs. The use of pre columns (also known as loading columns) is also a convenient way to remove the majority of baseline material between runs. Optimal results are achieved when preparing similar intermediates, since the purifications and impurity profiles are likely similar. In this case a column can be flushed and re-used multiple times. If the column is not to be re-used right away, it can be flushed with air and re-equilibrated at a later time. However, this can lead to gaps and air pockets if the column is not packed well, leading to decreased performance. Even if multiple re-uses are not possible, it should be remembered that even a single re-use of each column, on average, represents a reduction of solid waste and associated costs of purchase and disposal by half.[19]

## 8.6 Recycling of Uncontaminated Cardboard and Plastic

It is commonplace at most institutions to have a recycling program that will collect and recycle most of the easily recyclable items from the laboratory that are not contaminated, such as packaging materials, plastics from pipette tips, needle caps, computer cables, equipment, bottles, cans, *etc.* In fact many of the plastic items encountered in the laboratory are recyclable even though they are not stamped with the recycling code that we see on larger items. The plastic caps from disposable syringe needles for instance are polypropylene and can be recycled, even though they are not stamped presumably due to their small size. They should not be used to recap needles for safety reasons. Various small items from the end caps of columns to the boxes that LCMS vials come in, and almost any plastic used in packaging and shipping, can be recycled as long as it is not contaminated. Checking with your EHS organization or your local recycling agency will help you to determine which items are actually acceptable for recycling in your geographical area as some facilities do not handle all grades of materials. Networking with other laboratories to share shipping costs can also be helpful when smaller amounts of materials are being considered for recycling. While some of these items may be small in size, their quantity can still be significant based on broad use over time by a large group of people, such that the impact from recycling can be enormous.

## 8.7 Summary

In regards to recycling and reusing many laboratory materials, chemicals, and reagents, a large change has occurred and continues to grow across the pharmaceutical industry. This is by no means a completed journey. Everyday many new ideas and options are available to further explore and improve upon the current status. Focusing on the major areas discussed above will work towards revealing other opportunities to implement the ideas of recycling and re-use in the laboratory. While some changes may seem minor at first, if they can be expanded across the industry the impacts will be significant. One less test tube, one less bottle of solvent, one less silica column, one recycled catalyst, everyday, for every person across the industry will bring about the kind of grand change that the 12 principles of green chemistry were written to accomplish.

## References

1. P. T. Anastas and J. C. Warner, *Green Chemistry: Theory and Practice*, Oxford University Press, New York, 1998.
2. R. K. Henderson, J. Kindervater and J. B. Manley, *Lessons Learned through Measuring Green Chemistry Performance: The Pharmaceutical Experience*, American Chemical Society Green Chemistry Institute Pharmaceutical Roundtable, 2008, http://portal.acs.org/portal/PublicWebSite/greenchemistry/industriainnovation/roundtable/CTP_005585 (accessed December 17, 2014).
3. *Guidance for the Pharmaceutical Industry: Manufacturing, Processing or Holding Active Pharmaceutical Ingredients*, US Department of Health and Human Services, Food and Drug Administration, Atlanta, GA, March 1998. http://www.fda.gov/downloads/Drugs/Guidances/ucm070289.pdf (accessed April 2014).
4. http://www.veoliaes.com/en/services/small-business/recycling/solvents.html (accessed April 2014).
5. *Azeotropic Data—III*, ed. L. H. Horsley, Advances in Chemistry, American Chemical Society, Washington, DC, 1973, p. 116.
6. http://www.epa.gov/osw/hazard/recycling/regulations.htm (accessed December 17, 2014).
7. *Transition Metals for Organic Synthesis: Building Blocks and Fine Chemicals*, ed. M. Beller and C. Bolm, Wiley-VCH, Verlag GmbH, Weinheim, Germany, 2nd edn, 2004.
8. *CRC Handbook of Chemistry and Physics*, ed. D. R. Lide, CRC Press, Boca Raton, Florida, vol. 85, 2004, Section 14, pp. 1–47.
9. S. D. Ramgren, L. Hie, Y. Ye and N. K. Garg, *Org. Lett.*, 2013, **15**(15), 3950.
10. A. Molnar, in *Palladium Catalyzed Reactions: Practical Aspects and Future Developments*, ed. A. Molnar, Wiley-VCH, Weinheim, Germany, 2013, vol. 9, pp. 333–386.

11. *Recoverable and Recyclable Catalysts*, ed. M. Benaglia, John Wiley & Sons Ltd, West Sussex, United Kingdom, 2009.
12. L. Yin and J. Liebscher, *Chem. Rev.*, 2007, **107**, 133.
13. http://thalesnano.com/image/data/brochures/HC_brochure.pdf (accessed December 17, 2014).
14. http://www.strem.com/uploads/resources/documents/heterogeneouscatalysts.pdf (accessed December 17, 2014).
15. D. Astles, *Recycling Precious Metals using Functionalised Silica Scavengers and Catalysts*, presented at the Royal Society of Chemistry Symposium, The Sustainability Challenge, Munich, Germany, 2008.
16. Personal communication, A. Redman, L. Shuster, GlaxoSmithKline, 2014.
17. W. Vogel, *Glass Chemistry*, Springer-Verlag, Berlin and Heidelberg GmbH & Co. K, 2nd revised edn, 1994.
18. D. T. Richter, 14th Annual Green Chemistry and Engineering Conference, Washington D.C., 2010.
19. M. C. Bryan, B. Dillon, L. G. Hamann, G. J. Hughes, M. E. Kopach, E. A. Peterson, M. Pourashraf, I. Raheem, P. F. Richardson, D. T. Richter and H. F. Sneddon, *J. Med. Chem.*, 2013, **56**(15), 6007.

CHAPTER 9

# The Need for a Green Electronic Lab Notebook

ALEX M. CLARK* AND SEAN EKINS

Molecular Materials Informatics, Inc., 302-1900 rue St. Jacques Ouest, Montréal, Canada, H3J 2S1
*Email: aclark@molmatinf.com

## 9.1 Introduction

The field of chemical informatics (usually called *cheminformatics*) is broadly defined to cover the software methodologies that can be applied to chemical problems by representing structures and data in machine-readable form. The study of cheminformatics has its theoretical origins in library and information science, but has become most prominent under the aegis of the pharmaceutical industry.[1] Since the 1980s, numerous university research groups and small-to-medium sized companies have specialized in the field. There has been tremendous energy invested in the process of gathering chemical structures and associated experimental data, making it available for convenient searching and retrieval, mining it to develop and refine hypotheses, and deriving new insight into the relationships between structure and properties or biological activity.[2]

Cheminformatics has largely evolved within the boundaries of the needs of the pharmaceutical domain, which provided almost all of the industry funding and demand. Most of the techniques are predicated on ways to describe the atoms and bonds of the kinds of organic molecules that make up small molecule drugs and the building blocks of biopolymers. Molecular structures can be represented with atom coordinates suitable for generating

RSC Drug Discovery Series No. 46
Green Chemistry Strategies for Drug Discovery
Edited by Emily A. Peterson and Julie B. Manley

Published by the Royal Society of Chemistry, www.rsc.org

2D sketches, or as 3D conformations that are models for actual molecular geometry.

Structures are often grouped together in collections with associated data such as name, database identifiers, physical properties, and assay measurements. The structure definitions are often assembled into more highly marked up forms, such as chemical reactions or scaffold/substituent breakdowns. By collecting significant quantities of structure–activity data, cheminformaticians are able to apply a diverse and powerful toolkit of data mining, analysis, and prediction techniques, and so guide the efforts of drug discovery researchers. Methods such as substructure/similarity searching,[3] pharmacophores,[4] fingerprint-based clustering,[5] Bayesian inference,[6] quantitative structure–activity relationship (QSAR) models,[7] and scaffold matrix plots[8] have become mainstream in the realm of drug discovery.

Over the last decade, the amount of chemical structure data available publicly has grown to enormous quantities due to various curation efforts and a partial trend toward openness.[9] These collections of chemical structure representations are commonly annotated with any experimental properties that are on hand, which include those of interest to all kinds of chemists, such as melting point, boiling point, density, vapor pressure, *etc.*, as well as data that are of particular interest to medicinal chemists working in drug discovery, such as water solubility, water/octanol coefficient ($\log P$), and activity or toxicity measurement in various biological environments.

While access to numerous sources of data is a welcome development, it has not been the panacea that might be expected.[10] In recent years there has been increasingly vocal attention paid to data quality, for there are serious problems with almost all content sources. First among these is the general lack of provenance. Scientists typically publish the data in whatever form is convenient for manuscript creation and reading by other scientists, and this is seldom a format that is compatible with cheminformatics requirements. Much of the available data has therefore to be re-entered by a human operator, or extracted using a software algorithm, with little or no tracking of the origins. To make matters worse, most of the large public databases incorporate input from one another, and so errors get laundered and propagated. The fact that the original scientist who published the data is not part of the loop is a critical concern, since there is no way to know which data is valid, or who is responsible for fixing it. Estimates of error rates would suggest that even the most well respected public databases have an incidence of flawed data that is likely to be at least 5%,[11] and many chemical data experts privately suspect that it is much higher. Given that these are essentially transcription errors, the standard quality level is quite shocking, and is undoubtedly a major limiting factor holding back the field.

In addition to propagation of mistakes, publicly available cheminformatics data must grapple with numerous thorny issues associated with correctly representing a molecular species as a diagram. The study of cheminformatics could be described as a collection of simple rules that are generally applicable and easy to work with, mixed with an incredibly troublesome set of edge cases. Issues such as stereocenter resolution, tautomerization, protonation state, and aromaticity equivalence are some of the means by which two molecular diagrams might represent the same chemical, or two different chemicals. Creating a mapping between the known scientific observations about a molecule and an algorithmically accessible data-structure can never achieve perfection, but there is a consequential difference between *good* and *great*.

These issues have resulted in numerous publications focusing on issues within the organic subset to which drugs belong. Description of nonorganics is much less frequently discussed, for which the kinds of bonding arrangements and valence rules are far more diverse.[12] In terms of systematic efforts to describe these compounds, this is essentially an uncharted frontier by comparison, and becomes a poignant limitation for the drug discovery industry when it comes to representing chemical reactions. While there are only a handful of actual drugs with exotic nonorganic bond types, there are few total syntheses that have been carried out without using at least one inorganic or organometallic compound as a catalyst or stoichiometric reagent.[13]

While the drug discovery industry routinely employs cheminformatics methods to assist in its quest for small organic molecules with drug potential, as well as in a molecular book-keeping capacity for archiving existing knowledge, an ancillary mission that is common to the entire chemical industry is to carry out its research in a way that causes minimal damage to the environment. Besides lifesaving drugs, synthetic chemicals are used in essentially every facet of modern industry.[14] However, very often such chemicals, whether they be deliberate endpoints or unwanted byproducts, do not have a benign and quick degradation pathway. They can remain in the environment in which they are used, or spread to contaminate other areas, such as air and water, or accumulate in the food chain. It has been reported that bioactive metabolites or drugs themselves can accumulate in the environment and enter our water supply.[15] Green chemistry in its entirety is summarized by: "the utilization of a set of principles that reduces or eliminates the use or generation of hazardous substances in the design, manufacture and application of chemical products".[16] There have been recent efforts to associate the green chemistry properties of molecules (such as stability, aquatic toxicity, *etc.*) with the molecular features that enables prediction as well as the computational design of chemicals.[17] Given that the pharmaceutical industry has as its goal the improvement of the health of people throughout the world, it is increasingly becoming recognized that minimization of unintended environmental side effects is an important secondary objective.[18]

There are also significant external pressures from regulations and regulators, such as California's Green Chemistry legislation, as well as market pressure driven by increasing consumer awareness, to identify alternatives to chemicals of concern. From a business standpoint, it makes sense to start moving away from these chemicals and materials as soon as cost-effective alternatives are available. It is important in today's commercial environment to be perceived as green, and achieving this in reality means having an understanding of the chemicals in a product and their effect on the environment. Preferably companies should consider environmental impact, human health, ecotoxicity, and social metrics for various endpoints. It would be ideal if one could bring these metrics earlier into the planning process, preferably before the molecule is synthesized for the first time.

One of the primary goals of green chemistry is to encourage all kinds of synthetic chemists to consider greener alternatives when they plan out their chemical reactions. In an ideal world, all chemists would be able to digest all of the green chemistry literature and be able to make informed decisions as to the environmental consequences of their reaction pathways, and how these would be magnified in case the reaction pathway needs to be carried out on a larger scale. This of course being unrealistic, one of the appropriate questions to ask would be: How can cheminformatics help chemists explore greener reaction options that have been reported by their colleagues, and do so in a way that provides negligible additional burden on their already demanding schedules?

The remainder of this chapter will introduce some of the technologies that can be used to support this goal, and describe some of the limitations that need to be addressed to ensure that it is ultimately effective. These technologies will be framed in the context of blueprints for a hypothetical lab notebook product, which seeks to deliver seamless access to green chemistry design and optimization tools, in the form of databases, algorithms, and user interfaces.

## 9.2 Methods

Of all the ways in which software can be used to help chemists apply green chemistry to their synthetic procedures, this chapter will focus on two main categories for which structure-based algorithms and databases can be applied to routine workflows involving an electronic laboratory notebook (ELN) system. The market for ELNs is large and diverse,[19] and while many products are designed for scientists in general, others are focused on specific disciplines: a subset of these operate on chemical reactions as the central theme.

Reaction-centric ELNs are able to provide a variety of supporting cheminformatics methods, because they make use of domain-specific data-structures that are richly marked up for the benefit of software. These methods are divided into two categories: those that apply to chemical structures that represent individual compounds, and those that apply to the overall reaction description.

Cheminformatics methods are relevant to all stages of a synthetic procedure, starting from the moment that the chemist has begun to consider any of the reactants or products, through the drawing and balancing of the reaction, to recording of the measurements, to archiving the resulting data in a form suitable for later retrieval, analysis, and data mining.

### 9.2.1 Chemical Structures

An individual chemical structure can be drawn or algorithmically constructed using one of hundreds of different software applications on various different platforms, including personal computers, web applications, and mobile apps. Chemical structures are typically encountered in one of three broad categories of file formats:

- *Cheminformatics formats*: dedicated to capturing the atoms, bonds, and annotations that describe the nature of the chemical species in a way that can be unambiguously interpreted by a software algorithm.[20]
- *Chemistry formats*: dedicated to capturing the chemistry in a way that maps directly to the creation of a custom drawn diagram, intended primarily for manuscript creation.[21]
- *Dead formats*: print-ready graphics that contain no chemistry metadata.[22]

For cheminformatics purposes, the preference for avoiding *dead formats* is clear: graphics formats such as SVG, EPS, WMF, PNG, or PDF provide only the instructions necessary for a computer to render a picture in a human-readable way. If the data were created using chemistry aware software, whatever machine-readable content previously existed has most likely been lost, and can only be recreated by manual curation, or extremely error prone techniques such as optical structure recognition (OSR).[23]

The distinction between *cheminformatics* formats and *chemistry* formats is quite subtle; it is a matter of machines-first *versus* humans-first. A well designed *cheminformatics* format is a core data-structure that can capture the underlying chemistry in a concise and meaningful way, whereas a *chemistry* format is designed to facilitate the creation of graphics, by chemists for chemists. These two approaches have much overlap; they both consist primarily of atoms and bonds. A cheminformatics format restricts itself exclusively to well-defined concepts, at the expense of providing ways for the author to customize the appearance. A *chemistry* format provides a large number of annotation options that are usually meaningful to chemists who are examining the diagram with knowledge of context, but do not necessarily have any clear meaning to cheminformatics software. This is a significant problem because creators of chemical data are often not aware of these distinctions. Diagrams that are meaningful to chemists are often *almost* meaningful to software, which is another way of saying that the data is *wrong*, for purposes of interpretation by an algorithm. It only takes one missing or misinterpreted atom or bond property to imply a chemical species that was not intended by the scientist who drew the diagram.

Figure 9.1 illustrates several common examples of structures drawn in a way that is informative to chemists, but broken for cheminformatics purposes: (a) the use of partial resonance to indicate delocalized negative charge: the dotted line is a drawing object rather than being defined as a distinct property of the molecule, and the negative charge is not associated

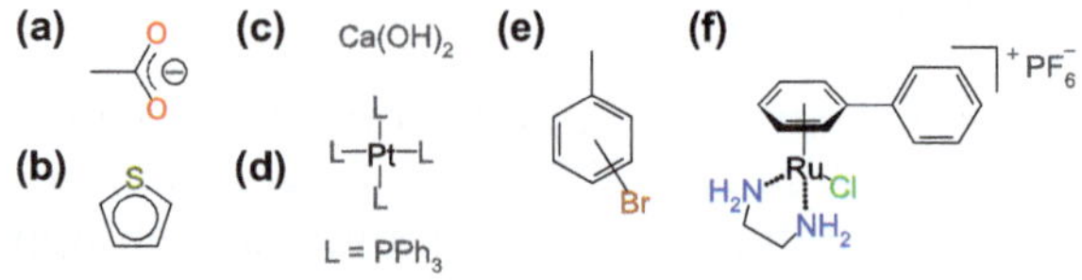


**Figure 9.1** Examples of chemical structures drawn for clarity of communication between chemists, but which cannot be correctly understood by chemical software algorithms.

with a molecular entity; (b) use of a ring to indicate aromaticity: to a cheminformatics algorithm, this is interpreted as *tetrahydro*thiophene + a meaningless drawing object; (c) a partially collapsed molecular formula that all chemists understand, but provides no connectivity information, which is required for cheminformatics techniques; (d) an inline abbreviation that is defined as freeform text below the structure definition, which is uninterpretable to cheminformatics software; (e) a mixture of aromatic substitution compounds, which a cheminformatics algorithm would interpret as toluene + bromomethane; (f) an η6-coordinated arene ligand, which would be interpreted as a methyl ligand and an uncoordinated biphenyl molecule; also, an overall positive charge notation which is not associated with the metal complex, and the use of wedge bonds to visually indicate orientation, rather than encoding of chirality.

For chemists to be able to successfully apply cheminformatics to support their research, it is necessary to either make use of software that operates directly on cheminformatics formats, or to use a specific subset of the functionality that is provided by chemistry drawing software. It is essential to avoid the use of any drawing objects that do not have a well-defined meaning to an algorithm.

Whenever a software application has on hand a user-specified chemical that is represented in a cheminformatics format, several options for further research become available. Many of these involve cross referencing the structure with structure-centric databases to lookup important green chemistry properties, while others involve the calculation of properties based on various kinds of models, or suggesting possible alternative options.

#### *9.2.1.1 Green Solvents*

Organic solvents are a major contributor to the environmental issues caused by chemical reactions, even at the laboratory scale. Many solvents present disposal risks due to volatility and flammability, and aromatic or chlorinated solvents are intrinsically problematic.[24] Relatively benign solvents, and even water, can also be a disposal problem if they cannot be separated from malign byproducts, such as dissolved heavy metals. When a laboratory reaction is being considered for scale up, the purchase price of solvents can become a major concern, and safety issues are correspondingly magnified.

For these reasons it is worth considering solvent options early in the design of a reaction step. Some reactions can only be performed effectively with a narrow range of solvents, especially if one of the starting materials has a limited solubility profile, the product is difficult to separate, or a reactive intermediate is particularly sensitive. Other reactions have broad degrees of freedom, and can be optimized for various different common solvents with similar polarity and proticity.

Recently three similar guides have been published regarding the environmental properties of commonly used laboratory solvents. The American Chemical Society Green Chemistry Institute Pharmaceutical Roundtable produced a list of solvents with ratings for safety, health, and environmental consequences pertaining to air, water, and waste disposal.[25] GlaxoSmithKline, and more recently Sanofi, produced similar solvent selection guides, with ratings for waste, environmental impact, health, flammability/explosion, reactivity/stability, and lifecycle.[26]

While these indications are to some extent subjective, they are backed up by considerable experience and provide valuable guidelines for making decisions about choice of solvent. The synthetic chemist is entirely responsible for deciding whether an alternative solvent is likely to be effective, but the information about its environmental and safety profile is conveniently tabulated. Deployment of these information resources in a cheminformatics-friendly format that can be integrated into chemical software projects will be described later in this chapter.

A major area of future work is to be able to provide software that can *recommend* a greener solvent for a given reaction.[27] This is an incredibly difficult challenge, given the nuances involved in optimizing solvent selection, but given an adequately large corpus of related reaction attempts carried out in different solvents, and the inclusion of yields and negative results,[28] it may be possible to build an expert system that is useful for selected reaction domains. At the present time, the quantity and level of detail of reaction descriptions needs to be improved before such a project is viable. The existence of effective solubility calculation methods for a diverse variety of solvents could be a particularly useful for guiding scientists toward the most promising alternate choices (*i.e.* high solubility of reactants and waste products, low solubility of desired products). Most contemporary efforts to model solubility are applied to water, in large part due to the water solubility of organic compounds being a critical consideration in life sciences research, given that inadequate dissolution in water can present a significant challenge to drug candidate development.[29]

### *9.2.1.2 Environmentally Sustainable Feedstocks*

There is a growing body of knowledge regarding chemical compounds that can be prepared in an environmentally sustainable way, which typically means utilizing plant biomass, the waste streams of other industrial processes, or by even more exotic methods such as fixing carbon dioxide directly from the atmosphere. The canonical biotechnology example is preparation of ethanol by fermentation of naturally occurring sugars, but there are an increasingly large number of more complex commercially viable sources that can be used as starting materials for an organic synthesis.[30]

The number of suitable starting materials that can currently be produced from renewable sources on a commercial scale is miniscule compared to the hundreds of thousands of purchasable fine chemicals whose lineage can be traced back to petrochemical ingredients, toxic reagents, or high energy industrial processes. At the present time, information about such compounds is scattered, and has not been centralized into a conveniently accessible database, nor is it made available as part of large reaction planning databases and tools.

Such a resource would be of considerable value, because synthetic organic chemists frequently have to choose between multiple possible pathways, meaning there are multiple options for the primary starting materials. Choices are usually made based on the difficulty of the synthetic procedures, number of steps, ingredient cost & delivery time, and expected yield. When it comes time for a reaction to be scaled up, the choice of starting material becomes much more important; having already optimized the synthesis to make use of environmentally benign materials from an early stage can eliminate significant downstream hurdles. To facilitate this, it will be necessary to curate a list of known renewable feedstocks, and ensure that they are associated with chemical structures, so they can be effectively cross referenced by cheminformatics software. Merely making this information available as a textual list, *e.g.* pages of chemicals referenced by name in a PDF file, is not sufficient to drive adoption in an increasingly digital workplace.

#### *9.2.1.3 Hazardous Byproducts*

Several collections exist describing materials that present disposal issues, whether due to immediate safety concerns, the consequences of discharging such materials into the environment, or legal regulations.[31] For the most part, these documents tend not to be available as structure-searchable databases, missing a great opportunity to apply cheminformatics to synthesis planning.

Given the availability of a suitably curated collection, it would be straightforward to ensure that each time a chemist draws a chemical structure for inclusion into part of a workflow it could be automatically checked against the list, and flagged accordingly. Particularly in the case of a chemical reaction, where all of the reactants, products, reagents, solvents, and byproducts are explicitly indicated by structure, this would be a very effective way to ensure that the chemist has access to relevant information about how to handle and dispose of the chemicals involved, whether they be added explicitly, or generated during the course of the reaction.

This information would need to be considered entirely precautionary; not all chemistry can be accomplished using safe, inexpensive, and clean materials, but there are often several viable options, and always a multitude of factors to take into account. In the event that problematic materials are required, it is essential to be aware of the risks and appropriate mitigation strategies, and to have practical information about disposal issues that neither over- nor understates the steps involved.

#### 9.2.1.4 *Toxicity Prediction*

In addition to precurated databases of the relatively limited number of chemicals that have been well studied with regard to their environmental malignancy, there is a significant body of work that has gone into building models for predicting various forms of toxicity (see Chapter 10).[32] Chemicals that are particularly harmful, *e.g.* to humans *via* skin contact or ingestion, waterways or air pollution, often have common structural motifs or molecular properties. Beyond the obvious repeat offenders (*e.g.* mustard ligands, polyaromatics, chlorocarbons, *etc.*) there can be structural trends that are not necessarily easy to categorize, but can be teased out by computational model building. The model building often draws from the same techniques that were originally designed for predicting biological activities needed for drug discovery.[33]

Because several toxicity areas have been expressed as models,[34] their application falls within the domain of cheminformatics. Whenever a user expresses a structure using a digital format that is unambiguously meaningful to an algorithm, any number of models can be applied to estimate the extent or probability of a compound being toxic in some particular environment. This means that the compound does not need to already exist in a database, but it does raise issues as to the efficacy and domain of the model itself.[35] For toxicity areas that have relatively few studied examples, the applicability of a novel compound that has little similarity to a set of tested examples is often questionable.

### 9.2.2 Chemical Reactions

In cheminformatics, a reaction is a highly marked up collection of chemical species. As for individual structures, the categories of file formats follow the same pattern, being divided into *cheminformatics* formats, *chemistry* formats, and the *dead* formats useful only for graphics display. Unfortunately the *cheminformatics* formats in general use are much less well designed for reactions. There are relatively few software packages that are built to capture machine-readable chemical reactions at the point of entry, and the commonly used file formats are incomplete and inadequate for the task of capturing a complete description.

The first step toward capturing a reaction in a way that can be used for cheminformatics is to represent *all* of the reaction components as chemical structures, to the extent that this information is available. This includes the stoichiometric reactants, as well as nonstoichiometric reagents such as catalysts, solvents, or reactants used in excess. It also includes *all* of the formal products, whether they be major products, minor products, or unwanted byproducts that are disposed of. The reaction should be balanced; the stoichiometry should be included as part of the metadata for each reaction component. Additional metadata, such as chemical names, the specific role of each component, physical properties and measured quantities, can be layered on top of the scheme.

Figure 9.2(a) shows a three-step reaction sequence for the preparation of vanillin,[36] drawn in a conventional diagram style, but taking care to ensure that all reaction components are present and represented with chemical structures that are interpretable by cheminformatics algorithms. Figure 9.2(b)–(d) represent the same information, broken out into steps. Each component is explicitly segregated into its own block, which can be associated with higher order metadata. Nonstoichiometric components (such as catalysts and solvents) are indicated, as are unwanted byproducts.

Whenever possible, the scheme should include sufficient atom-to-atom mapping information to allow an algorithm to unambiguously determine the reaction transform. With a well-designed interface capable of partially automating the mapping pairs, the additional effort required to describe a

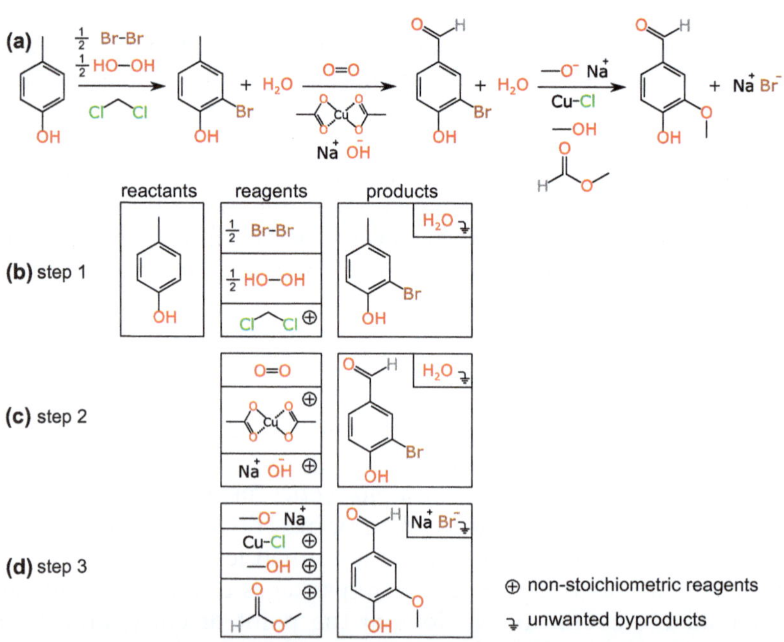


**Figure 9.2** (a) Example of a balanced reaction involving three distinct steps, rendered in a diagram format suitable for interpretation by chemists; (b), (c), and (d) each of the steps, with structures partitioned into components and classified as *reactant*, *reagent*, or *product*.

reaction would be a small fraction of the time taken to prepare the scheme. Once the reaction scheme has a correct atom-to-atom mapping, it is possible for software algorithms to classify and characterize the reaction. With this information, the data implies a specific transformation, which can be used to understand the nature of the chemistry being affected, and also to generalize by comparing the transformations of two different reactions. It can also be used as a template, *i.e.* the pertinent changes can be extracted from one reaction, and applied to another partially-specified reaction, either to fill in missing information or to conduct a search.

### *9.2.2.1 Green Metrics*

In the same way that medicinal chemists are increasingly focusing on improving ligand efficiency for a target,[37] chemists are almost always interested in evaluating their reactions in terms of metrics such as percentage yield and cost of ingredients. There are several ways to consider a chemical reaction for its green chemistry qualities, some of which are based on simple calculations, for example process mass intensity (PMI),[38] *E*-factor,[39] and atom economy.[40]

Figure 9.3(a) shows a simple reaction scheme for the acylation of salicylic acid to make aspirin. The reaction is carried out neat, with several equivalents of acetic anhydride serving as both reagent and solvent. The yield is given as 95%. Figure 9.3(b) shows the calculation of process mass intensity (PMI), by dividing the total masses of the reactants by the mass of the desired product, giving a value of 2.85. The ideal green reaction has a PMI of 1, which can be achieved by 100% yield with no byproducts, solvents, or excess reagents. Figure 9.3(c) shows the calculation of *E*-factor, which is closely related; the total amount of waste is divided by the product mass, giving a value of 1.85. The ideal value is 0, *i.e.* no waste. Note that because for most reaction schemes the yield is less than 100%, and not all of the byproducts are known, it is practical to calculate the waste as being the total mass of the input ingredients less the mass of the desirable products that were recovered. These metrics are most heavily influenced by use of solvent, since

**(a)**

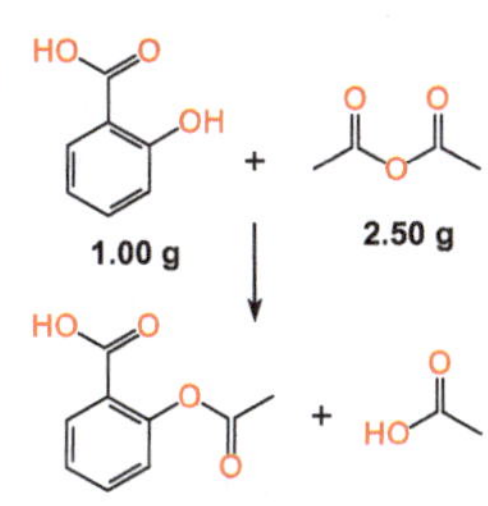


**(b)** process mass intensity $$= \frac{\sum \text{reactant mass}}{\sum \text{product mass}} = \frac{1.00 + 2.50}{1.23} = 2.85$$

**(c)** E-factor $$= \frac{\sum \text{waste mass}}{\sum \text{product mass}} = \frac{(1.00 + 2.50) - 1.23}{1.23} = 1.85$$

**(d)** atom economy $$= \frac{\sum \text{product molecular weight}}{\sum \text{reactant molecular weight}} = \frac{180.16}{138.12 + 102.09} = 75\%$$

**Figure 9.3** (a) Acylation of salicylic acid to make aspirin; (b) calculation of process mass intensity; (c) calculation of *E*-factor; (d) calculation of atom economy.

the combined mass of solvents typically outweighs that of excess reagents or unwanted byproducts, and it is usually disposed of rather than recovered.

Both the PMI and *E*-factor are useful ways to compare the efficacy of repeated experiments based on the same synthetic step. Like percentage yield, these values are useful ways to optimize the conditions, *e.g.* if the same yield can be accomplished when the quantity of solvent is reduced, or by using catalytic rather than stoichiometric reagents, these metrics both indicate an improvement. When comparing these metrics between different reactions, scientific judgment must be used; scaling according to molecular weight, the nature of the unwanted byproducts, and the relative difficulty of achieving favorably green conditions for a class of reaction must all be taken into account.

Figure 9.3(d) shows the calculation of the atom economy. Unlike the other two metrics, this is an intrinsic property of the reaction scheme, and is not dependent on the quantities measured during the experiment. Atom economy is an indication of the extent to which atoms are stoichiometrically discarded; the ideal value is 100%, which is the case when all of the reactant atoms are incorporated into the product. In this example the atom economy is 75%, since more than half of the molecular weight of the acetic anhydride reagent is discarded in the form of acetic acid. Paying attention to the atom economy during synthetic design favors methods that avoid bulky protecting groups and stoichiometric reagents that are composed of chemical functionality that is not part of the reaction payload.

What all of these metrics have in common is that the mathematics are trivial, but they cannot be calculated correctly unless the reaction scheme is complete. All components must be listed, with chemical structures and stoichiometry, and categorized correctly according to role. For PMI and *E*-factor, the quantities must all be available as mass. For atom economy, it is vital that the chemical structures are drawn in such a way that the molecular weight can be calculated correctly, which for many software packages is not the case for nonorganic structures.

The ability to have an electronic lab notebook always calculate these metrics, given that the underlying data-structure captures all of the necessary information correctly, has the potential to significantly improve the attitude of scientists towards green chemistry. Most synthetic chemists are accustomed to performing yield calculations, using a pencil and calculator if necessary, because this is considered to be an important metric of the reaction outcome. Expecting scientists to alter their work habits to manually calculate additional metrics is difficult. On the other hand, having software that is already useful for other reasons and provides an *always on* automatic calculation is far more likely to gain traction. By showing the metrics for previous and current reactions, these characteristics can easily be browsed and compared. When a chemist compares two reaction pathway options with different atom economy, or a repetition of a reaction with different PMI and *E*-factor numbers, it provides valuable feedback as to how the reaction might be refined.

### 9.2.2.2 Reaction Transforms

A *reaction transform* is a special instance of a chemical reaction description that contains enough information to indicate the parts of the molecules that are being changed. An example of a reaction transformation in chemistry vernacular is "acylation of a phenol derivative", possibly accompanied by a case study (*e.g.* the preparation of aspirin). This is understood to be a way to carry out a specific kind of change to a category of molecules with predictable results, and it is the applicability of these trends that makes up the basis of synthetic chemistry. A reaction transform expressed in the data-structures of cheminformatics allows an algorithm to start with a partial reaction description (*e.g.* a reactant or product) and fill out the entire reaction specification (or multiple possible reactions, in some cases). This is an invaluable tool for synthesis planning since the user can explore all the variations that are available given a pre-existing list of transforms. It can also be used to verify whether a fully specified reaction is compatible with the transform, which is useful for reaction naming.

In cheminformatics, the most popular methods for working with reaction transforms are based on the SMARTS molecular query format: reaction SMARTS and SMIRKS.[41] These SMILES-derived technologies are supported by several toolkits, and have the advantage of being line-based formats, which means their definitions can easily be encoded in text boxes, URLs, spreadsheets, and comma separated value files. Several useful tools have been constructed out of this technology, but there are important shortcomings. Firstly, these line-based formats contain no structural coordinates, so important drawing and orientation nuances are lost, which is a major drawback for practicing chemists. They also provide no metadata, and are limited to description of reactions and products, omitting reagents, catalysts, solvents, *etc.*, which makes these methods inadequate for many lab notebook purposes.

The most important information that needs to be provided in order to markup a reaction definition into a transform is atom mapping. While there are many published algorithms that can be used to guess atom-to-atom mapping, often involving some variation of the maximum common substructure algorithm,[42] obtaining a reliable answer by examining the unlabeled structure is not possible for a significant subset of reactions. Rearrangement reactions are particularly prone to needing special attention. For example, the Claisen rearrangement[43] illustrated in Figure 9.4(a) shows the reaction with the six atoms that make up the core transformation labelled according to the canonical definition of the reaction class. Figure 9.4(b) shows the numbering scheme that might be proposed by a well-designed maximum common subgraph algorithm. Even with some flexibility to allow mismatched bond types, this partial labelling scheme is not only incomplete, but also incorrect. Any attempt to elaborate on this starting point will derive a reaction transform that indicates the wrong chemistry.

For this reason, the software must at the very least allow the user to verify the atom mapping scheme, and, if necessary, correct it. Oftentimes it is only

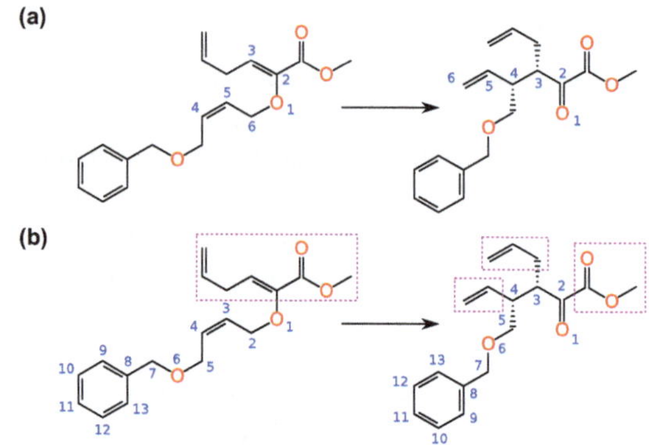


**Figure 9.4** Claisen rearrangement reaction: (a) correct atom-to-atom mapping of the atoms involved in the transform; (b) incorrect partial assignment that might be obtained from a maximum common subgraph algorithm.

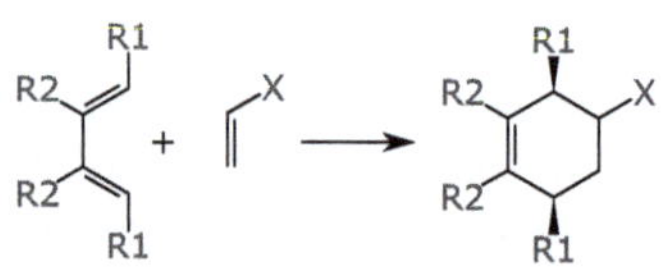


**Scheme 9.1** Diels–Alder reaction transformation, for which the use of placeholders R1 and R2 defines the symmetry requirement of the diene reactant.

necessary for the user to resolve one or two ambiguities for an algorithm to correctly assign the atom-to-atom mapping without error; even a single user-commanded atom mapping pair reduces the degrees of freedom exponentially.

Once the pertinent atoms have been mapped, the substituents that are not involved in the transform should be clipped and converted into placeholders. Atoms that change state, or are essential for the reaction description to be accurate, should be left as-is. Any possible substitution points should be annotated using labels, such as X, R, R1, *etc.*, that are consistent throughout the scheme. For example, Scheme 9.1 shows a Diels–Alder reaction template for which the diene is required to be symmetric due to the arrangement of the R1 and R2 labels. The dienophile denotes the substituent X which is required to be present, and identical, on both the reactant and product sides.

When there are multiple variations of a reaction, it must be decided whether to encode these variations within the same specification or create multiple transforms. The definitions of placeholders such as R or X could make use of query features to narrow down the possible matches (*e.g.* must be aromatic, aliphatic, halide, list of possibilities, *etc.*). In situations where the known cases are diverse, or the experimental conditions are different, creating multiple transforms is recommended. For example, Scheme 9.2 shows two closely related reactions: (a) Friedel–Crafts alkylation and (b) acylation. While it would be possible to describe both with a single reaction

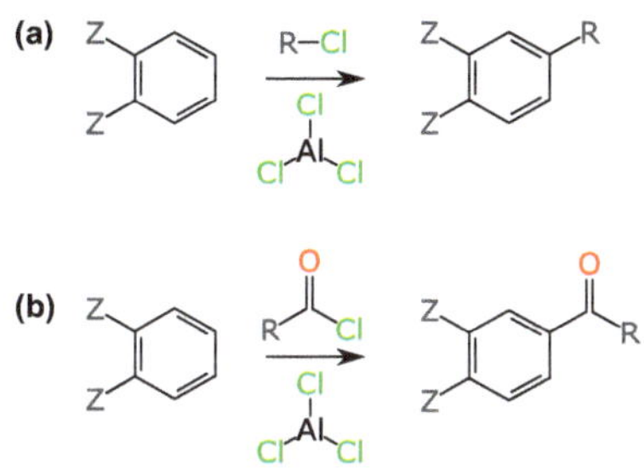


**Scheme 9.2** Representation of two closely related reactions as separate transforms: (a) Friedel–Crafts alkylation and (b) acylation.

transform, *i.e.* the R fragment being either *alkyl* or *acyl*, it makes sense to represent them separately since the reaction conditions differ.

Reaction transforms have quite diverse uses in cheminformatics (*e.g.* database querying,[44] reaction naming,[45] metabolism,[46] synthesis planning,[47] fragment generation[48]). For green chemistry purposes, an effort to curate the literature of greener reactions, each expressed as a transform with corresponding experimental details, would be an invaluable resource. Unrestricted access to such data and the algorithms necessary to apply these transforms would allow a chemist to scan the covered green chemistry literature with a single search query. This is in contrast to requiring chemists to be personally current on contemporary methods, or to hunt through existing collections of reactions and making a judgment call as to how "green" each of them is. The curated green reactions would ideally be linked to literature articles, describing their rationale and why they are more environmentally appropriate than alternative options.

## 9.2.3 Implementation

At the time of writing, the methods described in this chapter are mostly hypothetical, in development, or in some cases available as subset functionality in larger products. This section describes how the prerequisite functionality can be brought together to meet these needs.

### *9.2.3.1 Experiment Data-structures*

At its minimum, the form required to represent an experiment based on a chemical reaction must describe each of the chemical identities individually, and mark them up with properties such as *agent type*, *reaction step,* and *stoichiometry*. Additional properties such as name and quantity can be added as second order content. This can be conceptualized as a hierarchy (Figure 9.5).

The first hierarchical level is the *step*. Chemists often draw reactions as multiple steps, for several different reasons. If there is a well-defined intermediate, it is informative to indicate it as part of the scheme, in order that the beholder gains a more complete understanding of the process involved.

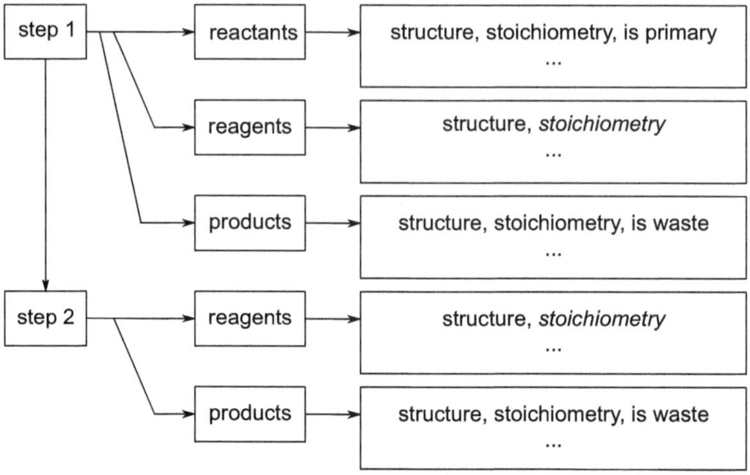


**Figure 9.5** Hierarchical representation of the data structure required to represent a multistep reaction as a collection of components.

Many reactions involve more than one step telescoped into the same flask, *e.g.* after the first reaction is complete, a second set of reagents is added to the mix, and only the final product is separated. Chemists also frequently draw entire synthetic campaigns as a many-step reaction; for cheminformatics purposes, though, it is often preferable to divide reactions into distinct laboratory exercises.

Within each step is contained some number of *reactants, reagents,* and *products*. Reactants and reagents are both used to describe chemical entities that are added to the reaction vessel, but by convention a reactant is always stoichiometric. Reagents, which are often drawn by chemists as species above or below the reaction arrow, are usually nonstoichiometric species. These include catalysts, solvents, and various kinds of cofactors. A reagent that is formally nonstoichiometric is not contemplated during the balancing of the reaction, and so contributes no atoms to the product. By this definition, reactants that are used in excess are still considered stoichiometric, even though their quantities are disproportionate.

Each reactant can be further annotated by whether it is *primary* or not. A primary reactant is an input material that is considered to be the basis for yield calculations, and is often referred to as a starting material or limiting reagent. Each product also has a flag that can be used to indicate that it is *waste, i.e.* a product that is unwanted and discarded. To ensure that a reaction is fully balanced, it is necessary to indicate all waste products.

For step 2 and beyond, there is no category for reactants; the reactions for subsequent steps are defined to be all of the *non-waste products* of the previous step. This allows a strong correspondence between the data-structure and the natural way that chemists draw chemical reactions. It also introduces the need to be able to indicate whether these additional reagents are *stoichiometric.*

#### 9.2.3.2 Balancing Reactions

To fully capture the properties of a reaction, particularly with regard to calculating green chemistry metrics, the reaction must be balanced, meaning that the total atom counts for the *reaction* and *product* sides for each step must be equal. This means that all stoichiometric components must be represented as correctly drawn chemical structures and with stoichiometry indicated. The stoichiometry value can be represented as integers, fractions, or decimals (*e.g.* 2, 2/3, 1.5), as long as the atom counts cancel out to within reasonable numerical precision. This should ideally apply to each step of the reaction. In some cases it is possible for an algorithm to automatically select stoichiometry values for the components in order to minimize the overall atom imbalance, although this approach can give misleading results when not all of the byproducts have been specified.[49]

As long as the chemical structures are represented in a format that guarantees the correct molecular formula, or the molecular formula is encoded in a separate field to overcome the limitations of common structure formats with unusual bonding types, the calculation is straightforward. Multiplying each atom count by the stoichiometry produces a tally on either side of each step, and these can be cancelled out. If there are any leftover atoms on either side, then the reaction is not balanced.

In some cases, the balancing of the reaction may not be possible if the reaction pathway is unknown or the nature of the byproducts has not been established. It is nonetheless good practice to describe balanced reactions whenever possible, for various reasons. A balanced reaction is less likely to contain mistakes due to the self-consistency check. It is also a more thorough classification that provides a lot of metadata about possible reaction mechanisms, and including the formal byproducts in the reaction scheme can provide valuable insight into safety issues and the environmental consequences of disposal.

#### 9.2.3.3 Quantity Calculations

The most common ways in which reaction ingredients are measured are by *mass* and *volume* (either of neat substance or as part of a solution). Quantities are often recorded as *moles,* which are readily interconvertible with mass, since the format requires that all components are represented by a structure that implies the correct molecular formula, and hence molecular weight. By providing *concentration* or *density,* the mass and volume also become interchangeable.

Since the format requires balanced reactions with specified stoichiometry, once the molar mass of a single stoichiometric component is known, it is possible to use the stoichiometry ratio to cross-populate the molar mass of all other stoichiometric components. Using a reaction description for which these correspondences are reliable is very powerful in terms of providing a helpful user interface, since in many cases it is only necessary to provide a single value (*e.g.* the anticipated or measured mass of the primary reactant) and have all the quantities of all other stoichiometric components inferred automatically.

To calculate inferred quantities, each component initially defines the following properties, each with standardized units:

- stoichiometry,
- molecular weight (g $mol^{-1}$, calculated from structure),
- mass (g),
- volume (mL),
- moles (mol),
- density (g $mL^{-1}$),
- concentration (mol $L^{-1}$),
- yield (%, products only).

The *density* and *concentration* properties are used to interconvert between *mass* and *volume*. The *molecular weight* is used to interconvert between *mass* and *moles*. The quantity properties (*mass, volume*, *moles,* and *yield*) have one of three state annotations: *actual, virtual, unknown,* or conflict. At the beginning of the calculation, each of these is set to *actual* if there is a user-entered value, and *unknown* otherwise. All calculated quantities are labelled as *virtual.* A user-provided quantity can be either a measurement, if the experiment is underway, or an intended quantity that will be replaced later, if the experiment is still in the planning phase. For the calculation to be useful, there must be at least one user-provided value that can be converted into *moles*.

The relationships between the stoichiometric components for a well-defined multistep reaction are determined by the following algorithm:

1. Iterate over all components:
   - *mass* = *moles* × *molecular weight:* if two of these are defined, calculate the third;
   - if *concentration* is defined, then *volume* = *moles* × *concentration:* if two of these are defined, calculate the third;
   - if *density* is defined, then *mass* = *volume* × *density:* if two of these are defined, calculate the third;
   - in each case, if all three properties are defined, and the numbers do not fit the formula, mark the quantities as having status of *conflict.*
2. If any properties were changed, goto 1.
3. Determine *reference moles* as a ratio: for each reactant (step 1 only) that is labelled primary, calculate *total moles* divided by *total stoichiometry;* if any are undefined, or the total is 0, stop.
4. Iterate over all *stoichiometric product* components:
   - *yield* = (100% × *moles*)/(*reference moles* × *stoichiometry*): if *yield* or *moles* is defined, but not both, calculate the other;

  - if *yield* and *moles* are both defined, but do not fit the formula, mark them both as having status of *conflict;*
  - if *moles* and *yield* are both undefined, calculate the *moles* as if the yield is 100%.
5. If any properties were changed, goto 1.
6. Iterate over all *stoichiometric* components:
  - *moles* = *stoichiometry* × *reference moles*: if *moles* is undefined, it can be calculated.
7. If any properties were changed, goto 1; otherwise, stop.

This iterative process involves filling in missing values (as *virtual* quantities) until no more can be determined. From the perspective of a synthetic workflow, the chemist starts by defining the complete and balanced chemical reaction. Preliminary values may be entered to trigger the calculation of estimated quantities of stoichiometric reagents, which is often useful prior knowledge.

Once the experiment begins, each measured quantity is entered for the corresponding component, along with density or concentration if applicable. The software automatically updates all inferred values with each change. A user-specified value always overrides a calculation. Until the reaction is complete, the quantities for the product are filled out as if the reaction achieved a 100% yield, *i.e.* the theoretical maximum is indicated. Once the reaction is complete, and the product has been isolated and measured, the quantity can be entered, and the percentage yield is calculated by comparing expected *versus* obtained.

### 9.2.4 Example Products

The electronic lab notebook field is a highly active market, which includes a diverse variety of products.[50] Many of these are designed for general purpose data collection, while a smaller subset focus on the specifics of chemical reactions, for example the Accelrys[51] and PerkinElmer[52] ELN products were originally chemistry-specific, before branching out. These products have the benefit of familiarity by the end user and some aspects of green chemistry support can be incorporated into the standard product or customized at an additional cost, depending on the product. However, to be able to unlock the full capabilities that are possible using cheminformatics methods, the core ideals of structure and reaction representation need to be built in from the ground up. Many of the methods and requisite data collections described in this chapter are not currently available as products, but there has been incremental progress to this effect. For disclosure, the authors were involved in the development of the following products, but this does not preclude the availability of other products.

#### *9.2.4.1 Green Solvents and Lab Solvents Apps*

The ACS GCI Pharmaceutical Roundtable's green solvent guide was released in 2011, and contains valuable reference information regarding the

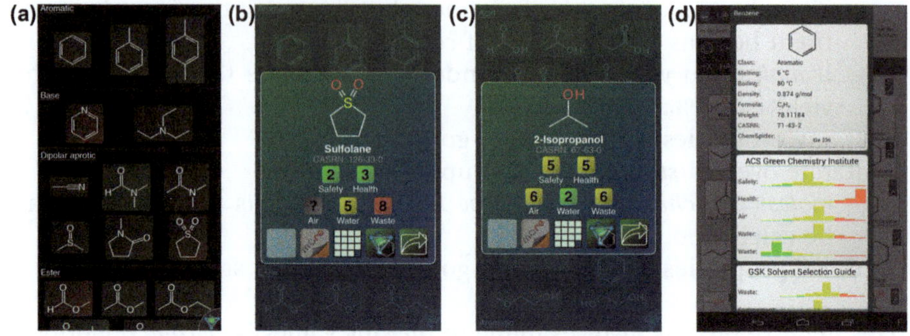


**Figure 9.6** (a)–(c) The Green Solvents app and (d) the Lab Solvents app.

environmental properties of many common solvents. The information was originally released as a PDF file, where all the information was provided in a table, with solvents referred to by name only. To help increase the visibility of this information and encourage its uptake, the data were "appified".

First of all, the compound names were mapped to chemical structures, and correlated with online databases such as ChemSpider.[53] An app for iOS-based devices (iPhone, iPod, and iPad) was created, called *Green Solvents*.[54] Figure 9.6(a) shows the main page, which is a simple interface allowing the grouped list of solvent structures to be panned up or down by swiping. Figure 9.6(b) and (c) show the results of tapping on a compound structure, which further reveals the name and environmental ratings. Later, a similar app entitled *Lab Solvents*[55] was created for Android-based devices (Figure 9.6d). This app draws upon the same source data, but also merges it with the content from the GlaxoSmithKline solvent selection guide.[56]

While these two apps are very simple, they have served to improve the visibility of green chemistry. The apps are free, and the curated source data has been made openly available. The modest effort required to markup the records into cheminformatics formats (structures, names, links, *etc.*) has proven to be useful for reuse in more advanced products.

### 9.2.4.2 *Yield101*

The *Yield101* app,[57] for iOS-based mobile devices, was designed for educational use. The primary workflow involves specifying a single step chemical reaction, using many of the core principles described in this chapter, and providing automatic calculation of quantities. Figure 9.7 shows the definition for the bromination of *p*-cresol.

In this case, the stoichiometry has been defined, and several of the quantities and properties have been filled in. The primary use case for this app is to benefit students in a teaching laboratory environment. By operating on chemical structures which are easy to visualize, and performing the

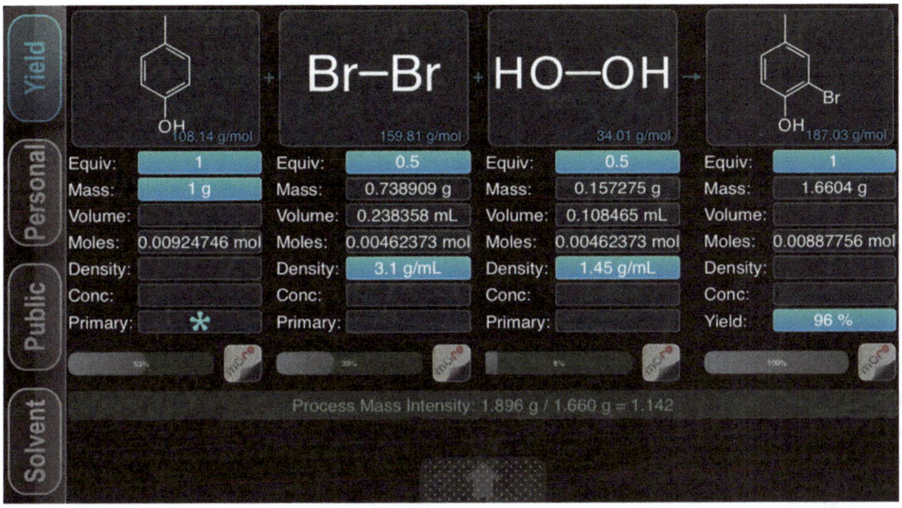


**Figure 9.7** The Yield101 app, showing a single reaction step with provided quantities.

quantity calculations and unit conversions automatically, this frees the students to focus more on the chemistry, and the hands-on practice of operating in an actual laboratory.

Despite its relative simplicity, the *Yield101* app implements many of the cheminformatics techniques that are necessary for a reaction lab notebook. While it is limited to describing a single reaction step, its structure-centric approach makes use of a data representation that is consistent with the ideals necessary for a complete and digitally pure reaction description. The app automatically calculates the process mass intensity (PMI) metric whenever the necessary quantity information has been filled in. It also provides easy access to the list of solvents that was curated for the *Green Solvents* app.

### *9.2.4.3 Open Drug Discovery Teams*

A third recently developed mobile app connected to green chemistry is the Open Drug Discovery Teams (ODDT) project,[58] which defines several topics, including precompetitive areas which include green chemistry. Each topic is built up by harvesting source documents from networks such as Twitter and Google Alerts newsfeeds. The green chemistry topic, for example, collects relevant material by monitoring the hash tag #greenchemistry. Each link to a relevant article is collected, and users of the app are able to browse the incoming content and vote articles up or down. Users can ensure that relevant content is picked up by the service by using their own Twitter account and matching up links with hash tags. The service also recognizes chemical content (*e.g.* molecules, reactions, *etc.*) and is able to display these intelligently, as well as launch the content using other compatible apps installed on the device.

The app is designed to provide a convenient and centralized location for timely newsfeeds, with a crowd curation component for prioritizing the most relevant content. For example, the raw cheminformatics content used for the Green Solvents app is included in this topic *via* a tweeted link.

### 9.2.4.4 Green Lab Notebook

At the time of writing, a more advanced project has recently been released in the form of a mobile app, with the objective of implementing all of the ideas covered in this chapter, called the *Green Lab Notebook* (GLN).[59] The scope of the project is to allow users to create a collection of experiments, each of which is a single- or multistep reaction, captured using a data-structure that provides all of the capabilities described earlier. In addition to providing editing and browsing tools for defining experiments, it will also include convenient access to a variety of supporting green chemistry data sources, such as environmentally rated solvents, sustainable feedstocks, known hazardous compounds, toxicity calculations, and green reaction transforms.

The GLN is being implemented initially as a mobile app for iOS-based devices, supported by hosted web-services to carry out more extensive cheminformatics operations and database lookups. This product will complement several cutting edge cheminformatics apps that are already available for these next generation computing devices.[60] While initially being deployed as a mobile app, the functionality and technology involved is equally applicable to conventional desktop applications, as well as the web platform.

Figure 9.8 shows several screenshots of the user interface for which the three-step synthesis of vanillin from *p*-cresol is shown. Figure 9.8(a) shows the main editing area where each of the reaction components can be specified. Figure 9.8(b) shows the information blocks that the app presents in order to allow the user to explicitly specify quantity information, and to view quantities automatically calculated by the app. Figure 9.8(c) shows the summary block that shows green chemistry metrics for the products of each of the three steps. For the benefit of understanding, each of the values is shown with its full working.

One of the most ambitious features intended for the GLN project is the use of reaction transforms, as described earlier in the chapter. The primary data object is the *experiment,* which typically corresponds to a specific laboratory exercise. However, by converting the reaction definition into a transform, by ensuring that atom mapping is complete, and expressing bystander fragments as R-groups, these partial definitions will be searchable and will be able to be applied directly to proposed reactants or products. Combined with a curated reference database of the transforms of known green reactions, this has the potential to make a significant amount of green chemistry knowledge available to working chemists, as well as to simplify many of the tedious steps of chemical reaction data entry.

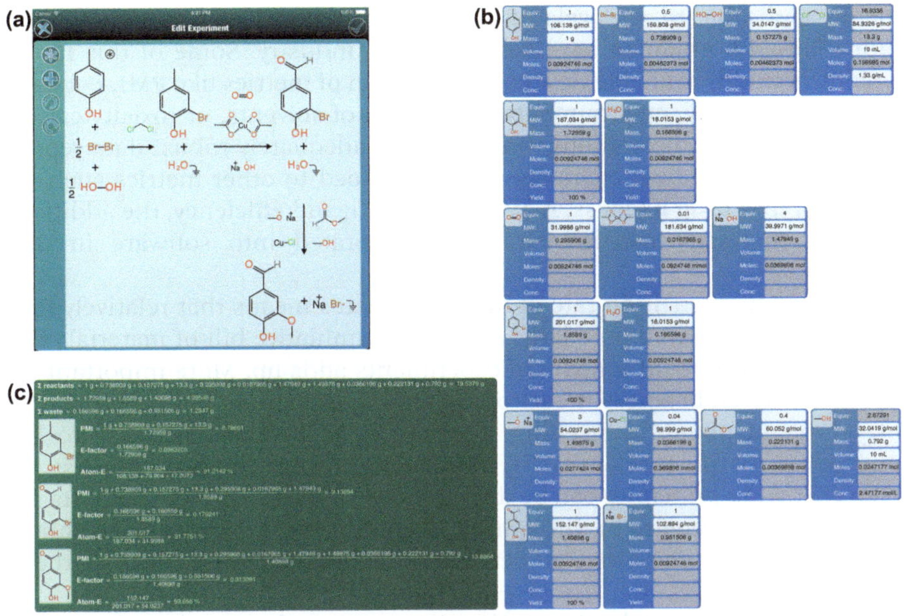

**Figure 9.8** The Green Lab Notebook app: (a) editing screen for a multistep reaction; (b) quantity blocks for each reaction component; (c) green chemistry metrics calculated for each non-waste product.

It is hoped that, in the near future, there will be a number of electronic notebooks that are attuned to the needs of synthetic chemists, and record the information in sufficient detail to provide similar functionality.[61]

## 9.3 Conclusion

Green chemistry as a discipline has generated a large amount of knowledge that is spread out across dedicated journals (such as the eponymous *Green Chemistry*, and *ACS Sustainable Chemistry and Engineering*) as well as being intercalated into journals devoted to other topics. There has been relatively little effort made to apply the discipline of cheminformatics to green chemistry,[62] and this has become a pertinent issue since the quantity of data has grown to the point where software is necessary to keep track of it. The contemporary body of cheminformatics methodologies that has been developed with drug discovery researchers as the primary beneficiaries is highly applicable to handling green chemistry data, but there are some areas in which these techniques need to be improved, particularly with regard to providing a more rigorous definition of chemical reactions and their constituent structures.

By designing new electronic lab notebook functionality with the explicit needs of green chemistry in mind, the opportunity exists to spread the principles of green chemistry further into the chemistry realm. This is equally applicable to all levels of career progression, from undergraduate studies through graduate level academia and industry. Some of this functionality is readily available now (*e.g.* calculation of metrics like PMI, *E*-factor and atom economy, environmental properties of solvents, or toxicity calculations), and is predicated only on having an adequately robust data representation. As medicinal chemists are accustomed to other metrics such as the rule of five, drug likeness, ADME/Tox, and ligand efficiency, the addition of green chemistry metrics can be incorporated into software in the same way.

While the high attrition rate of drug candidates means that relatively few chemicals are prepared on a large scale, the cumulative bill of materials for such a large number of individual experiments adds up. More important to the patent-driven business model of the pharmaceutical industry is that any delay brought about by needing to redesign a synthetic pathway to be appropriate for process chemistry represents a huge opportunity cost. If green chemistry principles can be designed into a synthesis at an early stage, with negligible decrease in throughput, there is a major economic incentive to ensure that all chemists are actively incorporating these ideas in their reaction planning.

Future green lab notebook functionality requires the curation of data resources such as environmentally sustainable feedstocks, chemical structures of hazardous byproducts, and reaction transform descriptions for reactions with highly favorable green chemistry characteristics. The advancement of cheminformatics software for green chemistry has the opportunity to march in lock-step with the increasing prevalence of electronic lab notebooks. It is important to incorporate green chemistry principles as a matter of priority since there is still time to influence the techniques that are considered to be best practices within the ELN marketplace.

# References


1. *Chemoinformatics: A Textbook*, ed. J. Gasteiger and T. Engel, Wiley-VCH, Weinheim, 2003.
2. (a) T. Engel, *J. Chem. Inf. Model.*, 2006, **46**, 2267; (b) *Handbook of Chemoinformatics Algorithms*, ed. J.-L. Loup and A. Bender, Chapman and Hall/CRC Press, Boca Raton, 2010; (c) P. Willett, *Annu. Rev. Inf. Sci. Technol.*, 2009, **43**, 1; (d) D. K. Agrafiotis, D. Bandyopadhyay, J. K. Wegner and H. van Vlijmen, *J. Chem. Inf. Model.*, 2007, **47**, 1279.
3. J. M. Barnard, *J. Chem. Inf. Comput. Sci.*, 1993, **33**, 532.
4. (a) S. Ekins, J. Olechno and A. J. Williams, *PLoS One*, **8**(5), e62325, DOI: 10.1371/journal.pone.0062325; (b) S. Ekins, C. Chang, S. Mani, M. D. Krasowski, E. J. Reschly, M. Iyer, V. Kholodovych, N. Ai, W. J. Welsh, M. Sinz, P. W. Swaan, R. Patel and K. Bachmann, *Mol. Pharmacol.*, 2007, **72**, 592.
5. J. W. Raymond, C. J. Blankley and P. Willett, *J. Mol. Graphics Model.*, 2003, **21**, 421.

6. (a) S. Ekins, A. J. Williams and J. J. Xu, *Drug. Metab. Dispos.*, 2010, **38**, 2302; (b) S. Ekins, *J. Pharmacol. Toxicol. Methods*, 2014, **69**, 115.
7. A. Tropsha, *Mol. Inf.*, 2010, **29**, 476.
8. A. M. Clark and P. Labute, *J. Med. Chem.*, 2009, **52**, 469.
9. A. M. Clark, A. J. Williams and S. Ekins, *J. Cheminf.*, 2015, manuscript accepted.
10. A. J. Williams and S. Ekins, *Drug Discovery Today*, 2011, **16**, 747–750.
11. P. Tiikkainen, L. Bellis, Y. Light and L. Franke, *J. Chem. Inf. Model.*, 2013, **53**, 2499.
12. A. M. Clark, *J. Chem. Inf. Model.*, 2011, **52**, 3149.
13. C.-J. Li and B. M. Trost, *Proc. Natl. Acad. Sci. U. S. A.*, 2008, **105**, 13197.
14. K. Kummerer, *Green Chem.*, 2007, **9**, 899.
15. D. Fatta-Kassinos, I. K. Kalavrouziotis, P. H. Koukoulakis and M. I. Vasquez, *Sci. Total. Environ.*, 2011, **409**, 3555.
16. P. T. Anastas and J. C. Warner, *Green Chemistry: Theory and Practice*, Oxford University Press Inc., New York, 1998.
17. (a) A. M. Voutchkova, T. G. Osimitz and P. T. Anastas, *Chem. Rev.*, 2010, **110**, 5845; (b) A. M. Voutchkova, *Green Chem.*, 2011, **13**, 2373.
18. P. J. Dunn, S. H. Galvin and K. Hettenbach, *Green Chem.*, 2004, **6**, 43.
19. M. Rubacha, A. K. Rattan and S. C. Hosselet, *J. Lab. Automat.*, 2011, **16**, 90.
20. (a) A. Dalby, J. G. Nourse, W. D. Hounshell, A. K. I. Gushurst, D. L. Grier, B. A. Leland and J. Laufer, *J. Chem. Inf. Comput. Sci.*, 1992, **32**, 244; (b) *SketchEl* molecule file format: http://molmatinf.com/fmtsketchel.html (accessed April 2014).
21. Software packages such as *ChemDoodle, ChemDraw*, *Accelrys Draw* and others make use of file formats designed for presentation of chemical structures: http://www.cambridgesoft.com/services/documentation/sdk/chemdraw/cdx (accessed April 2014).
22. A. M. Clark, *Mol. Inf.*, 2013, **32**, 291.
23. I. V. Filippov and M. C. Nicklaus, *J. Chem. Inf. Model.*, 2009, **49**, 740.
24. K. Alfonsi, J. Colberg, P. J. Dunn, T. Fevig, S. Jennings, T. A. Johnson, H. P. Kleine, C. Knight, M. A. Nagy, D. A. Perry and M. Stefaniak, *Green Chem.*, 2008, **10**, 31.
25. (a) D. J. C. Constable, P. J. Dunn, J. D. Hayler, G. R. Humphrey, J. L. Leazer, R. J. Linderman, K. Lorenz, J. Manley, B. A. Pearlman, A. Wells, A. Zaksh and T. Y. Zhang, *Green Chem.*, 2007, **9**, 411; (b) https://www.acs.org/content/dam/acsorg/greenchemistry/industriainnovation/roundtable/solvent-selection-guide.pdf (accessed April 2014); (c) D. Prat, O. Pardigon, H.-W. Flemming, S. Letestu, V. Ducandas, P. Isnard, E. Guntrum, T. Senac, S. Ruisseau, P. Cruciani and P. Hosek, *Org. Proc. Res. Dev.*, 2013, **17**, 1517.
26. (a) R. K. Henderson, C. Jiménez-González, D. J. C. Constable, S. R. Alston, G. G. A. Inglis, G. Fisher, J. Sherwood, S. P. Binksa and A. D. Curzons, *Green Chem.*, 2011, **13**, 854; (b) http://www.rsc.org/suppdata/gc/c0/c0gc00918k/c0gc00918k.pdf (accessed April 2014); (c) D. Prat, O. Pardigon, H.-W. Flemming, S. Letestu, V. Ducandas,

P. Isnard, E. Guntrum, T. Senac, S. Ruisseau, P. Cruciani and P. Hosek, *Org. Process Res. Dev.*, 2013, **17**, 1517.
27. L. Moity, V. Molinier, A. Benazzouz, R. Barone, P. Marionc and J.-M. Aubry, *Green Chem.*, 2014, **16**, 146.
28. It is important to distinguish between a negative result and a failure: the former is a carefully considered scientific claim that a given experiment does *not* proceed in the indicated way, and should ideally be accompanied by evidence, such as the characterisation of an unintended reaction product. Recording every failed experiment as a negative result is likely to be unproductive, resulting in a poor signal to noise ratio for machine learning purposes.
29. V. J. Stella and K. W. Nti-Addae, *Adv. Drug Delivery Rev.*, 2007, **59**, 677.
30. *Renewable Raw Materials: New Feedstocks for the Chemical Industry*, ed. R. Ulber, D. Sell and T. Hirth, Wiley-VCH, Weinheim, 2011.
31. (a) Environmental Protection Agency resources: http://www.epa.gov/osw/hazard (accessed April 2014).
32. (a) J. Kostal, A. Voutchkova-Kostal, B. Weeks, J. B. Zimmerman and P. T. Anastas, *Chem. Rev. Toxicol.*, 2012, **25**, 2780; (b) S. Ekins, *J. Pharmacol. Toxicol. Methods*, 2014, **69**, 115; (c) J. Kostal, A. Voutchkova-Kostal, J. B. Zimmerman and P. T. Anastas, *Proc. Natl. Acad. Sci. U. S. A.*, 2014, early edition (http://www.pnas.org/cgi/doi/10.1073/pnas.1314991111).
33. (a) P. Rydberg, D. E. Gloriam, J. Zaretzki, C. Breneman and L. Olsen, *ACS Med. Chem. Lett.*, 2010, **1**, 96; (b) J. Kirchmair, M. J. Williamson, J. D. Tyzack, L. Tan, P. J. Bond, A. Bender and R. C. Glen, *J. Chem. Inf. Model.*, 2012, **52**, 617; (c) T. J. Ritchie and S. J. Macdonald, *Drug Discov. Today*, 2014, **19**, 489.
34. S. Ekins, S. Andreyev, A. Ryabov, E. Kirillov, E. A. Rakhmatulin, S. Sorokina, A. Bugrim and T. Nikolskaya, *Drug Metab. Dispos.*, 2006, **34**, 495.
35. R. Liu and A. Wallqvist, *J. Chem. Inf. Model.*, 2014, **54**, 793.
36. J.-A. Jiang, C. Chen, Y. Guo, D.-H. Liao, X.-D. Pan and Y.-F. Ji, *Green Chem.*, 2014, **16**, 2807.
37. A. L. Hopkins, G. M. Keserü, P. D. Leeson, D. C. Rees and C. H. Reynolds, *Nat. Rev. Drug Discov.*, 2014, **13**, 105.
38. C. Jimenez-Gonzalez, C. S. Ponder, Q. B. Broxterman and J. B. Manley, *Org. Process Res. Dev.*, 2011, **15**, 912.
39. R. A. Sheldon, *Pure Appl. Chem.*, 2000, **72**, 1233.
40. B. M. Trost, *Angew. Chem., Int. Ed.*, 1995, **34**, 259.
41. D. Weininger, A. Weininger and J. L. Weininger, *J. Chem. Inf. Comput. Sci.*, 1989, **29**, 97.
42. J. D. Crabtree and D. P. Mehta, *J. Exp. Algor.*, 2009, **13**, 15.
43. J. Becker, L. Butt, V. von Kiedrowski, E. Mischler, F. Quentin and M. Hiersemann, *J. Org. Chem.*, 2014, **79**, 3040.
44. D. D. Ridley, *J. Chem. Inf. Comput. Sci.*, 2000, **40**, 1077.
45. H. Kraut, J. Eiblmaier, G. Grethe, P. Löw, H. Matuszczyk and H. Saller, *J. Chem. Inf. Model.*, 2013, **53**, 2884.
46. L. Riddler and M. Wagener, *Chem. Med. Chem.*, 2008, **3**, 821.

47. J. Law, Z. Zsoldos, A. Simon, D. Reid, Y. Liu, S. Y. Khew, A. P. Johnson, S. Major, R. A. Wade and H. Y. Ando, *J. Chem. Inf. Model.*, 2009, **49**, 593.
48. C. Steinmann, M. W. Ibsen, A. S. Hansen and J. H. Jensen, *PLoS One*,7(9), e44480, DOI: 10.1371/journal.pone.0044480.
49. A. M. Clark, M. Bronzetti and S. M. Muskal, White paper: Reaction101 and Yield101: Two mobile apps for chemistry with pedagogical value (http://molmatinf.com/docs/101Apps_WhitePaper.pdf, accessed November 2014).
50. P. Du and J. A. Kofman, *J. Lab. Automat.*, 2007, **12**, 157.
51. http://accelrys.com (accessed April 2014).
52. http://www.cambridgesoft.com (accessed April 2014).
53. H. E. Pence and A. J. Williams, *J. Chem. Educ.*, 2010, **83**, 1123.
54. http://itunes.apple.com/app/green-solvents/id446670983 (accessed April 2014).
55. http://play.google.com/store/apps/details?id=com.mmi.android.labsolvents (accessed April 2014).
56. S. Ekins, A. M. Clark and A. J. Williams, *ACS Sustainable Chem. Eng.*, 2013, **1**, 8.
57. S. Ekins, A. M. Clark and A. J. Williams, *Chem.-Bio. Inf. J.*, 2013, **13**, 1.
58. S. Ekins, A. M. Clark and A. J. Williams, *Mol. Inf.*, 2012, **31**, 585.
59. (a) A. M. Clark, *J. Cheminf.*, 2014, **6**(Suppl. 1), O1; (b) http://itunes.apple.com/app/green-lab-notebook/id908173898 (accessed November 2014).
60. (a) A. M. Clark, *J. Cheminf.*, 2010, **2**, 8; (b) A. J. Williams, S. Ekins, A. M. Clark, J. J. Jack and R. L. Apodaca, *Drug Discovery Today*, 2011, **16**, 928; (c) A. M. Clark, S. Ekins and A. J. Williams, *Mol. Inf.*, 2012, **31**, 569; (d) S. Ekins, A. M. Clark and A. J. Williams, *Mol. Inf.*, 2012, **31**, 585; (e) S. Ekins and A. M. Clark, *J. Cheminform.*, 2013, **5**, 13.
61. (a) M. E. Kopach and E. A. Reiff, *Future Med. Chem.*, 2012, **4**, 1395; (b) J. Zeng, M. Hillman and M. Arnold, *Bioanalysis*, **13**, 1501.
62. (a) R. N. Das and K. Roy, *Chemosphere*, 2014, **104**, 170–176; (b) K. Roy and R. N. Das, *J. Hazard. Mater.*, 2013, **254–255**, 166; (c) R. N. Das and K. Roy, *Mol. Divers.*, 2013, **17**, 151.

CHAPTER 10

# Toxicology for Chemical Safety and Sustainability

NICHOLAS D. ANASTAS

Poseidon's Trident, LLC, 83 Sassamon Avenue, Milton, MA 02186, USA
Email: mnanastas@comcast.net

## 10.1 Toxicology and Sustainable Chemistry

### 10.1.1 Principles of Toxicology

The critical aspects of toxicology necessary to inform safer chemical design are presented here. Several exceptional textbooks are available that serve as outstanding resources for those who want to investigate this fascinating field further.[1–5] This chapter provides a foundation upon which to build a conceptual dossier and a core skill set for characterizing toxicity in pursuit of designing safer chemicals.

Toxicology is the study of adverse effects on living systems resulting from chemical, biological, or physical agents.[5] This definition implies the enormity of the scope, the scientific complexity, and the depth of understanding required to effectively practice toxicology. Toxicology is both an art and a science similar to organic synthesis and medicine. The science of modern toxicology has gone beyond the traditional practice of a primarily descriptive discipline based on observations in whole animals to the current practice of using knowledge of the mechanisms of toxicity to describe, predict, and, ultimately, mitigate toxicity through molecular design. The art is in the interpretation of the results and ranking a toxicant relative to other compounds. The choice of toxicology topics to emphasize in this chapter is the

RSC Drug Discovery Series No. 46
Green Chemistry Strategies for Drug Discovery
Edited by Emily A. Peterson and Julie B. Manley

Published by the Royal Society of Chemistry, www.rsc.org

result of an assessment by the author of the concepts most useful to molecular designers involved in the practice of green chemical design. Important definitions used throughout this chapter are as follows:

(1) *Toxicity* is a relative property of a molecule's potential to cause harm.
(2) A *toxicant* is any agent capable of producing adverse responses in an organism.
(3) A *toxin* is a toxicant of natural origin, for example a natural product from a plant or a toxin from a venomous animal.
(4) A *xenobiotic* is a compound that is foreign to the organism.

Often times toxicant, toxin, and xenobiotic are used interchangeably, however incorrectly.

The central maxim of toxicology is that there is a quantitative relationship between the dose of a toxicant, toxin, or xenobiotic in an organism and the biological response it produces. This fundamental association is called the dose–response relationship and is essential to both toxicology and to pharmacology. The definitive determinant of toxicity is a function of the concentration of the ultimate toxicant at the target site for a long enough period of time, which is governed by the time course of action (toxicokinetics) and the response to the interaction at the target site (toxicodynamics).

*Toxicokinetics* describes the processes associated with the time course of a xenobiotic along its pathway to its receptor site or sites. Toxicokinetics is related to pharmacokinetics (the study of pharmaceutical kinetics within the body) but concerns the processes of toxicity instead of the action of a drug. Thus, the well-known principles of absorption, distribution, metabolism, and excretion, commonly referred to by the acronym ADME, also apply to the study of toxicokinetics.[6] All of these factors have a role in determining the amount of toxicant reaching the target site as well as the length of time the xenobiotic remains in the organism (Figure 10.1).

### *10.1.1.1 Absorption*

Absorption is the process of chemical, biological, and physical agents crossing biological membranes. The major sites of absorption are the gastrointestinal (GI) tract, the lungs, the skin (dermal absorption), and the gills of aquatic organisms. Absorption across each of these anatomic structures is dictated by the properties of the compound and the properties of the membrane itself. Non-polar, neutral, organic chemicals can transverse biological membranes through passive diffusion because of the lipid nature of

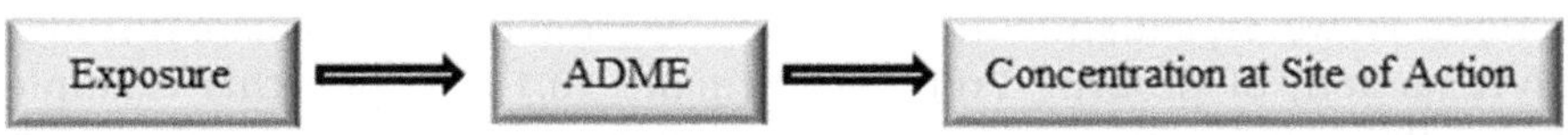


**Figure 10.1** Factors influencing the concentration of a compound at the target site.

biological membranes. Absorption from the gastrointestinal tract is one of the most common and most well-described routes of exposure due mainly to the study of pharmaceuticals.[7] For further reading on absorption from the respiratory tract and the dermis, see ref. 5.

#### *10.1.1.2 Metabolism*

Metabolism, or biotransformation, is the process of chemical transformation of a toxicant into a metabolite that likely possesses a different toxicity profile than the parent compound. Biotransformation affects both endogenous chemicals and xenobiotics. Metabolism can result in transformation into a product that is less toxic, more toxic, or equi-toxic, but in general, more water soluble and more easily excreted compounds result from metabolism.

Chemical modification can alter biological effects through toxication, also called bioactivation, which refers to the situation where the metabolic process results in a metabolite that is more toxic than the parent. If the metabolite demonstrates lower toxicity than the parent compound, the metabolic process is termed detoxication. These processes can involve both enzymatic and non-enzymatic processes. Excellent detailed reviews of metabolism can be found in comprehensive texts.[8]

The liver is the primary organ of metabolism in mammals for both exogenous and endogenous chemicals. The enzymes associated with biotransformation have broad substrate specificity that allows a few enzymes to interact with a myriad of xenobiotics. This attribute is a part of the adaptable defense mechanism present in all organisms. The synthesis of these enzymes is inducible, meaning that exposure to the compound itself or other compounds can cause an increase in the production of metabolic enzymes to deal with exposures to potential toxic compounds.

Metabolism is divided into two major phases based on the general reactions associated with each category. Phase I metabolism prepares compounds for excretion by revealing or adding a functional group that makes a compound more water soluble, and therefore more easily excreted. Phase I enzymes are located in almost all tissues; however, the greatest concentration is in the liver. Other sites include the gut, skin, and lung. The major classes of Phase I reactions are hydrolysis, reduction, and oxidation. Hydrolysis reactions include the cholinesterase and pseudocholinesterase enzymes involved with the metabolism of pesticides and other ester containing molecules. Not all biotransformations are detoxicating, in fact some transformations can lead to a toxic response or, alternatively, be harnessed for therapeutic effect. For example, the antibiotic sulfanilamide has been used safely for many years to treat streptococcal and pneumococcal infections. However, sulfanilamide is generated from the metabolism (through azo-reduction) of prodrug Prontosil (Scheme 10.1). Despite this example, most nitro-containing compounds can cause various adverse effects including methemoglobinemia and are potentially mutagenic. Metabolites of

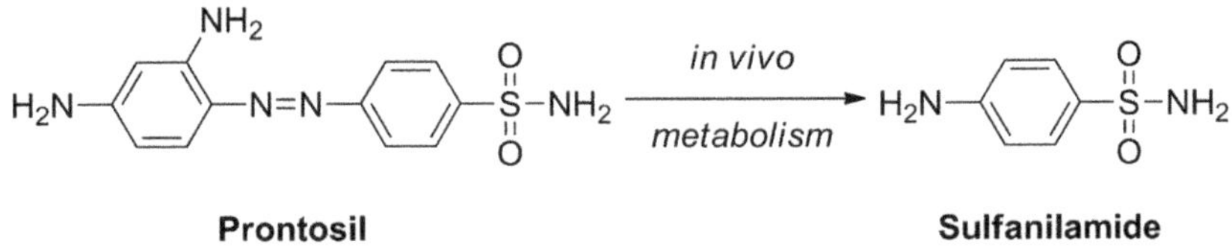


**Scheme 10.1** Azo-reduction of Prontosil to sulfanilamide.

nitroaromatic compounds have been shown to covalently bind to DNA. Reduction of the nitro group to the same reactive nitroso toxicophore as would be formed by the oxidation of the corresponding aromatic amine would likely have similar toxic consequences.[9]

The cytochrome P450 (CYP) family of enzymes are the major catalysts that are responsible for various oxygenation reactions associated with the Phase I metabolism and biotransformation of many xenobiotics.[10] Cytochrome P450 enzymes are located in the endoplasmic reticulum of the microsomal fractions of cells and have high catalytic versatility. The main function of this group of isozymes is to insert one atom of oxygen into a substrate, thereby increasing hydrophilicity. These enzymes are heme-containing proteins containing a reduced iron species essential for transferring electrons and work in concert with coenzymes NADPH and NADPH reductase.

Phase II metabolism is associated with synthetic conjugation reactions including glucuronidation, sulfation, methylation, and conjugation with glutathione. Often, the Phase I metabolites are used as substrates for Phase II reactions. Phase II products are water-soluble and therefore more easily excreted by the kidney.[11]

### *10.1.1.3 Excretion*

Excretion is the removal of a chemical from the organism through any of the available processes including respiration, or urinary, biliary, and fecal excretion. Toxicants or xenobiotics can be eliminated from organisms by several routes depending on the anatomy and physiology of an organism. For water-soluble compounds, or compounds that have been metabolized to water soluble compounds, the kidney is the main organ of excretion in most organisms. Compounds with molecular weights less than 60 kilodaltons (kDa) are filtered by the glomerulus. Volatile compounds are excreted primarily through exhalation. Lipid soluble compounds are excreted through bile and feces.

The applied dose of a chemical refers to the amount of a chemical that comes in contact with an organism or "receptor". The journey to the site of biological action is governed by four primary processes that control the amount of toxicant that will ultimately reach the site of biological action. The effective dose or biologically effective dose is the amount of a chemical that reaches the site of action. The effective dose is determined by (1) the extent and rate of uptake at the site of exposure, (2) the rate of distribution to the tissues, (3) the extent of biotransformation, and (4) how quickly and efficiently the compound is eliminated from the organism.

## 10.1.2 Disposition of Toxicants in Organisms

### *10.1.2.1 The Dose–Response Relationship*

The dose–response relationship describes the correlation between an increase in the dose of a chemical and the resulting increased response, which can be either beneficial or adverse. Though the relationship is elegant in its simplicity, it remains a formidable assignment to fully characterize the complex and often subtle nature of toxic responses. Toxicity is a function of dose, exposure, and time.[12] Consequences of the interaction of a molecule with a biological target will propagate through molecular, biochemical, cellular, and organism levels of organization, ultimately resulting in a biological consequence. This consequence can be detrimental in the case of toxicity, or beneficial in the case of therapeutic compounds.

The term dose refers to the total amount of a substance to which an organism is exposed. Dose is commonly expressed as mass of substance per weight of the organism per time (*e.g.*, mg $kg^{-1}$ $d^{-1}$). The entire dose is not necessarily absorbed and distributed to its site of action. The *external* or *applied dose* is the amount of a chemical at the interface between the environment and the organism. The *biologically effective* or *internal dose* is the amount of toxicant actually reaching the target. The total dose can be calculated if the duration and extent of exposure are known.

In both toxicology and pharmacology it is customary to plot the dose as the independent variable on the *x*-axis and the response as the independent variable on the *y*-axis. When the dose is plotted arithmetically, a hyperbolic curve is generated showing the increased response with increased dose. When the dose is log-transformed and plotted against response, a line segment is obtained making the statistical manipulation easier to evaluate.[13] An advantage to the log dose–response plot is a much more straightforward interpretation of differences in potency among a group of toxicants acting through similar modes or mechanisms of action. Chemicals producing the same maximal effect but at a lower dose will occupy a position farther to the left on the plot of the dose–response curve indicating greater potency.

Most chemicals follow a threshold response, that is, the probability of a response is essentially zero below a certain dose or concentration. This can also be defined as less than an observable response for a population. The position of the dose–response curve provides information on the amount of a chemical that is necessary to elicit a maximal response. Responses are normally distributed reflecting the variation within a population of responses. Those responding at lower doses or concentrations reflect sensitive individuals or hyper-responders. Individuals that are more resistant to the effects are hypo-responders, whereas most of the members of the exposed population respond to similar doses reflected in the median or average response.

Selective toxicity is succinctly defined as agents that produce injury to the undesirable entity (uneconomic species) without causing harm to the desired entity (economic species).[14] The concept of selectivity can be applied across species as in the case of pesticides that are designed to eliminate pests and not harm humans or non-target species or can be applied to situations within a single organism as in the case of anti-neoplastic chemicals that are designed to target vulnerable features of cancer cells while not damaging non-cancer cells.

Chemicals are used in commerce and industry, as medicines, and may be naturally occurring or synthetic. Under certain circumstances of exposure, these chemicals present a hazard to humans and the environment. The nature and severity of these hazards is determined by the physicochemical properties of the agent that determine its interaction with its ultimate target (intrinsic hazard) and its ability to come into contact with receptors (exposure). Determining whether a substance is toxic or non-toxic requires a set of metrics representing both assessment and measurement endpoints for clearly defined adverse outcomes.

Pharmaceuticals are an emerging class of environmental contaminants. Over the past 20 years or so, interest in the presence of pharmaceuticals in the environment, often referred to as PIE, has increased significantly. Concern over PIE has resulted in enhanced analysis of environmental media, research into potential health effects from exposure to these active compounds, and modeling environmental fate and transport. The concentrations of the pharmaceuticals and personal care products (PPCPs) reported to date are significantly below therapeutic levels.[15] However, this does not mean that we can dismiss this issue without considering potential effects on non-human organisms, both aquatic and terrestrial, as well as the potential adverse effects of longer-term exposure to mixtures acting through similar mechanisms, especially for endocrine disrupting chemicals. Excretion of prescribed medications and disposal of PPCPs into wastewater is thought to constitute the main route of entry into the environment.[16]

Drug discovery chemists must be aware of the full spectrum of consequences of the molecule they produce including therapeutic, off-target effect, human and ecotoxicity, and environmental fate.

### *10.1.2.2 Exposure*

Toxic responses cannot occur unless an organism is exposed to a chemical and the ultimate toxicant reaches its site of action. The amount of exposure is determined by the toxicokinetic principles outlined above.

Exposure periods are generally classified into four general categories: acute, subacute, subchronic, and chronic (Table 10.1). These well-developed time duration categories are approximate and usually apply to well-designed animal studies but rarely apply to actual exposure scenarios in residential and occupational settings. These intervals can be adjusted to satisfy experimental conditions or regulatory requirements. Acute effects occur

**Table 10.1** Typical exposure categories and time durations.

| Exposure category | Time duration |
|---|---|
| Acute | Less than 24 h |
| Subacute | Less than one month |
| Subchronic | Greater than one to three months |
| Chronic | Greater than three months; up to lifetime |

immediately upon exposure or within a very short period of time post exposure. These adverse effects can result from a single exposure or from multiple exposures within a very short period, for example the dermal corrosivity of strong acids, inhalation toxicity of carbon monoxide, or single ingestion of high doses of arsenic. Chronic effects manifest after repeated exposures, several months to the organism's entire lifetime. For the same chemical, acute effects are often vastly different from chronic effects.

#### *10.1.2.3 Spectrum of Toxic Effects*

Virtually every chemical is toxic at sufficiently elevated dose, and for an appropriately sufficient amount of time and route of exposure. Chemicals have a spectrum of undesired effects depending on the dose, frequency and duration of exposure, the intrinsic toxicity of the molecule, and the influence of protective or adaptive mechanisms. The dose needed to produce a particular deleterious effect among chemicals can range more than eight orders or magnitude. No chemical demonstrates a single, well-defined, and exclusive adverse effect. For example, acute exposures to volatile anesthetics result in dizziness and anesthesia whereas chronic exposure to lower concentrations can result in liver and kidney damage.

The potential for reversible toxicity is important for characterizing the significance of a toxicant. A compound that demonstrates irreversible effects, for example, permanent corrosive tissue damage or covalent binding to macromolecules, is of more concern than those compounds that demonstrate adverse effects that are reversible upon cessation of the exposure.

Toxicants can adversely affect a limited anatomical or physiological space or manifest toxicity throughout the system (*i.e.*, systemically). These effects are not limited exclusively to a particular toxicant or class of toxicant. Xenobiotics can manifest both types of characteristics. Some toxicants are so reactive that the damage is manifested directly at the site of exposure, for example, strong acids and bases and strong oxidizers or reducers are capable of causing immediate and irreversible necrosis of skin. Most other toxicants must be absorbed and transported to their site of action.

#### *10.1.2.4 Non-organ System Toxicity*

Not every compound interacts with an anatomical receptor or tissue that we refer to as organs. Many xenobiotics act through non-target system related toxic pathways. Three important categories on non-organ system related toxicity are carcinogenesis, reproductive, and developmental deficits.

### *10.1.2.5 Carcinogenesis*

Carcinogenesis is a multi-stage process associated with the induction of neoplasms, or new unregulated cell growths that leads to a family of disease states commonly termed cancers. Many carcinogenic molecules are compounds that undergo bioactivation to form electrophiles that can react with cellular nucleophiles like DNA and proteins. Covalent modification of DNA or proteins can lead to a myriad of adverse effects including disruption of critical enzyme pathways and destruction of structural cellular components.[17]

### *10.1.2.6 Reproductive and Developmental Toxicity*

The processes associated with reproduction and development in all organisms and plants are extremely complex, awe inspiring events requiring flawless execution of elegant combinations of timing and of process fidelity. Any errors at critical stages in the process can lead to devastating consequences including physical malformations, increased reproductive failures, physiological deficits, and death.

Reproductive toxicology is the study of the adverse effects on the male and female reproductive systems resulting from exposure to biological, chemical, or physical agents. The endocrine system is integral to reproduction and other physiological processes.

The endocrine system is a group of specialized organs, tissues that function to regulate many of the activities of the organs, and tissues and is the primary system responsible for homeostasis.[18]

Exposure to certain xenobiotics is associated with adverse endocrine effects. These chemicals are commonly referred to as endocrine disrupting chemicals (EDCs). Endocrine disruption can interfere at all levels of physiological organization and with all endocrine organs, for example, mimicking natural hormones, blocking endogenous receptors, or directly affecting the system itself.

Estrogen active, or estrogenic, compounds have received a great deal of attention from the research community and from the public media because of their putative role in breast and ovarian cancers as well as with disrupting reproduction and development in aquatic communities. Many compounds are suspected endocrine disruptors including pesticides, industrial chemicals (*e.g.*, nonylphenol), plasticizers (*e.g.*, phthalates), and some pharmaceuticals (*e.g.*, diethylstilbestrol, DES). Currently no comprehensive list of potential endocrine disrupting chemicals and their associated putative toxicophores is available; however, excellent work is being done by the United States Environmental Protection Agency (USEPA) to classify such toxicants.[19]

Endocrine disruption can affect any part of the endocrine system, including the organs themselves or as part of the biochemical pathways.

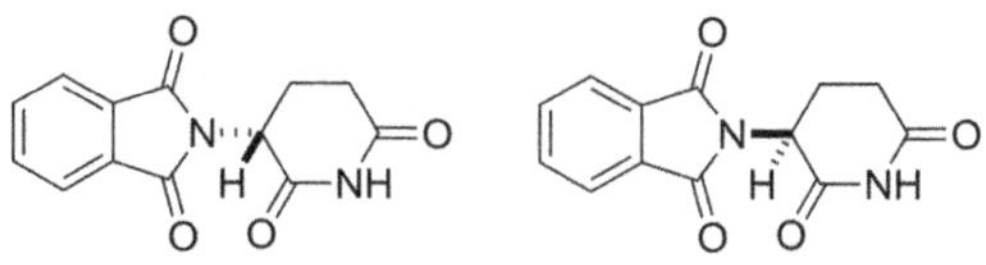


**Figure 10.2** Structures of (*R*)- and (*S*)-thalidomide.

There are several endocrine organs and systems; however, male and female reproduction and the thyroid system are the most widely studied. Adverse effects can occur through agonists that mimic endocrine activity or as antagonists when they block the natural processes of an organism.

Developmental toxicology is the study of adverse effects in a developing organism. Teratology is a sub-discipline of developmental toxicology that focuses on the specific time period between conception and birth. Many teratological effects are quite obvious, for example, cleft palate and missing limbs. The timing of exposure is of paramount concern especially for developmental toxicity. Critical exposure periods of susceptibility correlate with the timing of organ development and are quite precise. Any perturbations of the normal timing significantly increase the likelihood of adverse developmental effects including teratogenesis. Guidelines have been developed to assess developmental risks.[20]

One example of a teratogenic chemical is thalidomide. Thalidomide is among the most notorious examples of how a slight change in molecular structure can influence biological effects. Thalidomide was a drug prescribed to pregnant women to mitigate morning sickness. Within a year of its introduction onto the market, reports of severe limb malformations in newborns were disclosed. The drug was removed from the market soon after the incidents of these adverse effects appeared.

Thalidomide exerts its teratogenic effects by interfering with organogenesis between days 24 and 33 of development.[21] The exact mechanism is not known; however, intercalation into DNA is one of the leading hypotheses among the more than 30 proposed. Thalidomide contains a stereogenic center and thus racemic thalidomide exists as an equal mixture of two enantiomers ((*R*) and (*S*), Figure 10.2). Research has indicated that only the (*S*)-enantiomer results in toxicity. Understanding the relationship between the structural requirements and exposure limitations necessary for developmental toxicity has enabled pharmacologists to identify new clinical uses for thalidomide including treatment of leprosy, AIDS, and against some aggressive forms of cancers.[22]

### *10.1.2.7 Immunotoxicology*

The immune system is a highly complex and well-regulated system of cells, tissues, organs, and molecular mediators that respond to both endogenous and exogenous challenges. Concern for the potential toxicity to the immune system arises from the central role the immune system has in maintaining and protecting health. Immunotoxicology is a blend of toxicology and immunology defined as adverse changes in structure or function resulting from exposure to a toxicant.[23] Immunotoxicity can manifest as hypersensitivity or allergic reactions (*e.g.*, dermatitis, inflammation), immunodeficiency (*e.g.*, HIV/AIDS), and autoimmunity (*e.g.*, celiac disease). Substances that provoke immune responses are termed antigens. Antigens can be beneficial, for example, neutralizing infectious agents, or the responses can be adverse and result in anaphylaxis, or immune deficiency.

### 10.1.3 Linking Toxicology with Sustainability

Toxicology is composed of a diverse collection of sub-disciplines each focused on a specialized area of investigation. Modern toxicology evolved from the related field of pharmacology, a mature science that investigates the effects of chemicals on living systems for the purpose of therapeutics and other medicinal endpoints. Both disciplines are based on similar fundamental principles. Green toxicologists, as described below, use the principles of toxicology and chemistry to identify opportunities to design molecules with reduced hazard by establishing design rules.

The structure–hazard relationship forms the nexus between molecular design and toxicology. Because chemistry studies the properties and transformations of matter and toxicology is tasked with understanding the effects of chemicals on human health and the environment, understanding both chemistry and toxicology is unquestionably necessary to design safer chemicals.[24]

Several sub-disciplines of toxicology have evolved to fill specific research and regulatory needs and to meet science and policy objectives as part of green chemistry. The goal of design for hazard reduction (or benign-by-design) is to minimize, through molecular manipulation, the intrinsic toxicity associated with exposure to a chemical. The success of these efforts relies on cooperative efforts among toxicologists, synthetic chemists, and environmental scientists. This section focuses on describing the principles and practice of designing less toxic and sustainable chemistries to avoid both physical hazards and global hazards.

The evolution of toxicology from a primarily descriptive discipline into a well-developed predictive science relies on the newest approaches of molecular toxicology including the incorporation of toxicogenomics and other tools focused at the genetic level of organization to uncover toxic mechanisms of action at the biochemical, cellular tissue, and systems levels.[25] Systems biology has been an essential tool in framing the picture of toxicity in a more holistic way by describing and predicting adverse outcomes. A framework for designing safer chemicals is described in Section 10.4 providing a useful tool for identifying opportunities for safer chemical design.[26,27] Designing safer chemicals requires incorporating toxicology into the design phase of the molecular design process.

## 10.2 Green Chemistry

Green chemistry is the design of chemical products and processes that reduce or eliminate the use and generation of hazardous substances.[28] It is a philosophy, a pollution prevention strategy, and a design protocol that provides a holistic, conceptual framework for designing for reduced hazard. Green chemistry is applicable to each lifecycle stage providing strategies at one or several points. For example, using bio-renewable feed-stocks, incorporating atom economy, maximizing energy efficiency, minimizing water use, designing less toxic chemicals, and designing for rapid and efficient breakdown in the environment are all included in the guiding principles of green chemistry. Advancements in green chemistry have enabled success in addressing global climate change, energy production, improved water and food supply, and a reduction in the release of toxic chemicals to the environment.[29]

One area that has not received as much focus as other research efforts is designing chemicals with reduced toxicity, or Green Chemistry Principle number four. Multiple reasons exist for this deficit, including the lack of synthetic chemists trained in toxicology, the lack of hazard data of sufficient quantity and quality to inform the design of safer chemicals, and the complexity of elucidating adverse outcome pathways and the underlying structure–hazard relationship.

### 10.2.1 Expanding Green Chemistry Principle Number Four

The principles of toxicology must be incorporated into the chemistry curriculum as a core learning requirement to train the next generation of scientists in a holistic, trans-disciplinary manner. This training does not need to be as rigorous as that provided to professional toxicologists, but must be robust enough to enable a chemist to determine when to seek the advice of a trained toxicologist. Ideally, chemists and toxicologists work together as early in the process of designing safer chemicals as possible.

Chemists who are trained in both toxicology and synthetic methods will be able to recognize potential trouble within a molecule and apply appropriate design tactics to make a safer compound. The education and training of this new type of trans-disciplinary scientist requires a paradigm shift in the philosophy of the traditional role a chemist plays in designing chemicals and materials. The benefits of this multi-disciplinary education will result in synthetic chemists that design better drugs, since they will be versed in potential toxicity mechanisms that could lead to attrition of drugs from the clinics due to toxicity issues. Using the information from structural alerts as part of the strategy to design safer molecules provides an important strategy for chemists to link familiar molecular attributes with the associated biological endpoints. Gathering a comprehensive database supporting structure–hazard relationships is a daunting task; however, significant progress has been made in assigning molecular requirements for reactivity through a weight of evidence approach.

### 10.2.2 Green Toxicology

Synergy of these two disciplines has resulted in an emerging area of study termed "green toxicology" that seeks to formalize the early cooperation among chemists and toxicologists to design less hazardous products and processes and to guide research protocols. **Green toxicology** is the *application of the principles of toxicology to chemicals with the specific intent of deriving design protocols for hazard reduction.*[30] This is a step on a continuum linking chemistry, medicinal chemistry, toxicology, and finally green toxicology. The art and science of toxicology is applied to design for hazard reduction, a practice that is consistent with other sub-disciplines of toxicology (*e.g.*, clinical, forensic, aquatic, ecological, *etc.*). Green toxicology is essentially an expansion of green chemistry principle number four (design for reduced toxicity) to incorporate the latest toxicological tools, specifically computational and mechanistic protocols.

The following section outlines the principles of toxicology that can be applied to synthetic molecular design to construct less hazardous molecules *a priori*. Green toxicology also uses the principles of environmental fate and transport, and environmental or ecological toxicology to evaluate the behavior of chemicals in the environment by incorporating this information into evaluating biological, chemical, and physical hazards to wildlife and to plants.

## 10.3 The Role of Modern (Twenty-first Century) Toxicology in Designing Safer Chemicals

Toxicology testing is currently undergoing a paradigm shift from classical methods that rely on toxicodynamic outcomes from animal studies to testing that is focused on assessing perturbations to key biological pathways resulting in disruption of homeostasis that may be permanent or temporary.[31] The use of traditional animal toxicity testing approaches is expensive, slow, and rarely provides data that enables the articulation of a mode or mechanism of toxicity. Replacing existing methods with emerging twenty-first century methods, referred to as Tox 21, reduces the number of animals used, which has benefits scientifically, ethically, and financially.[32] The objectives presented here aim to advance a paradigm shift in toxicology that uses the tools of computational toxicology to catalog modes and mechanisms of action. The use of emerging toxicity assays can lead to more rapid data generation examining a wider array of tools using complementary biologically endpoints (Table 10.2).

**Table 10.2** Examples of emerging *in vitro* toxicity assays.

| *In vitro* assay type or name | Assay description or endpoint |
|---|---|
| Ames assay | Mutagenicity in *Salmonella typhimurium* |
| BSEP (bile salt export pump) | Ability to excrete bile salts |
| CYP (cytochrome p450) superfamily of monooxygenases | Binding to one or multiple isoforms of the cytochrome enzyme |
| hAR | Binding to the androgen receptors |
| hER | Binding to the estrogen receptors |
| hERG | Binding to the ether-a-go-go receptor |
| HSP (heat shock protein) | Binding to this co-chaperone molecule |
| PPAR (perioxisome proliferator activated receptor) | Binding to receptors and subsequent stimulation of β oxidation of lipids |
| PRX (peroxiredoxin activity) | A measure of potential oxidative damage from peroxides and hyper-peroxides |
| TNF (tumor necrosis factor) | Associated with apoptosis |

### 10.3.1 Tools for Designing Safer Chemicals

Systems toxicology is the integration of disparate data generated from classical toxicology, molecular biology, computational chemistry, computer science, and mathematics to describe responses to a toxic insult. Deriving meaningful information from the wealth of information generated from this approach requires an integrated strategy designed to identify pathways of toxicity with reasonable certainty.[33] The past decade has seen an evolution of a common language that enables productive conversations between toxicologists and chemists in the common pursuit of designing chemicals possessing reduced hazard. These tools include computational toxicology (also referred to as *in silico*), *in vitro*, and whole animal (*in vivo*) testing. Computational toxicology is the application of mathematical and computer models and molecular and biological and chemicals approaches to explore quantitative and qualitative relationships between chemical exposure and adverse health outcomes.[34] This emerging research integrates recent technological advances made in molecular biology, so-called "omics" technology, chemistry, and computer science to describe the effects of chemicals on biological pathways. The tools supporting these efforts include high throughput screening (HTS) and high throughput content methods HCS, QSAR, and others mentioned earlier.

### 10.3.2 The Role of Pharmaceutical and Medicinal Chemistry

It can be confidently stated that safer chemical design began in the pharmaceutical industry.[35] Pharmaceutical companies traditionally rely on the use of animal testing for developing medicines and for predicting toxicity of therapeutic targets as required by the Food and Drug Administration (FDA). From the beginning, the pharmaceutical industry has led the advancement of safer chemical design by developing novel and more accurate *in vitro* and *in vivo* techniques. These advancements have reduced the use of animal testing, which results in financial benefits, improved animal welfare, and helps to prevent late stage compound attrition by identifying toxicity early in the drug discovery process.

The primary goal of most pharmaceutical companies is to develop therapeutic compounds for disease modification. At the same time a medicinal chemist is designing for therapeutic functionality and efficacy, they must design a compound for minimal toxicity. In other words, the quest for the ultimate pharmacophore includes a simultaneous search for any potential toxicophores in a putative marketable molecule while ensuring that the medicine works. This task is not trivial, considering the number of variables that must be optimized to create a successful drug and the requirement for novel intellectual property means that often the potential for toxicity is hard to predict for a particular scaffold. Every year, millions of potential therapeutic candidates are screened for efficacy and toxicity, and, of those, only one or two will be approved by the Federal Drug Administration. *In silico* methods have been developed and utilized quite successfully in the pharmaceutical industry to characterize potential toxicity associated with drug candidates.[36] The structural alerts and tools provided by the extensive research in the pharmaceutical industry can help chemists in other disciplines design safer chemicals. Furthermore, the tactics used to increase bioavailability for pharmaceuticals can be consulted for how to make industrial chemicals less bioavailable.

## 10.4 Hierarchy of Knowledge

Safer chemical design through structural modification generates new molecules, compounds, or materials requiring hazard characterization. The approaches used for assessing toxicity for new chemicals share a commonality with the tools for assessing the toxicity of existing chemicals. Safer chemical design through structural modification generates new molecules, compounds, or materials requiring some level of hazard characterization. Not all available data are equally useful when initiating a molecular design project. A tiered system of data usability for safer chemical design is required. Over the past decade or so, several hazard reduction design framework protocols for industrial chemicals have been put forward.[37,38] All of these protocols incorporate a holistic, systematic approach that highlights the overall value of understanding the mechanism of toxicity as completely as possible.

### 10.4.1 Mechanistic Data: Tier 1

Mechanistic toxicology focuses on elucidating and describing the molecular events from exposure to the events that lead to the disruption of biological targets and describes the resulting adverse outcomes on living systems. A *mechanism* of action is defined as a detailed description of the key molecular events associated with a toxic response. A *mode of action* is a generic description of the key events and processes, starting with the interaction of an agent with a cell, through functional and anatomical changes, resulting in toxicity.[39] Ideally, the steps in the pathway of toxic response are identified and are connected to the manifestation of toxicity. This connection is extremely challenging and few mechanisms have been described in detail. Significant progress has been made in the past several decades identifying mechanisms and modes of action. However, there is much work to be done. Chemists are familiar with reaction mechanisms as they pertain to synthetic reactions, but are not often familiar with those same fundamental mechanisms applied to biological situations.[40]

Determining individual steps involved with the manifestation of an adverse response provides a starting point for documenting opportunities to control the structure–toxicity relationship, which is a critical step in the process for designing safer chemicals.

The manifestation of toxicity reflects a complex sequence of connected events. Elucidating these mechanisms forms the nexus between synthetic chemistry and toxicology, and develops the basis of green toxicology and safer chemical design.

Mechanisms are rarely single events occurring in isolation. In reality, toxic mechanisms are complex multifaceted events involving feedback and repair mechanisms, transport to the site of action, molecular transformation, reaction with target molecules, and ultimately excretion. Modes of action can be divided into two general categories: non-specific (narcosis) and specific modes of action. Narcosis, also called baseline toxicity, is a generalized depression of biological activity resulting from the presence of toxicants. The exact mechanism is not known; however, there have been several theories advanced to ascribe a more precise characterization of the process. Some examples of chemicals that act through baseline narcosis include ethanol and general anesthetics.[41] Narcosis has been further divided into polar and non-polar modes depending on the influence of physical chemistry parameters including molecular weight and geometry.[42]

Specific toxicity reflects the interaction of compounds with identifiable biological targets. Examples of specific endpoints include oxidative uncouplers, DNA alkylators, acetylcholinesterase inhibitors, and central nervous system (CNS) active compounds.

The progress in the rapidly expanding, multidisciplinary field of toxicogenomics has provided essential insight into the action of xenobiotics at the level of gene expression.

Toxicogenomics holds the promise of revealing the effects of toxicants on the genome and the expression of these alterations at the biochemical, cellular tissue, and organism level. Tens of thousands of genes exist in the human genome and not all of them are expressed. The differential expression of these genes is responsible for normal function as well as for the responses to exposures to a xenobiotic.

### 10.4.2 Quantitative Structure–Activity Relationships (QSAR): Tier 2

The toxicity–structure relationship is a fundamental physicochemical property of a molecule that lends itself to modeling and quantitative structure–activity relationship (QSAR) modeling. A more in-depth treatment and discussion of the history of QSAR can be found in the work done by Selassie and others.[43] QSAR models can be used for establishing quantitative relationships between structure and activity or structure and property, to predict the potential activity for compounds of unknown toxicity and for designing safer chemicals.

Molecular descriptors are any parameter used in the development of either a structure–activity relationship (SAR) or QSAR to model any type of property under investigation, for example, linking molecular structure, molecular weight, polar surface area, or $\log P$ with some type of biological attribute or response (*e.g.*, toxicity, $LC_{50}$, carcinogenicity, *etc.*). QSARs were among the first applications of *in silico* toxicology.[44] These relationships can provide initial screening data to prioritize further testing, or provide support as part of a weight of evidence approache for supporting mechanistic determinations for adverse outcome pathways. A plethora of potential property–response combinations exist and many have been developed with a high degree of repeatability.[45] Some of the more common properties are listed in Table 10.3.

A QSAR is only as good as the quality of the data used to develop the model. There are potentially significant limitations in using QSARs to describe structure–activity relationships and in applying these relationships to predicting the responses of untested molecules to the same or similar physicochemical properties. One must understand the underlying mechanisms and modes of action as completely as possible as well as all other factors associated with modifying toxicity for the chemical under study. Modifying factors among and between species include anatomical and

**Table 10.3** Examples of molecular descriptors.

| | |
|---|---|
| Physicochemical | Lipohilicity (as $\log P$)<br>Water solubility<br>Henry's law constant<br>Rates of reaction<br>Molecular weight (important for biliary clearance) |
| Topological parameters | Molecular connectivity<br>Molecular volume and surface area<br>Stereochemistry<br>Molecular volume/surface area<br>Structural fragments (alerts)<br>TPSA |
| Electronic and steric parameters | Hammett constants<br>Taft constants<br>$pK_a$<br>Dipole moment<br>Frontier molecular orbitals |

physiological differences, referred to as interspecies, and then there are the intra-species differences accounting for variability in the populations associated with genetic polymorphism, age, gender, and disease state. A structure–activity relationship is usually derived using molecules of similar molecular structure, for example, congeners of polychlorinated biphenyls or substituted halogenated aromatics. Extrapolating these results to predict the toxicity of compounds with different molecular attributes is not often successful.

More quantitative treatment of structure–activity relationships using *n*-octanol as the preferred lipophilic solvent was performed by Hansch and others.[46,47] These studies established the octanol–water partition coefficient ($K_p$) as the standard for characterizing lipophilicity in biological systems. Hansch and others performed the seminal work in investigating the influence of partition coefficients with biological action.[48] This parameter is more commonly expressed as $\log P$ {*i.e.*, $\log P_{oct/wat} = \log([solute]_{oct}/[solute]_{water})$} and occupies a central role for predicting the behavior of non-polar organic molecules. Log $P$ is one of the most important of the molecular descriptors used to date and reflects the influence that the degree of lipophilicity has on the movement of compounds across biological membranes. Compounds with low lipid solubility do not pass as readily through membranes as do compounds that are more lipid soluble, but they may have higher aqueous solubility. In addition, it is known in the pharmaceutical industry that compounds with high $\log P$ tend to have poorer toxicological outcomes because they tend to have more non-specific interactions with off-target receptors leading to off-target toxicity.

### 10.4.3 Kinetics and Dynamics: Tier 3

Advancements in mechanistic toxicology and identifying the factors affecting toxicokinetics and toxicodynamics have provided opportunities to exploit molecular soft spots to design safer chemicals. The use of toxicogenomics is central to the emerging field of computational toxicology providing information on transcript, protein, metabolomic, and lipid profiling with conventional toxicology methods to investigate the interactions between gene expression and toxicant exposure.[49] Traditional techniques include determining absorption, distribution, metabolic, and excretion rates as well as investigating key metabolic pathways leading to ultimate toxicants or metabolism leading to detoxication.

### 10.4.4 Bioavailability: Tier 4

Often the most straightforward and initial consideration for screening structural changes for hazard reduction is limiting the ability of a compound or material to reach its ultimate target site of action. As described previously, several physicochemical factors influence bioavailability, including ionization potential (*i.e.*, $pK_a$), molecular weight, lipophilicity, and water solubility among many others. The pharmaceutical industry has developed clever relationships (such as Lipinski's "Rule of 5") that link the optimal molecular properties required for a putative therapeutic compound to achieve oral availability.[50] Designing molecules for reduced toxicity can use these guidelines in reverse to minimize the potential for a molecule to reach its site of action.

Cholestyramine is a polymer containing a quaternary ammonium moiety that is prescribed to patients to absorb bile acids including cholesterol. It is not absorbed from the gastrointestinal tract because if its molecular size but also because of the positively charged ammonium group. Charged molecules are generally not absorbed well through most common routes of exposure of administration.

This same concept can be demonstrated with the neuromuscular blocking agent curare, which is actually a mixture of various constituents depending on the botanical source. The active component in curare is *d*-tubcurarine, an alkaloid that contains a quaternary ammonium feature. Absorption occurs efficiently in muscle where the pH is approximately 7. However, the pH of the gut is approximately 1.5, much more acidic, ensuring that the majority of curare is in a positively charged state, therefore limiting absorption into the systemic circulation and, thereby, mitigating toxicity.

## 10.5 Articulating and Compiling Design Rules and Guidelines

Computational toxicology represents a paradigm shift in how toxicology research is advancing the assessment and design of chemicals in the twenty-first century.[51] The United States Environmental Protection Agency (USEPA) defines computational toxicology as the "integration of modern computing and information technology with molecular biology to improve agency prioritization requirements and risk assessment of chemicals".[52] The primary difference between traditional toxicology testing and computational toxicology techniques is one of scope in that the numbers of chemicals studied simultaneously, as well as the number of endpoints and biological pathways examined, are in the hundreds to thousands.

Some excellent illustrative examples exist to show how a holistic, systems-based approach can advance the development of design guidelines for making safer, healthier, and sustainable chemicals, compounds, and materials. DeVito and coworkers used computational toxicology to design safer nitriles using a mechanistic and QSAR approach as described earlier. Understanding of the mechanistic nature and QSAR of a class of toxic chemicals can lead to a statement of design rules that inform the design of safer chemicals. Nitriles are a functional group (CN) contained in chemicals that are widely used for various applications including as solvents, in medicines, and in other industrial applications. Nitriles occur naturally in both plants and animals and are also synthesized. Their ubiquitous nature and volume of use mean that the number of individuals potentially exposed to nitriles is significant; therefore, evaluating and reducing the risk associated with exposure to this class of compounds is warranted.

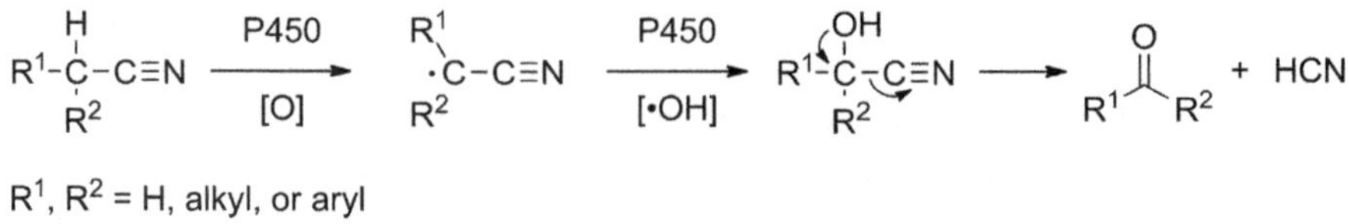


**Scheme 10.2** Mechanism of cyanide release from nitriles.

Depending on how a nitrile is incorporated into a chemical, the toxicity of nitriles can be similar to the toxicity of cyanide intoxication, implying that that the cyanide moiety from the molecule is the ultimate toxicant. DeVito and others have developed a mechanistic-based model for predicting the acute toxicity of nitriles based on the rate of hydrogen atom abstraction by CYP 450 and hydrogen radical stability as the primary variables.[53] An evaluation of the toxicity, as measured by the $LD_{50}$, compared with the structural characteristics of selected nitriles, revealed that the critical mechanistic step is the rate of α-hydrogen abstraction (Scheme 10.2).[54] A greater rate led to a greater acute toxicity. From this evaluation, structural modifications for reduced hazard, or design rules, were derived. Among the molecular attributes that are associated with lower acute toxicity were (1) steric hindrance around the α-hydrogen to restrict cytochrome P450 enzyme access, (2) groups that reduced the stability of the α radical, and (3) avoidance of hetero-containing groups on the α carbon (Scheme 10.2).

### 10.5.1 Design for Reduced Oral Absorption

For chemicals that are not intended to be used as pharmaceuticals, reducing the potential for absorption could be a viable method to improve the safety of these compounds. The process of oral absorption in humans is dependent upon several physicochemical characteristics of the molecule, the absorptive surface of the membrane, and the surrounding conditions (*e.g.*, pH). Absorption is often a first and necessary step in a complete pathway to toxicity, therefore, any molecular modifications that reduce the potential for absorption will likely reduce or eliminate toxicity.

Lipinski and colleagues examined the influence of a selected set of physicochemical properties on the extent of oral absorption for a group of pharmaceuticals to determine whether a relationship could be established for predicting the success of new drugs as part of the drug discovery process, providing a set of guidelines for deciding whether an unknown molecule is "drug-like" referring to the likelihood of the compound being absorbed orally.[50] Compounds have a higher chance of being well absorbed if they possess the following characteristics:

- Log $P < 5$,
- molecular weight of $< 500$ g $mol^{-1}$,
- fewer than five hydrogen bond donors,
- fewer than ten hydrogen bond acceptors.

### 10.5.2 Reverse Design

For disciplines outside the realm of drug discovery, these guidelines can be used to design safer molecules by exploiting molecular features that cause a decrease in absorption of a potentially toxic molecule. This idea of "reverse design" applies the inverse of Lipinski's rules to reduce the likelihood of a toxicant reaching a site of action. As these design guidelines are identified, they must be collected, sorted, and compiled into resources that are useful to synthetic chemists. Useful references include ADME information,[55] the so-called "3/75" rule with regard to toxicity for drug molecules,[56] and multiple-parameter optimization (MPO) for CNS drug-like properties[57] as well as for brain penetration.[58]

## 10.6 Conclusions

We face a new frontier in toxicology and safer chemical design, where as citizens and scientists, environmentalists, and engineers, we have the opportunity to act upstream of pollution thereby decreasing the downstream pressure from the sources of inherently toxic pollution by advocating for, and applying, the principles of green chemistry and green toxicology.

The desire for designing safer chemicals has been articulated for many years and has been incorporated into pharmaceutical and industrial chemistry research strategies. The first step in designing a safer chemical is to establish the relationships between molecular structure, functionality, and adverse biological (toxicological) outcomes. Chemists are familiar with the properties that are required for functionality, for example, in the preparation of dyes, solvents, surfactants, pharmaceuticals, and other commercially important products. These chemists may not be as familiar with the process of evaluating the structure–hazard relationship for potential toxicity.

The same chemical principles apply to designing for functionality as for designing for reduced hazard. Instead of functionality, the spectrum of toxic effects must be considered in the design phase and must also be considered design flaws. There are many potential challenges to designing safer chemicals. Comprehensive hazard evaluation is an extremely complex undertaking reflecting the inherent complexity of biological systems. Some of these challenges include:

1. The lack of specificity and selectivity for highly reactive chemicals. Highly reactive chemicals present unique challenges to green toxicologists because of the unpredictability of the potential interactions with biological targets. Highly reactive chemicals react quickly and indiscriminately, leading them to be called promiscuous molecules.
2. Structural diversity is vast and as a result the prediction of toxicity is made more challenging, especially based on two-dimensional inspection.
3. The fate of a chemical entity is dependent on several processes within the body.

The complexity of designing safer and sustainable chemicals and materials requires a trans-disciplinary approach. Uncertainty in the nature of biological systems, the behavior of molecules within these systems, and limited data are significant limiting components of this process. Extrapolating vast amounts of information from *in vitro* test systems to predict *in vivo* outcomes in whole organisms, especially humans, must include a discussion of the associated limitations and uncertainties.[59] Constructing large-scale data repositories requires a robust strategy for interpreting and controlling these data being generated by HTS and thousands of data points per sample. Major challenges must be overcome to develop mechanisms to offer these structure–toxicity relationship data among various research and business organizations, given that companies may have concerns regarding sharing of proprietary data, to further the mission of understanding adverse outcome pathways at a molecular level of organization. Ideally the grand plan is to ultimately describe the entire "toxome," or, in other words, account for all of the potential pathways of toxicity.[60,61]

## References

1. *Cassarett and Doull's Toxicology: The Basic Science of Poisons*, ed. C. D. Klaassen, McGraw-Hill, New York, 8th edn, 2010.
2. *Principles and Methods of Toxicology*, ed. A. Wallace Hayes, CRC Press, Philadelphia, 4th edn, 2014.
3. *Fundamentals of Aquatic Toxicology: Effects, Environmental Fate and Risk Assessment*, ed. G. Rand, CRC Press, Philadelphia, 2nd edn, 1995, p. 1148.
4. *A Textbook of Modern Toxicology*, ed. E. A. Hodgson, John Wiley and Sons, Hoboken, NJ, 4th edn, 2010, p. 672.
5. D. L. Eaton and S. G. Gilbert, in *Casarett and Doull's Toxicology: The Basic Science of Poisons*, ed. C. D. Klaassen, McGraw Hill, New York, 7th edn, 2008.
6. (a) D. J. Burkett, *Pharmacokinetics Made Easy*, McGraw-Hill, New York, 2002; (b) *Handbook of Pesticide Toxicology: Principles*, ed. R. Krieger, Elsevier Science, San Diego, 2nd edn, 2001.
7. Y. Kwon, *Handbook of Essential Pharmacokinetics, Pharmacodynamics and Drug Metabolism for Industrial Scientists*, Kluwer Academic Press, New York, 2001.
8. A. Parkinson and B. W. Ogilvie, in *Cassarett and Doull's Toxicology: The Basic Science of Poisons*, ed. C. D. Klaassen, McGraw-Hill, New York, 7th edn, 2008, ch. 6, pp. 161–304.
9. G. F. Smith, *Prog. Med. Chem.*, 2011, **50**, 1.
10. F. P. Guengerich and J. S. Macdonald, *AAAP J.*, 2007, **8**, 1.
11. W. B. Jacoby and D. M. Zeigler, *J. Biol. Chem.*, 1990, **265**, 20715.
12. *Handbook of Pesticide Toxicology: Principles*, ed. R. Krieger, Elsevier Science, San Diego, 2nd edn, 2001.

13. L. Shargel and A. D. Yu, *Applied Biopharmaceutics and Pharmacokinetics*, Appleton and Lane, Norwalk CT, 3rd edn, 1993.
14. A. Albert, *Selective Toxicity: The Physico-chemical Basis of Therapy*, Chapman and Hall, London, 6th edn, 1981.
15. B. A. Boxall, M. A. Rudd, B. W. Brooks, D. J. Caldwell, K. Choi, S. Hickmann, E. Innes, K. Ostapyk, J. P. Staveley, T. Verslycke, G. T. Ankley, K. F. Beazley, S. E. Belanger, J. P. Berninger, P. Carriquiriborde, A. Coors, P. C. DeLeo, S. D. Dyer, J. F. Ericson, F. Gagne, J. P. Giesy, T. Gouin, L. Hallstrom, M. V. Karlsson, D. G. J. Larsson, J. M. Lazorchak, F. Mastrocco, A. McLaughlin, M. E. McMaster, R. D. Meyerhoff, R. Moore, J. L. Parrott, J. R. Snape, R. Murray-Smith, M. R. Servos, P. K. Sibley, J. O. Straub, N. D. Szabo, E. Topp, G. R. Tetreault, V. L. Trudeau and G. Van Der Kraak, *Environ. Health Perspect.*, 2012, **120**, 1221.
16. G. T. Ankley, B. W. Brooks, D. B. Huggett and J. P. Sumpter, *Environ. Sci. Technol.*, 2007, **41**, 8211.
17. J. S. Walsh and G. T. Miwa, *Ann. Rev. Pharmacol. Toxicol.*, 2011, **51**, 145.
18. P. E. Molina, Endocrine Physiology, *Lange Physiology Series*, McGraw-Hill, 2006, vol. 2, p. 304.
19. USEPA 2014, http://www.epa.gov/endo/ (accessed 12/6/14).
20. USEPA, *Guidelines for Developmental Risk Assessment*, US Environmental Protection Agency, Risk Assessment Forum, Washington DC, EPA /600/ FR-91/001 1991.
21. T. D. Stephans, C. J. W. Bunde and B. J. Filmore, *Biochem. Pharmacol.*, 2000, **59**, 1489.
22. H. A. Fine, W. D. Figg, K. Jaeckle, P. Y. Wen, A. P. Kyritsis, J. S. Loeffler, V. A. Levin, P. M. Black, R. Kaplan, J. M. Pluda and W. K. Yung, *J. Clin. Oncol.*, 2000, **18**, 708.
23. M. I. Luster, P. P. Simeonova and D. R. Gormolec, Immunotoxicology, in *Encyclopedia of Life Sciences*, John Wiley & Sons, 2001.
24. J. Bragg, Structural Alerts for Toxicity, in *Burger's Medicinal Chemistry, Drug Discovery and Development*, ed. D. J. Abraham and D. P. Rotella, John Wiley & Sons, Hoboken, NJ, 7th edn, 2010, pp. 301–334.
25. Z. Gregus, in *Cassarett and Doull's Toxicology: The Basic Science of Poisons*, ed. C. D. Klaassen, McGraw-Hill, New York, 7th edn, 2008, vol. 3, pp. 49–122.
26. N. Anastas, in *Green Chemistry Metrics*, ed. A. Lapkin and D. Constable, Blackwell, London, 2008, ch. 2, pp. 27–40.
27. N. D. Anastas and J. C. Warner, *Chem. Health Saf.*, 2005, 3.
28. P. T. Anastas, and J. C. Warner, *Green Chemistry Theory and Practice*, Oxford University Press, UK, 1998.
29. P. T. Anastas and M. Kirchoff, *Acc. Chem. Res.*, 2002, **35**, 686.
30. N. Anastas, in *Green Techniques for Organic Synthesis and Medicinal Chemistry*, ed. W. Zhang and B. Cue, John Wiley and Sons, UK, 2012, vol. 1, pp. 3–31.
31. T. Hartung, *Nature*, 2009, **460**, 208.

32. National Research Council, *Toxicity in the 21st Century: A Vision and a Strategy, National Research Council Committee on Toxicity Testing and Assessment of Environmental Agents*, National Academy Press, Washington, DC, 2005.
33. T. Hartung, E. Van Vliet, J. Jaworska, L. Bonilla, N. Skinner and R. Thomas, *ALTEX*, 2012, **29**, 119.
34. R. Kavlock, G. T. Ankley, T. Collette, E. Francis, K. Hammerstrom, J. Fowle, H. Tilson, G. Toth, P. Schmieder, G. D. Veith, E. Weber, D. C. Wolf and D. Young, *Reprod. Toxicol.*, 2005, **19**, 265.
35. K. L. Chapman, H. Holzgrete, L. E. Black, M. Brown, G. Chellman, C. Copeman, J. Couch, S. Creton, S. Gehen, A. Hoberman, L. B. Kinter, S. Madden, C. Mattis, H. A. Stemple and S. Wilson, *Res. Toxicol. Pharmacol.*, 2013, **66**, 88.
36. L. G. Valerio, *Toxicol. Appl. Pharmacol.*, 2009, **241**, 356.
37. S. Devito and R. Garret, *Designing Safer Chemicals*, American Chemical Society, Washington, DC, vol. 640, 1996, pp. 194–223.
38. A. M. Voutchkova, T. G. Osimitz and P. T. Anastas, *Chem. Rev.*, 2010, **110**, 5845.
39. USEPA 2005. Guidelines for Carcinogen Risk Assessment, Risk Assessment Forum, Washington, DC, EPA/630/P-03/001F, March 2005.
40. *Goodman & Gilmans: The Pharmacological Basis of Therapeutics*, ed. J. G. Hardman and L. E. Limbird, McGraw-Hill, New York, 10th edn, 2001.
41. G. D. Veith and S. J. Broderius, *Environ. Health Perspect.*, 1990, **87**, 207.
42. D. W. Roberts and J. F. Costello, *QSAR Comb. Sci.*, 2003, **22**, 226.
43. C. D. Selassie, S. B. Mekapati and R. P. Verma, *Curr. Top. Med. Chem.*, 2002, **2**, 1357.
44. I. Rusyn and G. P. Daston, *Environ. Health Perspect.*, 2010, **118**, 1047.
45. C. Hansch, P. P. Maloney, T. Fujita and Q. M. Muir, *Nature*, 1962, **194**, 178.
46. S. P. Bradbury, C. L. Russom, G. T. Ankley, T. W. Schultz and J. D. Walker, *Environ. Toxicol. Chem.*, 2003, **22**, 1789.
47. C. Hansch and T. Fujita, *J. Am. Chem. Soc.*, 1964, **86**, 1616.
48. (a) C. A. Hansch, R. Steward, S. M. Anderson and D. Bentley, *J. Med. Chem.*, 1968, **11**, 1; (b) C. Hansch, R. M. Muir, T. Fujita, P. P. Maloney, F. Geiger and M. Streich, *J. Am. Chem. Soc.*, 1963, **85**, 2817.
49. M. D. Waters and J. M. Fostel, *Nat. Rev. Genet.*, 2004, **5**, 936.
50. C. S. Lipinski, F. Lombardo, B. W. Dominy and P. J. Feeney, *Adv. Drug Deliv. Rev.*, 1997, **23**, 3.
51. R. J. Kavlock, G. Ankley, J. Blancato, M. Breen, R. Conolly, D. Dix, K. Houck, E. Hubal, R. Judson, J. Rabinowitz, A. Richard, R. W. Setzer, I. Shah, D. Villeneuve and E. Weber, *Toxicol. Sci.*, 2008, **103**, 14.
52. USEPA 2003, A Framework for Computational Toxicology Research Program, Washington, DC, EPA 600/R-03/65.
53. S. C. DeVito, in *Designing Safer Chemicals*, ed. S. C. DeVito and R. Garrett, American Chemical Society, Washington, DC, 1996, vol. 640, pp. 194–223.

54. J. Grogan, S. C. Devito, R. S. Pearlman and K. R. Korzekwa, *Chem Res. Toxicol.*, 1992, **5**, 548.
55. M. P. Gleeson, *J. Med. Chem.*, 2008, **51**, 817.
56. J. D. Hughes, J. Bragg, D. A. Prince, S. Bailey, G. A. DeCrescenzo, R. V. Devraj, E. Ellsworth, Y. M. Fobian, M. E. Gibbs, R. W. Gilles, N. Greene, E. Huang, T. Krieger-Burke, J. Loesel, T. Wager, L. Whiteley and Y. Zhang, *Bioorg. Med. Chem. Lett.*, 2008, **18**, 4872.
57. T. T. Wagner, X. Hou, P. R. Verhoest and A. Villalobos, *ACS Chem. Neurosci.*, 2010, **1**, 435.
58. L. Di, H. Rong and B. Feng, *J. Med. Chem.*, 2013, **56**, 2.
59. J. D. Pliel, M. A. Williams and J. R. Sobus, *Toxicol. Lett.*, 2012, **215**, 201.
60. T. Hartung and M. McBride, *ALTEX*, 2011, **28**, 83.
61. M. Bouhifd, H. T. Hodgberg, A. Kleesang, A. Maertens, L. Zhao and T. Hartung, *Basic Clin. Pharmacol. Toxicol.*, 2014, **115**, 24.

CHAPTER 11

# *Environmental Regulations and the Green Chemist*

DAVID TAYLOR

wca, Brunel House, Volunteer Way, Faringdon, Oxfordshire SN7 7YR, UK
Email: david.taylor@wca-consulting.com

## 11.1 Introduction

I suspect that the link between the medicinal chemist working at the bench and the growing volumes of environmental regulation will seem extremely tenuous to most of the readers of this book; however, I would urge you to persevere for three reasons.

Firstly, green chemists, like all other scientists, are subject to a host of health, safety, and environmental regulations at their place of work.

Secondly, despite the commonly held viewpoint that the only regulations that matter to the pharmaceutical industry are those derived from the medicines regulators such as the US Food and Drug Administration (FDA) and the European Medicines Agency (EMA), this is not the case and increasingly pharmaceuticals are coming to the attention of environmental regulators across the world.

Finally, decisions taken by medicinal chemists in the laboratory can have profound implications in the decades which follow should the drug on which they are working succeed in its clinical trials and finally enter the market. For example, one drug company had a promising candidate drug under investigation which their medicinal chemists had synthesized using a simple chlorofluorocarbon as a starting material. However, this substance was subject to an international convention banning its manufacture and

RSC Drug Discovery Series No. 46
Green Chemistry Strategies for Drug Discovery
Edited by Emily A. Peterson and Julie B. Manley

Published by the Royal Society of Chemistry, www.rsc.org

use, and consequently taking the molecule into full scale production using this process route would have been fraught with difficulty, both in terms of obtaining consent for residual discharges, but, more importantly, identifying a supplier that was able and willing to provide the volumes of material required. Fortunately, the candidate drug failed in its early clinical trials and the potential problems were avoided. Consequently, all medicinal chemists should have a basic understanding of the environmental regulations that will impact on the drugs that they are developing when they finally reach the market.

### 11.1.1 Regulation of Good Practice

The research, clinical trials, and manufacturing of pharmaceuticals are covered by rigorous compliance regulations with the objective of ensuring consistent high quality. These Good Laboratory Practice (GLP), Good Clinical Practice (GCP), and Good Manufacturing Practice (GMP) regulations are strictly enforced and subject to random inspection. In addition, no new medicinal entity (NME) can be introduced onto the market until the appropriate medicines regulator, such as the EMA or the FDA, has approved its safety, effectiveness, and quality.

When a medicine does finally receive approval for marketing, that authorization relates to both the medicine and the method by which it was manufactured. Simple product sampling techniques, as used in other industries, are insufficient to ensure the quality that is needed, and medicines regulators require manufacturers to follow strict Good Manufacturing Practice guidelines (GMP).[1] These involve a holistic approach to the whole manufacturing cycle. There is a requirement for extensive and rigorous qualification and validation of equipment and procedures, together with comprehensive documentation of every aspect of the process. Regulatory agencies undertake regular, often unannounced inspections and will expect to inspect any new manufacture prior to start-up. These manufacturing quality requirements are intended to ensure consistency between the medicine that was tested in the clinical trials and the product eventually used by the patient.

These GxP regulations have led to a perception within the industry that it is already very strictly regulated. However until the end of the twentieth century the vast majority of this regulation was focused entirely on human safety and little attention was paid by the medicines regulators to environmental issues. The last decade has seen a gradual change in some aspects of medicines regulation. For example, strict application of the GMP regulations, perversely, often inhibited the implementation of improvements to the environmental sustainability of manufacturing processes, since any significant changes trigger a requirement for further confirmatory clinical data. A more pragmatic approach is now being taken that enables some improvements to the manufacturing process to continue to be made. Similarly, although the medicines approval process remains dominated by considerations of human safety, it now increasingly includes an assessment of environmental risk.

### 11.1.2 The Development of Environmental Regulations

The human race has been concerned with protecting the environment for over 2000 years[2] with the earliest codified regulation appearing in England in the reign of Richard II.[3] However a significant body of environmental regulations only began to appear in the twentieth century. The number of environmental regulations in the developed world then began to increase exponentially in the 1970s after the first United Nations conference on the state of the environment.[4]

The growth of these regulations has been somewhat piecemeal with many laws introduced as reactions to specific incidents. For example, in the United Kingdom the Clean Air Act[5] resulted from the disastrous great London 'smog' of 1952; in the European Community the Detergents Directive[6] was introduced to deal with excess foam in rivers and wastewater treatment facilities; the 1980 CERCLA ('Superfund') Regulations in the USA were introduced as a consequence of the contaminated land issues at Love Canal.[7]

In addition there is increasing international and global cooperation with the beginnings of some global harmonization, for example, with the Globally Harmonized System (GHS) Packaging & Labelling initiative dealing with the transport of chemicals.[8]

Since the late twentieth century there have been attempts in the developed world to simplify and codify these environmental regulations into more coherent frameworks. This is intended to eliminate overlapping regulations, close gaps and loopholes while at the same time increase their efficiency and effectiveness. In the European Union the Water Framework Directive[9] and the REACH Regulation[10] are typical examples of this. In both cases the new law has brought together, expanded, and simplified several previously existing pieces of legislation to provide a greater degree of coherence.

It is perhaps not surprising that environmental regulations differ, both in scope and style, from country to country and region to region. This is partly because individual countries began the process at different times and approached this issue at different rates, depending on their own experience. In the 1970s and 1980s the United States was leading the development of environmental legislation, but since the 1990s it has been the European Union that has been in the lead. Major differences also exist in the approach taken to regulation in different parts of the world due to the existence of different value systems. This is perhaps most obvious in the different approaches currently taken in North America compared to Europe. This is often attributed to the endemic use in Europe of the Precautionary Principle, but in reality the difference relates to the different attitude to risk tolerance between the two regions. In the European Union, the risk averse value systems of the Nordic countries such as Denmark and Sweden and The Netherlands now dominate the regulatory arena and these values differ markedly to those of North America and in particular the USA where there is a higher risk tolerance.

### 11.1.3 Environmental Regulation and the Medicinal Chemist

At first sight there does not appear to be any reason why the medicinal chemist, working in a pharmaceutical R&D organization, should be at all concerned about the scope or impact of environmental regulations, other than those that might have some impact on the disposal of waste from the laboratory. However, the medicinal chemist is a key player in the eventual environmental consequences that may stem from the introduction of a new medical entity (NME) into the market. Decisions taken during lead optimization can, for example, play a critical role in determining the eventual rate of environmental degradation of the molecule. The synthetic route chosen to make the initial trial batches of the API and subsequently their scale-up to pilot plant level can determine the final manufacturing route. This influences the starting materials, intermediates, and solvents that may be involved and determines the overall process efficiency. All these decisions will have an impact on the final production of the active pharmaceutical ingredient (API). The medicinal chemist, as well as striving to implement "green chemistry", should also be aware of the overall regulatory environment that surrounds pharmaceuticals.

The remainder of this chapter summarizes the key types of regulation in three sections: the first considers those regulations related to the research and development process including the Health, Safety, and Environmental regulations that apply to work in the laboratory; the second examines the regulation of manufacturing processes; the final section looks at broader environmental quality regulations that are likely to have a significant impact on the product itself as it enters the environment after use by the patient.

## 11.2 Research & Development

### 11.2.1 Regulations in the Laboratory

Most of this chapter is concerned with the implications of environmental legislation on the output of medicinal chemists. This legislation has only an indirect impact on the chemist working in the laboratory. However, some aspects of environmental, health, and safety legislation have a direct impact and medicinal chemists need to maintain an awareness of these regulations and the regular changes that take place to them.

The laboratory is a workplace and most legislatures have introduced increasingly stringent health and safety regulations to try to minimize accidents and ill health in the workforce. In addition, employers have frequently gone beyond the basic requirements of the regulations by introducing their own more stringent operating procedures and codes of practice.

Medicinal chemists working in the laboratory have legal and ethical responsibilities for safety and the protection of human health and the environment. The examples given below are drawn mainly from the United Kingdom, but similar regulations and practices exist in many countries, although sometimes at different stages of development. The Royal Society of Chemistry has produced two very useful Guidance notes: *Individual Legal Responsibilities for Health and Safety*[11] and *Individual Legal and Ethical Responsibilities for Environmental Protection.*[12] These provide a broad overview of these two areas.

In the United Kingdom the introduction of the 1974 Health & Safety at Work Act set a new standard in this type of legislation.[13] The Act itself sets out broad principles to be followed by both employers and employees. For example, employers must *"provide and maintain a safe working environment without risk to their employees with adequate facilities and arrangements for their welfare"*. Employees also have responsibilities. They must *"take reasonable care both for their own health and safety and for that of other persons who may be affected by their acts or omissions at work. They must co-operate with their employer so far as is necessary to enable [the employer] to comply with the requirements of the Act"*.

The legislation is in the form of an enabling act. This allows for both the repeal and consolidation of relevant statutory provisions, and for new regulations to be made specifying how the general principles outlined in the Act should be applied in specific circumstances. Not surprisingly, a wide range of regulations have been made since the Act was first passed in 1974. These concern reporting of accidents, diseases and dangerous occurrences, first aid, safety signs, the notification and control of various hazardous activities, the control of substances hazardous to health, the classification, labelling and packaging of dangerous substances, *etc.*

In considering their health and safety responsibilities, it is important for medicinal chemists to understand where they sit in the company governance structure. Employers and employees usually have defined responsibilities in regulations, but what of the manager or team leader? A manager is clearly an employee with the responsibilities that follow. However, as an agent of the employer, the "manager" also effectively controls other employees and the work being done, and thus has additional responsibilities.

When the employer appoints persons to managerial positions they should be educated and trained in the relevant management systems in order to ensure the health and safety of persons in their charge. Responsibilities and duties increase with each level of management. Therefore, each promotion should be accompanied by appropriate training to ensure that managers continue to be able to discharge their health and safety duties effectively.

#### *11.2.1.1 Risk Assessment & Risk Management*

The main duty of an employer is to assess the risks to health and safety of its employees, and others such as contractors and visitors, arising out of their activities. These must cover the effective planning, organization, control, monitoring, and review of preventative and protective measures that have been introduced for health and safety purposes.

It might be thought that the significant risks to a medicinal chemist at work in the laboratory will come from exposure to chemicals. This will of course be a major part of risk management, but other issues will also be of potential concern, *e.g.* the use of compressed gases, the use of machinery such as centrifuges and hotplates, fire safety, and waste management. In addition, appropriate procedures need to be in place to deal with lone working, people with disabilities and women who are pregnant.

In the United Kingdom a major change to the regulation of substances in the workplace occurred with the introduction, under the Health and Safety at Work act 1974, of the Control of Substances Hazardous to Health (COSHH) Regulations in 2002.[14] This was an upgrade of a 1999 regulation and incorporated the provisions of several relevant European Union Directives.

The COSHH Regulations provide general requirements on employers to protect employees and other persons from the hazards of substances used at work by risk assessment, control of exposure, health surveillance, and incident planning. There are also duties on employees to take care of their own exposure to hazardous substances. In addition the regulations prohibit or restrict the use of a number of substances in the workplace in line with decisions made under the EU REACH Regulations.[10] These COSHH regulations can be considered to exemplify current best practice. More information is available from the United Kingdom Health and Safety Executive, and several the key requirements are described below:[15]

COSHH Regulation 6 requires that an employer should not carry out work liable to expose employees and non-employees, such as members of the public, to a substance hazardous to health without a risk assessment. This assessment must be reviewed regularly and use the information provided by the chemical supplier.

COSHH Regulation 7 requires that an employer prevent exposure to hazardous substances or, if this is not reasonably practicable, that he adequately controls exposure. As a consequence employers must take all reasonable steps to ensure that control measures, and any necessary equipment of facilities, are properly used or applied and, significantly, employees must use the control measures properly, return them after use, and report any defective equipment. This is not a top down one-sided regulation but involves shared responsibilities. Where necessary, workplace exposure or health monitoring may also be required.

COSHH Regulation 12 requires that all employees liable to exposure to hazardous substances are provided with suitable and sufficient information, instruction, and training. The increasing prevalence of regulatory systems like REACH,[10] described in more detail in Section 11.2.3.1, is contributing to a major improvement in the availability of information on the hazards, risks, and appropriate control systems for hazardous substances. This is, however, an area that can often be overlooked because the employer may make assumptions that, for example, new postgraduate employees will have received adequate training in these areas while at university.

COSHH Regulation 13 requires that employers prepare for possible accidents, incidents, and emergencies involving hazardous substances. These should include emergency procedures that are regularly tested. It is not sufficient to simply inform new employees that they should evacuate the building when the alarm sounds. Indeed initial testing of an emergency drill can be a salutary experience, often demonstrating unexpected flaws and/or ambiguities. When I worked in a GLP compliant facility we had a weekly test of the alarm system, using a different initiation point each time, and an unannounced full evacuation drill at about 5–6 week intervals. It required a number of iterations before this worked smoothly, but you do not want to identify problems for the first time during a real emergency. Similarly, it helps if the first time you use a fire extinguisher is not during a real fire! Consequently using extinguishers that have gone beyond their expiry date for practice sessions can be very valuable.

### *11.2.1.2 Environmental Impact*

As we have seen there is a wide range of health and safety issues for medicinal chemists to consider in their day to day operations, but almost none of these have consequences for the manufacture and sale of the final end product. Environmental issues are the reverse with very few having an impact on day to day laboratory operations but several having consequences for the end product as we shall see later in this chapter.

As far as laboratory activities are concerned the key environmental issue is the handling of waste and associated waste disposal. This is an area that is now highly regulated in both developed and developing countries, especially in regard to hazardous substances. In the European Union, for example, it is no longer acceptable to dispose of small electrical devices, such as hand held calculators, in the trash. Such items must be disposed in line with the Waste Electronic and Electrical Equipment (WEEE) Regulations.[16]

Simply disposing of all wastes into the site drainage system or discarding waste into the trash on an "out of sight, out of mind" principle is no longer viable. Regulatory constraints usually exist on what substances can be disposed of in various ways, *i.e.* to landfill or to sewer. Most laboratories are now connected to public sewers and the local sewage operator will have set conditions on what can be accepted into the sewer. Similar constraints exist on laboratory waste with segregation needed, *e.g.* for sharps and items containing hazardous substances such as fluorescent light tubes that contain small quantities of mercury.[17] Consequently regular contact needs to take place between staff working in laboratories and the site waste management team. This can also be a significant cost issue. In one R&D department, because of lack of training in waste management, it had become common practice for all waste to be disposed in the clinical waste.

This unnecessarily increased the waste disposal bill, which was reduced dramatically following training of laboratory staff.

Responsible laboratory waste management[18] should, in addition to complying with regulations, follow the principles of the waste hierarchy.[19] In particular, all chemical reactions should be periodically reviewed to ascertain whether safer, more efficient methods could be used, thereby minimizing the waste that has to be disposed of and at the same time ensuring that any residual waste can be disposed of safely. The following guiding principles should be considered to reduce waste, where possible at source.

In a laboratory context this could mean:

- ➢ Planning work carefully so as to minimize raw material consumption.
- ➢ Putting objects back into use. Cleaning and re-labelling reagent bottles for re-use, or return to supplier is a good example.
- ➢ Recovering value from waste by recycling, *e.g.* careful segregation of waste solvents for recycling or use as low-grade fuel.
- ➢ Incinerating non-recyclable combustibles (using approved methods to minimize atmospheric pollution) to reduce waste volume and toxicity.
- ➢ As a last resort, rendering the waste less environmentally harmful by an appropriate treatment and dispose of it to landfill – at an appropriate site. The amount of waste discharged to landfill should have the lowest practicable volume and the lowest achievable environmental toxicity.

### 11.2.2 Greening the Product

All medicinal chemists learn at their mother's knee that the three critical criteria for any NME are efficacy, safety, and quality. In other words the drug must work in the patient, its side effect profile must be such that the benefits of the therapy outweigh any risks, and it must be capable of being manufactured at a consistent high quality. Impact on the environment has previously not been an issue of concern. Consequently, the requirements for the evaluation of the environmental impact of active pharmaceutical ingredients until recently were extremely rudimentary and only required in the USA and Europe.

However, the emergence of the pharmaceuticals in the environment issue[20] has fundamentally changed this. In 2007, the European Medicines Evaluation Agency (EMEA) released a guidance document setting out a comprehensive methodology for the evaluation of the environmental risk assessment of medicines for human use.[21] This now applies, with some exceptions, to all new active ingredients entering the EU market and to any existing substance where a change in use patterns would lead to a significant increase in environmental exposure. The first pre-screening step in the assessment provides, with some caveats, for an exclusion from any further assessment for substances whose predicted environmental concentration is $<0.01\ \mu g\ L^{-1}$. However, in practice very few substances fall below this limit, which equates to a patient dose of $<2\ mg\ day^{-1}$.

Similar legislation is currently being drafted in other countries with implementation expected in the near future in Canada and Japan.[22] In 1997, the Food and Drug Administration relaxed its requirements for the environmental risk assessment of active ingredients as part of a deregulation initiative in the USA.[23] The new requirements provide a "categorical exclusion", *i.e.*, an exemption from environmental assessment for all active ingredients unless they exceeded a concentration of 1 $\mu g\ L^{-1}$ at the point of entry into the aquatic environment. In essence, provided that the applicant confirms that there are no "extraordinary circumstances" to prevent it, an environmental risk assessment for a product is only required when sales exceed 44 tons per year. It is possible, however, that the USA will introduce similar regulations to those being developed elsewhere over the next few years.

These new regulations mean that for the first time APIs will be subjected to the same sort of environmental evaluation previously applied only to pesticides and biocides. At present the consequences of this are uncertain. In the European Union the result of this environmental assessment cannot yet be used to prevent the granting of a marketing authorization even where the assessment indicates that an environmental impact is likely,[1] although the applicant is expected to limit the environmental impact:

> *"The environmental impact should be assessed and, on a case-by-case basis, specific arrangements to limit it should be envisaged. In any event this impact should not constitute a criterion for refusal of a marketing authorisation."*

The current guidance document does not give any indication of what might be expected under these circumstances other than to provide information:[21]

> *"When the possibility of environmental risks cannot be excluded, precautionary and safety measures may consist of:*
>
> ➢ *An indication of potential risks presented by the medicinal product for the environment.*
>
> ➢ *Product labelling, Summary Product Characteristics (SPC), Package Leaflet (PL) for patient use, product storage and disposal.*
>
> *Labelling should generally aim at minimising the quantity discharged into the environment by appropriate mitigation measures."*

However, although the medicines regulators seem not to be too concerned at present about APIs that might have an adverse environmental impact, the information may also be used by environmental regulators to exert controls on the discharge of the material into the environment from manufacturing facilities and sewage treatment facilities (see Section 11.3).

There is also a growing expectation in the literature that pharmaceutical companies, being aware of the concern over pharmaceutical residues in the environment, will now be pursuing a strategy of "green by design".[24] In other words endeavoring to create new APIs that are designed to have lower environmental impact and, in particular, low persistence. This somewhat naïve expectation is mainly due to a lack of understanding of the difficulties involved; however, work is in progress in some companies to see if, and how, this issue might be addressed in the future.[25]

### 11.2.3 Greening the Process

In addition to environmental regulations specifically related to the pharmaceutical product itself, a range of regulations also apply to the manufacturing process by which the product is made. These seek to minimize the overall environmental impact of the manufacturing facility both during normal operations and under abnormal conditions. Historically, such regulation has been concerned solely with direct emissions from the facility to air and water. However, in recent years there has been a trend towards a more integrated approach encompassing the broader life-cycle of the manufacturing process. Most of these regulations are not specific to pharmaceutical manufacture, but they nevertheless can act as a driver to improve process design.

Pharmaceuticals are produced in relatively small quantities, from a few kg per year for some anticancer drugs to a few hundred tons per year for more widely used medicines, and a few thousand tons per year for some analgesics. This is in contrast to many bulk chemicals where 1000 tons per day production is common. Unlike most "bulk" chemicals, however, most pharmaceuticals are very complex organic molecules that have to be constructed using multiple synthetic steps, often involving the isolation of intermediate products. As a consequence, process efficiency has historically been very low.[26]

Driven by cost and sustainability issues, the research pharmaceutical companies have in recent years become industry leaders in the introduction of green chemistry and technology techniques into their process design. Companies have developed sophisticated systems to ensure that potential environmental consequences, as well as health and safety considerations, are taken into account in the selection of reagents and solvents.[27] Sustainability metrics are routinely used to compare alternative process routes.[28] This has led to major improvements in efficiency in these complex syntheses and pharmaceutical companies regularly win US Presidential Green Chemistry Challenge Awards.[29] The major companies are also now collaborating at the American Chemical Society Green Chemistry Institute Pharmaceutical Roundtable and sponsoring research that should lead to even more sustainable synthesis routes.[30]

The medicinal chemist has a key role to play in the design of the initial synthetic route to produce the target molecule. This route may sometimes be substantially modified as the drug moves on towards eventual manufacture and sale, but decisions taken in these early stages often have a substantial impact on the final manufacturing route. The medicinal chemist should therefore maintain an awareness of the changing regulatory environment that surrounds this activity.

Changing a manufacturing process after the drug has received marketing approval can be very difficult. For example, when a new drug enters the final phase of clinical testing, its synthetic procedure is locked into place by regulators such as the US FDA to ensure consistency. No changes can then be made to the procedure without going through a formal re-approval process. This means companies can be reluctant to make changes, even when a simple one, such as switching a solvent, can save a lot of money and reduce environmental impact. This has now been recognized by both FDA and industry as a major inhibition to moving towards sustainability and recently discussions have been taking place to identify what measures can be taken to alleviate this problem while maintaining patient safety and drug efficacy.[31]

As noted above, the pharmaceutical industry has been an early adopter of the principles of green chemistry;[32] however, medicinal chemists no longer have a completely free hand in selecting the process chemicals that they would like to use, and there is an increasing number of regulations applying to the management and use of chemical substances, which can restrict the options available still further.

### *11.2.3.1 Regulations Controlling Chemical Use*

There are two types of regulation that impact the use of chemicals:

- ➢ Firstly, there are those regulations seeking to restrict which chemicals should be available for use. This type of regulation is accelerating and has potentially damaging consequences. As more and more substances are restricted, the breadth of the chemical palette shrinks. Since we can never know all the possible uses of a substance this poses a problem for society; impose draconian restrictions on use and lose any future potential benefits, or attempt to manage all the risks posed by a substance, recognizing that some risks are not yet known. Only the latter approach retains the opportunity for society to gain those, as yet unknown, benefits.
- ➢ Secondly, there are those regulations which seek to ensure that chemicals are managed safely in a way which minimizes any adverse effects on both human beings and the environment. Although these can sometimes become unnecessarily bureaucratic, such regulations make a significant contribution to health, safety, and well-being, and to the protection of the wider environment.

Regulations are nearly always enforced at the national (or even local) level; however they can be agreed at the global, international, regional, or national level. Three major global treaties now exist that have a bearing on chemical use. The earliest is the 1987 Montreal Protocol on Substances that Deplete the Ozone Layer[33] (a protocol to the Vienna Convention for the Protection of the Ozone Layer). This treaty is designed to protect the ozone layer by phasing out the production of numerous substances that are considered to be responsible for ozone depletion. This regulation might seem esoteric but one of its consequences is, for example, that carbon tetrachloride, along with several other substances, is no longer available for use even as a laboratory solvent or reagent. The second treaty is the 2001 Stockholm Convention on Persistent Organic Pollutants whose objective is to minimize global contamination from persistent man-made substances.[34] The signatories to this treaty agree to end the production and use of the designated chemicals. Initially 12 compounds were identified for control, predominantly old agrochemicals. Since 2001 a further ten compounds have been added and other substances are currently under consideration. The majority of the compounds currently scheduled for elimination under this treaty are heavily halogenated organic materials, but in principle the treaty can be applied to any substance that is persistent, bioaccumulative, toxic, and capable of natural long range transport. The final treaty is the 2013 Minamata Convention, which seeks to eliminate human health and environmental problems related to exposure to mercury and mercury compounds.[35] These three treaties, the provisions of which are applied by national regulators, all seek to eliminate the manufacture and use of specific chemicals. The chemicals affected are therefore removed entirely from the chemical inventory and are no longer available to the medicinal chemist, however useful they might be.

The first systematic approach to the regulation, rather than the elimination, of chemical use was the introduction in the United States in 1976 of the Toxic Substance Control Act. This was followed by a series of similar laws in many countries including Australia, Canada, China, and Japan. The principal objective of these regulations was to ensure that relevant information was made available concerning the hazardous properties of chemicals in order that users and regulators could assess any risks to human health or the environment resulting from their use. Most systems adopted a two stage approach. An inventory of "existing" substances currently on the market was assembled, and after a set date all substances introduced into the market that were not on the inventory were declared to be "new substances. New substances needed to have a minimum package of hazard data before manufacture or use, and a system was put in place to ensure that data on "existing" chemicals would be retrospectively provided. This latter aspect of the regulations was usually only partially successful. A very large number of substances were involved and the costs and resources needed to generate the necessary data were substantial. As a consequence slow progress was made.

In the European Union a similar series of Directives and Regulations eventually culminated in the introduction in 2006 of the **R**egistration **E**valuation **A**uthorization and Restriction of **Ch**emical substances (REACH) Regulation.[10] However, in addition to requiring the provision of hazard data the REACH Regulation also transfers the "burden of proof" that a chemical can be used safely from the regulator to the manufacturer and also enables the competent authorities to directly control the uses to which substances may be put. A 'Restriction' can be applied preventing the substance being used for specified purposes or a substance may require an 'Authorization' which only allows the substance to be used for specified purposes.

Substances used in human medicines, including excipients, are exempt from many of the provisions of REACH including Authorization but they are not exempt from the Restriction provisions in REACH Title VIII. This enables the European Commission in principle to restrict the manufacture, marketing and use of *any* substance including pharmaceuticals:

> *"When there is an unacceptable risk to human health or the environment, arising from the manufacture, use or placing on the market of substances, which needs to be addressed on a Community-wide basis, Annex XVII shall be amended in accordance with the procedure referred to in Article 133(4) by adopting new restrictions, or amending current restrictions in Annex XVII, for the manufacture, use or placing on the market of substances on their own, in preparations or in articles, pursuant to the procedure set out in Articles 69 to 73. Any such decision shall take into account the socio-economic impact of the restriction, including the availability of alternatives."*
>
> EC Regulation 1907/2006, Article. 68.

This provision is unlikely to be used in practice but, if necessary, it could be used to restrict sales of any pharmaceutical which, post launch, was found to be posing an unacceptable risk to the environment.

Although substances used "in human medicines" are exempt from most of the REACH regulation, substances used "in the manufacture of human medicines" are covered by the regulation. Thus, pharmaceutical companies need to comply with the registration provisions of REACH for all the substances used in their manufacturing processes. They also need to take into account the stringent and potentially very time consuming procedures that will apply to substances on Annex XIV, *i.e.* those requiring 'Authorization'.

One of the objectives of REACH is that "substances of very high concern" should, in principle, only be used where there is no satisfactory alternative and where the socioeconomic benefit outweighs the potential damage to human health and/or the environment. Substances of very high concern are defined in Article 57 of the REACH Regulation and include carcinogens, mutagens, and reprotoxins together with substances that are "persistent, bioaccumulative and toxic" (PBTs) or "very persistent and very bioaccumulative" (vPvBs). There is also a "catch all" criterion included:

**REACH Article 57f** *Substances – such as those having endocrine disrupting properties or those having persistent, bioaccumulative and toxic properties or very persistent and very bioaccumulative properties, which do not fulfil the criteria of points (d) or (e) – for which there is scientific evidence of probable serious effects to human health or the environment which give rise to an equivalent level* *of concern to those of other substances listed in points (a) to (e) and which are identified on a case-by-case basis in accordance with the procedure set out in Article 59.*

A pharmaceutical company that wishes to use, or continue to use, a substance in its manufacturing processes that has been included in Annex XIV will have to seek authorization. It should not be difficult to demonstrate a positive socioeconomic analysis for a human medicine, but before being granted an "authorization", the applicant will have to demonstrate that all alternative substances and process routes have been investigated. Initial experience with the authorization procedures indicate that gaining authorization can be time consuming and involve expenditures >$1M. Any authorization that is granted is also time limited, which may require a series of applications. A prudent company therefore will be rigorously applying the principles of green chemistry in its process design and pursuing an active avoidance policy with respect to substances that might be candidates for Annex XIV. Several NGOs have already provided lists of substances, such as the ChemSec SIN List, which they consider are candidates for Annex XIV.[36]

Following the agreement in the European Union of the pioneering REACH Regulation and its successful initial implementation, there has been interest in other parts of the world in adopting similar strategies. In 2006 Canada amended its 1999 Canadian Environmental Protection Act by introducing a comprehensive Chemicals Management Plan.[37] In 2010, the Chinese government introduced major amendments to its existing chemicals control laws. "Measures for the Environmental Management of New Chemical Substances", came into effect, under the Ministry of Environmental Protection (MEP) Order No. 7.[38] This regulation adopts several of the same principles and concepts of the European regulation, and thus has been dubbed "China REACH". In 2013 the government of South Korea introduced the Registration and Evaluation of Chemicals Act, commonly known as "K-REACH".[39] K-REACH is a chemicals management system, based on the EU counterpart, which aims to "protect human health and the environment from the risk of the use of chemical substances".

Active consideration is also being given in the United States to fundamentally reform the 1976 Toxic Substances Control Act, and bipartisan legislation, the Chemical Safety Improvement Act, was tabled in 2013.[40] This legislation would grant additional powers to the regulator: firstly to prioritize chemicals according to their risk to human health or the environment, and secondly to take action against those deemed to pose the greatest risk. In the meantime some states, in particular California, consider that changes are urgently required and have begun to take independent action. Individual states in the USA have frequently been in the lead in seeking to eliminate chemicals deemed to be hazardous with 18 individual states being responsible for the introduction of 80 laws in the twenty-first century. In 2008, the state governor authorized two of the six bills making up the California Green Chemistry Initiative. This controversial legislation required the state Department of Toxic Substances Control to assess chemicals and prioritize the most toxic for possible restrictions or bans, and create an online toxics information clearinghouse with information about the hazards of thousands of chemicals used in California. The regulations were expected to motivate manufacturers of consumer products containing chemicals of concern to seek safer alternatives. It was intended to put an end to chemical-by-chemical bans and remove harmful products at the design stage. The Safer Consumer Products Regulations have already begun to identify a list of priority products that contain substances of concern and for which actions need to be taken.[41] Other US states such as Massachusetts, Vermont, and Minnesota are now considering similar initiatives.

Although the detailed provisions vary, all these twenty-first century laws are leading towards a situation where the national (or local) regulator can severely restrict or completely prevent the manufacture and use of substances deemed to be hazardous, thus removing them from the inventory of chemicals available to the medicinal chemist.

### *11.2.3.2 Avoidance Strategies and Substitutions*

Regardless of their scientific validity, it is clear that over the next few years we will see the emergence of an increasing number of lists of "priority substances" that have been identified as being of "concern" to human health and/or the environment. There is likely to be a considerable overlap of the listed substances. Some of these lists, like those in the Stockholm Convention[34] or Annex XIV of the EU REACH Regulation, will have an immediate impact on use and availability of the listed substances, whereas others will gradually increase public pressure for action by the regulatory authorities.[42]

Bringing a new pharmaceutical to the market is a time consuming exercise regularly taking >10 years from initial laboratory synthesis to full scale production. It is therefore necessary that substances that are going to be essential in the final manufacturing route should remain available throughout the patent life of the drug. There is consequently a need for the medicinal chemist to maintain an awareness of developments in chemical control and to follow an avoidance strategy in relation to substances of potential concern.

Chemists adopting green chemistry principles already regularly consult tools such as solvent selection guides to choose the most appropriate substances to use in synthetic routes. A similar approach needs to be taken with substances of concern. For example AstraZeneca maintains a Substance Avoidance Database:[25]

**Substance Avoidance Database:** *This tool lists all substances on relevant regulatory lists from around the world to highlight substances that should be avoided when developing manufacturing processes. For the most problematic substances, alternative choices are offered for the process development scientists. The database is continually updated.*

The emphasis here is on "avoidance"; in other words, the medicinal chemist should be actively seeking to avoid substances that have been flagged up as potential concern where that is practicable. Sometimes it is simply not possible to avoid the use of one or more of these substances but in that case the fact should be flagged to the appropriate staff in the company as early as possible so that the consequences and risks can be evaluated in a timely manner.

If a substance of concern can be replaced by another, then this substitution needs to be done with care to avoid replacing a substance currently on an "arbitrary" list of concern with an unlisted substance that actually has a worse human and/or environmental profile. Substitution is far from being a simple process; there are many traps for the unwary and, in some circumstances, substitution can produce a worst outcome than the *status quo*.[43] For example, in the 1960s, chlorinated hydrocarbon insecticides were phased out because of the detrimental effects that they were causing on raptorial birds.[44] However, the substances that replaced them, the organophosphate insecticides, were subsequently found to have significant long-term toxic effects on humans.[45] They were, in turn, replaced by the synthetic pyrethroid insecticides, which were then found to be responsible for major environmental problems because of their very high toxicity to aquatic life.[46]

### 11.2.4 Pilot Scale Manufacture

As will be obvious from above, R&D activities themselves are, by and large, subject to only light touch regulations. The main regulatory aspects of the work of the medicinal chemist therefore are in anticipation of the consequences for the product and process when they move out of the R&D department and into manufacturing, distribution, and sales. However, a "grey" area can exist between R&D and manufacturing in relation to pilot plants, *i.e.* are pilot plants a part of R&D and thus likely to be exempt from regulations that affect manufacturing or are they actually small-scale manufacturing facilities? This is a particular issue for pharmaceuticals where a full-scale plant for some highly potent drugs may be no bigger than the pilot plant itself.

In the UK, the regulators' original position was that pharmaceutical pilot plants were part of the production facilities and were being used to manufacture APIs for "commercial purposes". They therefore came within the scope of the EU IPPC Directive and its requirements which were then being implemented in the UK as the Environmental Permitting Regulations.[47] After long discussions, it was finally agreed that this was inappropriate.[48]

## 11.3 Manufacturing

The medicines regulators like FDA and EMA are frequently considered within the pharmaceutical industry to be the only regulators that have any significant influence on pharmaceuticals. However, there are several environmental regulators that are also now beginning to exert an increasing influence.

In most parts of the world the local or national regulator(s) responsible for the environment also have powers to limit the amount of material that may be released from a manufacturing process into the atmosphere or local watercourses. In some cases, where the factory effluent is discharged into a sewerage system, this control may be exerted indirectly *via* limits on the eventual release by the sewerage system operator.

In the United States a comprehensive document was published by the Environmental Protection Agency in 1998 outlining Best Available Treatment options and the regulatory treatment of effluents arising from pharmaceutical manufacture.[49] However, the most comprehensive control system is currently that provided in the European Union by the Directive on Industrial Emissions and this is becoming a model for the development of similar legislation across the world.[50]

The purpose of the Directive is to achieve integrated prevention and control of pollution arising from those industries considered to pose the largest risk to the environment (these activities are listed in Annex I of the Directive). It lays down measures designed to prevent or, where that is not practicable, to reduce emissions to the air, water, and land from these activities, including measures concerning waste, in order to achieve a high level of protection of the environment taken as a whole. Research and development is excluded from the provisions of the Directive, but pharmaceutical manufacturing is specifically included in Annex 1 of the Directive under Section 4.5 "Production of pharmaceutical products including intermediates."

The eight general principles are laid out in Article 11, which requires the Competent Authorities in each Member State to ensure that installations are operated in such a way that:

(a) all the appropriate preventive measures are taken against pollution;
(b) the best available techniques are applied;
(c) no significant pollution is caused;
(d) the generation of waste is prevented in accordance with Directive 2008/98/EC;
(e) where waste is generated, it is, in order of priority and in accordance with Directive 2008/98/EC, prepared for re-use, recycled, recovered or, where that is technically and economically impossible, it is disposed of while avoiding or reducing any impact on the environment;
(f) energy is used efficiently;
(g) the necessary measures are taken to prevent accidents and limit their consequences;
(h) the necessary measures are taken upon definitive cessation of activities to avoid any pollution risk and return the site of operation to a satisfactory state defined in accordance with Article 22.

The key principle from the perspective of the pharmaceutical sector is (a) supported by (d) and (e). In essence to obtain a permit to operate a manufacturing facility, the operator needs to be able to demonstrate that "best available techniques" (BAT) are in use for all the products being manufactured. Specific guidance on what is considered to be BAT for the manufacture of pharmaceuticals is provided by the European Commission.[51] This needs careful consideration by the manufacturer. Article 12.1(g) of the Directive makes it clear that the application for a permit, a document which is in the public domain, must include: *"the proposed technology and other techniques for preventing or, where this is not possible, reducing emissions from the installation."* In other words a simple statement that this process *is* BAT is not sufficient; the applicant must document both the rationale and a description of the alternatives that have been rejected. In addition, unlike its predecessor the Integrated Pollution Prevention and Control (IPPC) Directive,[52] all new manufacturing facilities must, at a minimum, meet the BAT standards contained in the BAT Reference Document[51] while existing facilities must be improved to meet this standard. The "quick and dirty" process is no longer acceptable in the twenty-first century; the production process must be the best available and this poses a challenge for the R&D chemists.

Unlike the majority of "bulk" chemicals, most pharmaceuticals are complex organic molecules that have to be constructed using multiple synthetic steps, often involving the isolation and purification of intermediate products. As a consequence process efficiency has historically been very low.[26] In recent years, driven by both cost and sustainability issues, the research pharmaceutical companies have become industry leaders in the introduction of green chemistry and technology techniques into their process design. The implementation of environmental legislation such as this directive provides a further stimulus to medicinal chemists to take a holistic view of their process design activities, trying to consider the sustainability aspects as well as the purity of the end product.

The permit granted to the operator by the competent authority will set out a range of conditions that must be met. These normally include limitations on what may be discharged into air and water. These limits can relate either to integrated parameters such as pH or biological oxygen demand (BOD) or to specific substances such as copper. Until recently competent authorities had not set specific limits on individual pharmaceutical active ingredients, but as the Industrial Emissions Directive[50] (and its predecessor the 1996 IPPC directive)[52] has come into full effect in the last few years a number of specific limits on APIs have started to appear.[53] With the growing interest, and public concern, about the presence of pharmaceutical residues in drinking water the number of specific limits imposed on the release of active ingredients from manufacturing facilities can be expected to increase.

Potentially of greater significance is the provision (Article 18) in the Directive that requires measures in addition to BAT in order to prevent an environmental quality standard from being exceeded:

> *"Where an environmental quality standard requires stricter conditions than those achievable by the use of the best available techniques, additional measures shall be included in the permit, without prejudice to other measures which may be taken to comply with environmental quality standards".*

An increasing number of precautionary environmental water quality standards are being introduced by Member States as part of the implementation of the EU Water Framework Directive (see Section 11.4) and this is likely to provide additional pressure to reduce emissions from pharmaceutical manufacturing facilities.

The 2010 Industrial Emissions Directive[50] also incorporates parts of the previous Solvent Emissions Directive.[54] The aim of this Directive was to prevent or reduce the effects of volatile organic compounds (VOCs) on the environment (mainly *via* the atmosphere) and reduce the potential human health risks from solvent-based activities. The new provisions, Annex VII Part 1.8, specifically relate to pharmaceutical manufacture:

> *"The chemical synthesis, fermentation, extraction, formulation and finishing of pharmaceutical products and, where carried out at the same site, the manufacture of intermediate products".*

Most of the requirements of the directive related to emission and inventory control. For example, in the pharmaceutical sector, total plant emissions for new installations should not exceed 5% of solvent use (<15% for existing installations). However, one part of the directive has potentially negative consequences in terms of green chemistry.

Article 58 requires that installations which use substances or preparations containing volatile organic compounds that are classified as carcinogenic, mutagenic, or toxic to reproduction, and which carry specified risk phrases, have to take steps to replace them, as far as possible, with less harmful substances and preparations within the shortest possible time.

This is a hazard based substitution requirement that does not allow risk (exposure) to be taken into account. As a consequence it may lead to the mandatory substitution of low risk medium hazard solvents by low hazard ones that pose a higher risk. For example, a proposal was recently considered by the European Chemicals Bureau (document no. ECBI/74/95-Add 3) for the classification and labelling of ethanol as a mutagen under the Dangerous Substances Directive (67/548/EEC).[55] The consequence of such a classification would be that under this Directive ethanol would need to be replaced in pharmaceutical manufacturing processes despite it being probably the safest and greenest of all organic solvents.

Thus pharmaceutical manufacturing facilities are becoming subject to increasingly stringent environmental regulations and chemists involved in process design need to be aware of both current constraints and those that are currently being discussed. This is important because the lead time for the introduction of new regulations can be shorter than the lead time required to bring a new pharmaceutical to the market.

One subject under active discussion at present is a possible change to the GMP regulations to incorporate requirements about the quality of effluent discharges. This is a response to information that appeared in 2007.[56] Until 2007 there was a general assumption that discharges of active pharmaceutical ingredients (APIs) from manufacturing facilities would be trivial.

This was because it was known that these substances were extremely valuable and therefore it was assumed that substantial efforts would be made to avoid any loss of material. However, this confuses the value of the material in the market, which is indeed very high, with the cost of production, which is often relatively small. Larsson and his coworkers[56] looked at the discharge of APIs from manufacturing plants in India and found that in some cases effluents contained mg $L^{-1}$ concentrations of individual APIs. Since then several other similar emissions have been identified in China and Norway.[57,58] Thus, although the effluent discharges from many pharmaceutical production plants are stringently controlled, this is by no means universal. Since national environmental regulations are at different stages of development, one solution to this problem is seen as amending the GMP regulations that apply to all pharmaceutical production facilities, to incorporate appropriate effluent discharge controls. This was first suggested in 2002 by Velageleti and Burns and is now being championed within the European Union by the Swedish Government.[59,60] If approved this would incorporate mandatory emission standards to the aquatic environment for individual pharmaceuticals within the GMP requirements.

## 11.4 Products and Patients

In addition to regulations governing research and development and manufacturing processes a series of increasingly stringent regulations also apply to general environmental quality which, although not aimed specifically at the pharmaceutical industry, may nevertheless have a significant secondary impact.

There has been, for many years, an assumption that pharmaceuticals have no detrimental effects in the environment at the concentrations currently being detected. This assumption remains true when considering acute responses, but may not always be true in the case of chronic responses to long-term exposure. The effects of ethinylestradiol on fish feminization[61] and the devastating impact of diclofenac on the Asian Vulture[62] demonstrate that significant impacts, although rare, are certainly possible. Consequently, it would not be surprising if environmental regulators wanted reassurance that the risk benefit calculations relating to pharmaceuticals had properly taken environmental risk into account, and to seek to regulate where environmental damage was being caused by exposure to pharmaceutical residues.

In 2000 the European Union introduced a new holistic framework directive on water quality[9] which subsumed several previous Directives into a single piece of legislation, the Water Framework Directive. The objective of this Directive, and the Marine Strategy Framework Directive which followed in 2008, is that all community waters should be of "good ecological quality".[63] This may have implications for pharmaceutical companies in a number of ways. There is unlikely to be any significant impact on point source discharges from manufacturing facilities within most parts of the European Union, since these have been under stringent control for many years. Although as quality standards become more stringent, additional measures may be needed at some locations. However, the release of pharmaceutical products from domestic wastewater treatment plants as a consequence of patient use may now become an issue in some areas.

To meet the requirements of the Water Framework Directive, each Member State has to evaluate its own water quality, identify why some areas do not reach the required standard and then implement improvement plans. If a pharmaceutical residue is perceived to be contributing significantly to poor water quality, it is probable that an environmental quality standard will be set for it. Compliance with this standard will then be needed which can only be accomplished by improving wastewater treatment or restricting sales. Draft water quality standards have already been proposed in Germany for carbamazepine (0.5 μg $L^{-1}$), diclofenac (0.1 μg $L^{-1}$), erythromycin (0.02 μg $L^{-1}$), ibuprofen (7.1 μg $L^{-1}$), and metoprolol (7.3 μg $L^{-1}$) and in the United Kingdom for 17α-ethinyloestradiol (0.0001 μg $L^{-1}$).[64,65]

In addition to this activity within individual Member States, Article 16 of the Water Framework Directive requires the European Commission to identify a list of "priority pollutants" for control by standards set at Community level. All emissions of a subset of this list, the "priority hazardous pollutants", need to be reduced to zero by 2015. At the present time no pharmaceuticals appear on this list of 45 priority pollutants.[66] However, during the negotiations, the European Parliament and several Member States argued strongly to have carbamazepine, clotrimazole, estradiol, ethinyloestradiol, and diclofenac added to the list. Although this was unsuccessful, three of these pharmaceuticals (estradiol, ethinylestradiol, and diclofenac) have been included in a new "Watch List" (Directive 2013/39 Article 8b).[66] This is intended to monitor and collect information on emerging pollutants from a network of monitoring sites across Europe, to assess their suitability for future inclusion in the priority substances list. Work is already in progress on the next revision of the Priority Substances list which is due to be published in 2017 and this is likely to contain several pharmaceuticals, including some (or all) of those on the current "Watch List". The consequences for patient safety and pharmaceutical sales could be significant.

Many countries also have stringent regulations governing the quality of drinking water. Although these increasingly place limitations on a range of micro-contaminants there are no countries at present that explicitly include pharmaceuticals in these listings. In 2011 the World Health Organization (WHO) carried out a review of pharmaceuticals in drinking water and concluded that the current levels of exposure to pharmaceuticals in drinking-water also suggest that the development of formal guideline values for pharmaceuticals in the WHO Guidelines for Drinking-water Quality is unwarranted.[67]

In Australia the Environment Protection and Heritage Council, the Natural Resource Management Ministerial Council, and the Australian Health Ministers' Conference have published, in connection with proposals for some water reuse projects, draft drinking water standards for 87 individual pharmaceuticals.[68] These are based on human toxicological data and range from 0.01 μg $L^{-1}$ to 35 mg $L^{-1}$.

In the United States contaminant levels in drinking water are regulated under the 1974 Safe Drinking Water Act. The Environmental Protection

Agency regularly reviews these contaminants and in 2008 they released a third draft Contaminant Candidate List for public review and comment.[69] As part of the process to develop the list, the Agency evaluated pharmaceuticals and personal care products to identify those that had the potential to occur in drinking water provided by public utilities. EPA considered 287 chemicals identified as pharmaceuticals and personal care products and eventually concluded one antibiotic (erythromycin) and nine hormones (17 alpha-estradiol, 17 beta-estradiol, equilenin, equilin, estriol, estrone, ethinylestradiol, mestranol, and norethindrone) should be included on the Contaminant Candidate List because they are known or anticipated to occur in public water systems and may require regulation.[70]

In the 1998 European Union Directive on Drinking Water Quality 23 individual chemical parameters have specific limits together with two group limits for polycyclic aromatic hydrocarbons and pesticides.[71] In the case of pesticides no individual pesticide is permitted to exceed 0.1 $\mu g\ L^{-1}$ and pesticides in total should not exceed 0.5 $\mu g\ L^{-1}$. These limits were a compromise, the original demand was for a "zero" limit for pesticides in drinking water and these standards were set, in 1998, as the effective analytical detection limit. The application of the Directive to pharmaceuticals was discussed when it was reviewed in 2008 but no pharmaceuticals were added.[72]

The environmental concentrations of pharmaceuticals are predominantly due to current use by patients and subsequent discharge *via* domestic effluent treatment plants. Domestic effluent treatment plants were not specifically designed to eliminate micropollutants including pharmaceuticals and thus it is not surprising that removal rates vary dramatically (from 0% to 100% removal) both between treatment plants and between different pharmaceuticals within treatment plants. In most cases the current environmental concentrations do not exceed the Environmental Quality Standards (EQS) currently being proposed, but this is not always the case. Where an environmental quality standard is being exceeded, there will be an expectation from the environmental regulator that action will be taken to reduce inputs in order to correct the situation.

The scope for corrective action is, however, limited and problematic. There are only two ways to reduce environmental inputs: improve the efficiency of the effluent treatment plant or reduce the use of the pharmaceutical. The costs of upgrading effluent treatment plants can be extremely large. In the UK the cost of upgrading the 1360 effluent treatment plants to remove sufficient ethinylestradiol to enable 100% compliance with the proposed EQS has been estimated as *ca.* €35b.[61] On the other hand, substitution of ethinylestradiol or restricting its use would have severe social and political consequences. Ethinylestradiol may be exceptional, in that its potency is very high, making it necessary to treat effluents with a very high efficiency, and its use is essential. In more typical cases, control on use, for example, returning an "over the counter" medicine to "prescription only" or advising medical practitioners to make more use of alternative therapies, might be plausible, although public outrage might be expected.

## 11.5 A Look to the Future

Benjamin Franklin said that nothing in life was certain apart from death and taxes, to which he might now have added "and increasingly restrictive regulation".[73] This is a natural outcome for a society that is becoming more risk averse as it becomes wealthier.

The current trend is for an increasing number of chemicals to be classed as "chemicals of concern". In many instances this is based solely on hazard information. This is then followed by draconian controls up to and including cessation of manufacture and use. Medicinal chemists will therefore be faced with increasingly stringent controls on what they can use and the development and use of avoidance databases will become essential.

Environmental risk assessment of pharmaceuticals will become more stringent and exhaustive. This will be exacerbated by increasing concerns about endocrine disruption, worries about "unknown unknowns" with respect to nanomaterials and the implications of mixtures and breakdown products. We can also expect major advances in our understanding of the interaction of pharmaceuticals with biological systems other than mammals, enabling more intelligent testing strategies to be devised.

It is likely that Environmental Quality Standards will be developed for more and more pharmaceuticals. These will be applied in Europe by individual member states such as Germany, Denmark, and Sweden and are likely, in a small number of cases, to become requirements of European Union legislation. This in turn will lead to more pressure to reduce the sales of some products so that they are only used by patients for whom the drug is essential and for which no satisfactory alternative exists.

Advances in the development of "green pharmaceuticals" will also take place, albeit somewhat slowly.[74] The biggest change is likely to be derived from the development of biopharmaceuticals which are inherently degradable. Improvements in the environmental degradability of conventional pharmaceuticals will take longer and is predicated on achieving a greater understanding of the relationship between chemical structure and degradability.

Further developments in transparency of the potential environmental impact of pharmaceuticals can also be expected. The research pharmaceutical industry has made a good start with the voluntary development of the Swedish Environmental Classification Scheme and this has generated much interest in other countries.[20,75,76]

## References

1. *Off. J.*, 2004, **L136**, 34–57.
2. *The Holy Bible*, King James Version, 1712, Deuteronomy, **23**, 13.
3. *12 RIC II*, HMSO, 1388, **c10–13**.
4. *Report of the UN Conference on the human environment (Stockholm)*, UNEP, 1972.
5. *Clean Air Act*, HMSO, 1956, **c52**.
6. *Off. J.*, 1973, **L347**, 51–52.

7. *The Comprehensive Environmental Response, Compensation and Liability Act*, 1980, United States Government, **CFR42**, 96–510.
8. *A Guide to the Globally Harmonized System of Classification and Labelling of Chemicals (GHS)*, OSHA, 2006.
9. *Off. J.*, 2000, **L327**, 1–72.
10. *Off. J.*, 2006, **L396**, 1–849.
11. I. Wrightson, S. J. Cooper, M. Crookes, N. King, P. Lewis, J. Larner, D. H. Lohmann, C. Maxwell, D. M. Sanderson, P. Jackson, S. Whitehead, C. Grundy, C. Watts, S. Lipworth and C. Dempsey, *Individual Legal Responsibilities for Health and Safety*, Royal Society of Chemistry, London, 2013.
12. I. Wrightson, S. J. Cooper, M. Crookes, N. King, P. Lewis, J. Larner, D. H. Lohmann, C. Maxwell, D. M. Sanderson, P. Jackson, C. Watts and S. Lipworth, *Individual Legal and Ethical Responsibilities for Environmental Protection*, Royal Society of Chemistry, London, 2011.
13. *Health and Safety at Work Act*, HMSO, 1974, **c37**.
14. *The Control of Substances Hazardous to Health Regulations*, HMSO 2002, Statutory Instrument, 2002 No 2677.
15. Health & Safety Executive, COSHH essentials: Easy steps to control chemicals, *Control of Substances Hazardous to Health Regulations (HSG193)*, HSE Books, London, 2nd edn, 2003.
16. *Off. J.*, 2012, **L197**, 38–71.
17. Recycling Mercury-Containing Light Bulbs (Lamps), US Environmental Protection Agency, 2014. http://www.epa.gov/osw/hazard/wastetypes/universal/lamps/.
18. I. Wrightson, S. J. Cooper, M. Crookes, P. Jackson, N. King, P. Lewis, J. Larner, D. H. Lohmann, C. Maxwell, D. M. Sanderson, C. Watts, S. Lipworth, *Waste Management in Chemical Laboratories*, Royal Society of Chemistry, London, 2012.
19. W. Hansen, M. Christopher and M. Verbuecheln, *EU Waste Policies and Challenges for Local and Regional Authorities*, Ecologic, Berlin, 2002.
20. *Pharmaceuticals in the Environment, Current knowledge and needs assessment to reduce pressure and impact*, ed. B. Roig, IWA Publishing, London, 2010.
21. *Guideline on the environmental risk assessment of medicinal products for human use.* European Medicines Evaluation Agency, London, 2007, Guidance Note EMEA/CHMP/SWP/4447/00.
22. *An Environmental Assessment Regime for New Substances in Products Regulated under the Food and Drugs Act. Health Canada – Options Analysis Paper.* Health Canada, 2005.
23. *Guidance for Industry: Environmental Assessment of Human Drug and Biologics Applications.* US Center for Drug Evaluation and Research, 1998, CMC 6 (Rev 1).
24. K. Kummerer, *Green Chem.*, 2007, **9**, 899.
25. AstraZeneca, *Product Environmental Improvement*, 2014, http://www.astrazeneca.com/Responsibility/The-environment/Product-environmental-improvement.
26. R. A. Sheldon, *Chemtech*, 1994, **24**, 38.

27. A. D. Curzons, D. C. Constable and V. L. Cunningham, *Clean Products and Processes*, 1999, **1**, 82.
28. C. Jiménez-González, D. J. C. Constable and C. S. Pondera, *Chem. Soc. Rev.*, 2012, **41**, 1485.
29. *Presidential Green Chemistry Challenge*, US Environmental Protection Agency, 2008.
30. J. M. Crow, *Chemistry World*, 2008. http://www.rsc.org/chemistryworld/News/2008/July/09070801.asp.
31. D. K. Leahy, J. L. Tucker, I. Mergelsberg, P. J. Dunn, M. E. Kopach and V. C. Purohi, *Org. Process Res. Dev.*, 2013, **17**, 1099.
32. P. T. Anastas and J. C. Warner, *Green Chemistry: Theory and Practice*, Oxford University Press, New York, 1998.
33. UNEP, *The Montreal Protocol*, 1987, http://ozone.unep.org/new_site/en/montreal_protocol.php.
34. UNEP, *The Stockholm Convention*, 2001, Web: http://chm.pops.int/default.aspx.
35. UNEP, *The Minamata Convention*, 2014, Web: http://www.mercuryconvention.org/.
36. ChemSec, *The SIN List Database*, 2013, Web: http://www.chemsec.org/what-we-do/sin-list/sin-database.
37. *Chemicals Management Plan*, Government of Canada, 2006.
38. *The Provisions on Environmental Administration of New Chemical Substances,* Government of China, 2010.
39. *The Act on the Registration and Evaluation of Chemicals*, Government of Korea, 2013.
40. US Government, Chemical Safety Improvement Act, 2013, Web: https://www.govtrack.us/congress/bills/113/s1009.
41. *Safer Consumer Products Regulations*, California Department of Toxic Substances Control, 2014.
42. *REACH Authorisation List,* European Chemicals Agency, Helsinki, 2014.
43. *A Guide to Substitution*, UK Chemicals Stakeholder Forum, London, 2010.
44. N. C. Fox and J. W. Lock, *N. Z. J. Ecol.*, 1978, **1**, 118.
45. *Organophosphorous Cumulative Risk Assessment*, 2006, US EPA Office of Pesticide Programs.
46. A. Croxford, Sheep dip: Impacts on Aquatic Life, 2005, Regional Fisheries Ecology and Recreation Advisory Committee report no. T/RFERAC/05.
47. *The Environmental Permitting (England and Wales) Regulations*, HMSO, 2010, Statutory Instrument 2010 No 675.
48. *Regulatory position statement Permitting the production of phase 1, 2 and 3 clinical trial pharmaceuticals*, UK Environment Agency, 2011, MWRP RPS 048 Version: 1.
49. *Development document for final effluent limitations guidelines and standards for the pharmaceutical manufacturing point source category*, US Environmental Protection Agency, 1998, EPA-821-R-98-005.
50. *Off. J.*, 2010, **L334**, 17–119.
51. *Reference Document on Best Available Techniques for the Manufacture of Organic Fine Chemicals*, European Commission, 2006.

52. *Off. J.*, 1996, **L257**, 26–40.
53. *Integrated Pollution Control License 673*, Irish Environmental Protection Agency, 2004.
54. *Off. J.*, 1999, **L851**, 1–22.
55. *Off. J.*, 1967, **L196**, 1–98.
56. D. G. L. Larsson, C. de Pedro and N. Paxeus, *J. Hazard. Mater.*, 2007, **148**, 751.
57. D. Li, M. Yang, J. Hu, L. Ren, Y. Zhang and K. Li, *Environ. Toxicol. Chem.*, 2008, **27**, 80.
58. K. Thomas and K. H. Langford, *Green & Sustainable Pharmacy*, ed. K. Kummerer and M. Hempel, Springer, Berlin, 2010, 14, pp. 211–224.
59. R. Velagaleti, P. K. Burns, M. Gill and J. Prothro, *Environ. Health Perspect.*, 2002, **110**, 213.
60. *Platform to enable the initiation of a revision of EU legislation on Good Manufacturing Practice, GMP, in order for legislation also to comprehend environmental considerations*, Report from the Swedish Medical Products Agency, 2011.
61. R. Owen and S. Jobling, *Nature*, 2012, **485**, 441.
62. R. Cuthbert, M. A. Taggart, V. Prakash, M. Saini and D. Swarup, *PLoS ONE*, 2011, **6**(5), e19069.
63. *Off. J.*, 2008, **L164**, 19–40.
64. *ETOX: Information System Ecotoxicology and Environmental Quality Targets*, Umweltbundesampt, Berlin, 2008. http://webetox.uba.de/webETOX/public/search/ziel.do.
65. S. Jobling and S. Owen, *Late Lessons from Early Warnings II*, European Environment Agency, Copenhagen, 2013, pp. 278–307.
66. *Off. J.*, 2013, **L226**, 1–17.
67. *Pharmaceuticals in Drinking Water*, World Health Organization, 2011, WHO/HSE/WSH/11.05.
68. *Australian Guidelines for Water Recycling – Augmentation of Drinking Water Supplies*, 2007, ISBN 1 921173 20 3.
69. *Federal Register*, 2008, **73**, 9627–9654.
70. *Federal Register*, 2009, **74**, 51850.
71. *Off. J.*, 1998, **L330**, 32–54.
72. C. Jørgensen, B. H. Buchardt, J. Fawell and O. Hydes, *Preliminary draft final report on Establishment of a list of chemical parameters for the revision of the Drinking Water Directive*, European Commission, Brussels, 2008, ENV.D.2/ETU/2007/0077r.
73. B. Franklin, *in a letter to Jean-Baptiste Leroy, 1789*, in *The Works of Benjamin Franklin*, ed. J. Bigelow, Putnams & Sons, New York, 1904.
74. D. Taylor, *Green & Sustainable Pharmacy*, ed. K. Kummerer and M. Hempel, Springer, Berlin, 2010, 7, pp. 105–126.
75. B. Mattson, I. Näsman and J. Ström, *RAJ Pharma*, 2007, 153.
76. *Pharmaceuticals for Human Use: Options of Action for Reducing the Contamination of Water Bodies A Practical Guide*, ed. F. Keil, Institute for Social-Ecological Research (ISOE), Frankfurt, 2008.

CHAPTER 12

# *Patent Protection and Green Chemistry Innovation*

ANDREW D. COHEN*[a] AND ANDREW V. TRASK*[b]

[a] Patterson Belknap Webb & Tyler LLP, 1133 Avenue of the Americas, New York, NY 10036, USA; [b] Williams & Connolly LLP, 725 Twelfth Street NW, Washington, DC 20005, USA
*Email: acohen@pbwt.com; atrask@wc.com

The brief overview given in this chapter is no substitute for the advice and recommendation of a lawyer. The goal of this chapter, rather, is to provide the pharmaceutical scientist with a basic overview of the field, such that the scientist may communicate more knowledgeably with his or her lawyer when evaluating the potential patentability of a green chemistry invention.

## 12.1 An Overview of Patent Law and Incentives for Green Chemistry Innovation

This section provides a summary of the requirements for obtaining a patent, an overview of the U.S. patent system, and a discussion of incentives, both inside and outside the patent system, for green chemistry innovation.

### 12.1.1 Requirements for Patenting

Patents are permitted on a variety of inventions, including "any new and useful process, machine, manufacture, or composition of matter, or any new and useful improvement thereof."[1] In practice, most inventions fit into one

RSC Drug Discovery Series No. 46
Green Chemistry Strategies for Drug Discovery
Edited by Emily A. Peterson and Julie B. Manley

Published by the Royal Society of Chemistry, www.rsc.org

of these statutory categories of patentability.[2] In the pharmaceutical industry, it is common to seek patent protection on compositions of matter – *e.g.*, new molecular entities. But other aspects of the pharmaceutical development process may also involve patentable innovation. Targets for potential patent protection include new synthetic routes or reagents, salt forms and solid forms of an active pharmaceutical ingredient, pharmaceutical formulations, drug combinations, methods of using a drug to treat a particular disease, and so on.[3] Although most of those categories could entail opportunities for improvements in sustainability, this chapter will focus primarily on green chemistry improvements to the synthesis or solid form selection of pharmaceutical compounds. Both instances may present opportunities for innovation and patent protection even if the structure of a pharmaceutical compound has already been disclosed – provided the invention meets all the requirements of patentability. For instance, it is well established that a new method of synthesizing a known compound may be patentable.[4] And a new solid form of a known compound may, in some circumstances, be separately patentable.

Whatever type of invention is sought to be patented, it must satisfy the prerequisites for patentability. One such requirement is novelty.[5] If an invention previously has been disclosed to the public, such as through a publication or public use, it fails the novelty requirement. Lack of novelty also bars patentability if an invention has been "inherently" disclosed. For example, a previously unknown metabolite may not be patented if a method of administering its parent compound has previously been disclosed, and that method necessarily resulted in the creation of the metabolite in the patient's body.[6]

Another criterion for patenting is nonobviousness.[7] Even if an invention is novel, it may not be patented if, at the time the patent application was filed, it would have been obvious to someone possessing an ordinary level of skill in the field of the invention (who is sometimes called the "skilled artisan"). The obviousness inquiry involves evaluating several factors.[8] One factor is the differences between the claimed invention and the "prior art" – *i.e.*, the body of knowledge publicly available at the time the patent application was filed. In the case of improved synthetic methods, a court or patent examiner may evaluate the differences between the known method of synthesis and the improved method for which patent protection is sought. Such differences may include the number of synthetic steps or the particular reagents employed. Generally speaking, the greater the differences between the prior art and the claimed invention, the more this factor will weigh in favor of patentability.

Even if the differences between the claimed invention and the prior art appear slight, strong "objective evidence" may demonstrate nonobviousness. Courts have identified various categories of potentially relevant objective evidence. Among these, evidence of unexpected results can be especially probative of whether a pharmaceutical compound or process would have been obvious to the skilled artisan. As the case law explains, "that which would have been surprising to a person of ordinary skill in a particular art would not have been obvious."[9] A new method of synthesis, for example, may possess unexpectedly superior properties such as improved yield, reduced quantities of byproducts, or improved purity levels. Such findings may support the nonobviousness of a new synthetic process even if the evidence was not discovered until after the patent application was filed.[10] Moreover, a new solid form of a known pharmaceutical compound may possess properties including improved filtration or handling, isolation from more sustainable solvents, or improved crystallization yield, which may provide both environmental benefits and supporting evidence of nonobviousness.[11]

New solid forms may also provide benefits that lead less directly to a green chemistry result. Different solid forms of a pharmaceutical compound may possess different dissolution rates and thus different bioavailability. Improved bioavailability may mean that less compound is needed to elicit the same pharmaceutical effect, which in turn may mean that less compound must be manufactured, less waste will be generated, and less compound and its metabolites will be excreted. Improved physical properties may also lead to greater ease in formulating a drug product and, consequently, less waste. Each of these results carries implications for sustainability.

Apart from novelty and nonobviousness, a patent application must also adequately describe the invention and enable a person of ordinary skill to use it without having to engage in undue experimentation.[12] Although these and other requirements for patentability are not a focus of this chapter, each criterion for patentability must be satisfied before a patent may issue covering an invention.

### 12.1.2 The United States Patent System

Although by no means a comprehensive description of the U.S. patent system, the following overview aims to contextualize the ensuing discussion of green chemistry patents and court decisions involving issues of relevance to green chemistry.

Before a patent may issue on an invention, a patent application must be filed and reviewed by an examiner at the U.S. Patent and Trademark Office (USPTO). The application includes a written description, which provides details on the background and various aspects of the invention. The application concludes with claims, which serve to define the invention that the applicant seeks to patent. The examiner reviews the claims in view of the various requirements for patenting discussed above. If the examiner concludes that the claims do not meet one of the requirements for patenting, a written rejection is issued. The applicant (or the applicant's agent or attorney) may then respond to the rejection. A response may include arguments why the invention is patentable, a declaration by an inventor or another individual in support of the patentability of the claims, and/or an amendment of the claims. Because the back-and-forth between the examiner and the applicant may continue repeatedly, it is not uncommon for a patent to issue many years after the initial application was filed. Ultimately, if the applicant succeeds in overcoming the examiner's rejections, a patent will issue on the claimed invention.[13]

Once the patent issues, the patent owner may file suit in a federal district court[14] against a defendant who is allegedly infringing the patent. The defendant may allege in response that the patent is not infringed. In addition, or in the alternative, the defendant may allege that the patent is invalid. An invalidity challenge asserts that the patent does not meet one of the criteria for patentability discussed above. Thus, in some cases, the defendant may ask the court to repeat some of the same inquiries performed by the patent examiner prior to issuance of the patent. In deciding a case, a judge or jury finds the facts, and the court resolves legal issues by taking into account the patent statutes and relevant prior court decisions dealing with similar issues. If the court determines that the patent claims are not infringed or are invalid, then the defendant may avoid liability for patent infringement. If, instead, the court determines that the defendant is infringing a valid patent, the defendant may be subject to money damages for infringing conduct and/or an injunction against further infringement.

Special laws govern patent litigation between generic and brand pharmaceutical companies in connection with a generic's attempt to enter the market prior to the expiration of the brand's patent(s) covering a particular product. The Drug Price Competition and Patent Term Restoration Act of 1984, commonly known as the Hatch–Waxman Act, created this specialized form of patent litigation.[15] Its principal *quid pro quo* was the institution of an Abbreviated New Drug Application (ANDA) procedure, by which generic applicants could make use of a brand company's clinical data concerning safety and efficacy by demonstrating that the generic product is bioequivalent to the brand product. In exchange, brands received a new chemical entity (NCE) statutory exclusivity period, within which period the FDA could not approve a generic product. Additionally, the Hatch–Waxman Act set up a litigation process through which an ANDA filer could challenge a brand company's patent before launching its generic product on the market. The details of the Hatch–Waxman Act are beyond the scope of this chapter, but suffice it to say that under this regime patent litigation is now a significant aspect of a pharmaceutical company's business strategy, on both the brand and the generic sides.

The litigation process, whether under the Hatch–Waxman Act or otherwise, can be complex, costly, and time-consuming. Years may pass from the time that a patent infringement action is filed until the issuance of a district court's final judgment. Even after a judgment is obtained, the losing party may challenge that judgment on appeal, typically at the United States Court of Appeals for the Federal Circuit. Located in Washington, D.C., the Federal Circuit has specialized jurisdiction to review district court judgments in patent cases filed anywhere in the country. Although a decision by the Federal Circuit often marks the conclusion of a patent case, the losing party may petition the United States Supreme Court for further review.

The Supreme Court, however, has discretion over which cases it decides, and it typically reviews only a handful of patent cases per year.

### 12.1.3 Alternative Incentives for Green Chemistry Technology

Green chemistry principles offer certain inherent incentives for innovation, including both economic and moral. As to the former, many of the advancements in green chemistry result in lower waste or increased yield, often reducing costs. And as to the latter, the prospect of reducing the environmental impact of chemical processes is incentive itself for green innovations. To encourage innovation, various organizations and policy makers have established other means to incentivize innovations in green chemistry technologies.

Government organizations have offered certain incentives, such as prizes, to encourage innovation in green chemistry technologies. For example, in the United States, beginning approximately 20 years ago, the Environmental Protection Agency's (EPA's) Office of Chemical Safety and Pollution Prevention has sponsored the Presidential Green Chemistry Challenge Awards in partnership with the American Chemical Society Green Chemistry Institute. The award encompasses three focuses: greener synthetic pathways, greener reaction conditions, and the design of greener chemicals. Since the Award's inception in 1996, the EPA has received roughly 1500 nominations, and the 93 technologies that have won the award through 2013 have alone eliminated nearly 1 billion pounds of hazardous chemicals and solvents, saved 21 billion gallons of water, and eliminated environmental emission of nearly 8 billion pounds of carbon dioxide equivalents.[16] Similarly, in Japan, various chemical organizations have come together to form the Green & Sustainable Chemistry Network, which promotes innovation in green chemistry in part through the Green & Sustainable Chemistry Award. And in Europe, the European Association for Chemical and Molecular Sciences established the European Sustainable Chemistry Award. These public recognitions for advancements in green technology provide further incentives for innovation.

Similar programs in the private sector also foster such innovation. For example, Johnson & Johnson has established a program by which it incentivizes product teams to innovate greener products. The program, called Earthwards™, provides intracompany recognition for greener products and the teams that developed them.[17] Several products developed by Janssen – a Johnson & Johnson pharmaceutical company – have received Earthwards recognition for advancements in green chemistry that contributed to higher yield and lower waste. Pfizer, the first pharmaceutical company to appoint a full-time head of Green Chemistry, has established a number of internal incentives for green chemistry innovation.[18] Pfizer's Site Green Chemistry Award, first awarded in 2002, recognizes an individual employee or team for a significant improvement in any aspect of green chemistry technology. And Pfizer's Sustainability Science Award, ongoing since 2007, is given to three employees each year for advancements with positive environmental implications. Other pharmaceutical companies have developed similar awards to promote advancements in sustainability science.

The patent system itself also provides an incentive for innovation. In exchange for a public disclosure of how to make and use one's invention, an inventor is granted a temporally limited right to exclude others from making and using the invention. Beyond that, however, one way to incentivize green chemistry technology innovation is by offering access to quicker grants of patents. In late 2009, the USPTO established such a program, the Green Technology Pilot Program. Under the Pilot Program, the USPTO would accelerate the processing of patent applications directed to green technology, considerably shortening the approximately-three year time for examining a patent application.[19] Because a shorter waiting period for patent protection could provide significant strategic advantages, the USPTO hoped to incentivize innovation in green technology – including green chemistry – through this program.[20] Initially scheduled to last for one year, the program was successful enough to warrant an extension through the end of 2011,[21] and then by an additional three months through the end of March 2012. When the Pilot Program ultimately closed, the USPTO had granted over 3000 petitions for treatment under the program, and had issued over 1000 patents.[22] Indeed, the Pilot Program closed not because it was unsuccessful, but because the USPTO established a prioritized examination program extending to all technologies that offered examination of patent applications on an even faster track.[23]

## 12.2 Case Studies

As activity in the field of green chemistry grows, examples of the patentability of such technology continue to emerge. This section provides examples of patents covering green chemistry inventions with potential application to the pharmaceutical industry, as well as descriptions of some court decisions with potential implications for pharmaceutical green chemistry innovation.

### 12.2.1 Examples of Issued Patents Covering Green Chemistry Inventions

#### *12.2.1.1 A Greener Pinner Reaction*

U.S. Patent No. 6,806,380 ("the '380 patent") is entitled "Modified safe and efficient process for the environmentally friendly synthesis of imidoesters." Imidoesters are building blocks for the synthesis of heterocycles, which are themselves useful in the preparation of pharmaceutical compounds. Imidoesters are typically prepared *via* the Pinner reaction, involving bubbling an excess of hydrochloric acid gas through an anhydrous solution of a nitrile and an alcohol (Scheme 12.1). Of course, the use of excess hydrochloric acid

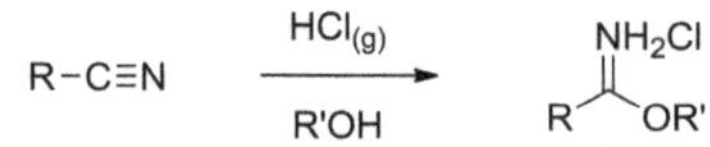


**Scheme 12.1** The Pinner reaction.

gas is neither safe nor environmentally friendly, leading the inventors of the '380 patent to develop a method of conducting the Pinner reaction using an anhydrous solution of hydrochloric acid.[24]

During prosecution, the patent examiner rejected the application, asserting that the invention was obvious. Specifically, though the examiner found no single prior art reference that modified the Pinner reaction to replace gaseous hydrochloric acid with a solution of hydrochloric acid, the examiner asserted that it would have been obvious to the skilled practitioner to substitute a solution of hydrochloric acid in the well-known Pinner reaction. To overcome this rejection, the applicants submitted a declaration from a scientist addressing long-felt need as objective evidence of non-obviousness. The scientist explained that at the time of the invention there existed a long-felt need to overcome the drawbacks of the traditional Pinner reaction, including the adverse environmental impact of using gaseous hydrochloric acid as well as the need for extensive purification of the byproduct. The examiner agreed that the "applicant has demonstrated that a modification of the Pinner process – along the lines of the instant invention – was not contemplated by other practitioners in the art as a practical solution to the drawbacks of using HCl (g) in said process." The issuance of the '380 patent highlights that a new process overcoming well-known drawbacks of an older process – including adverse environmental impacts – may be patentably nonobvious.

### *12.2.1.2 Improved Syntheses of Sitagliptin*

In 2006, Merck won the Presidential Green Chemistry Challenge for an improved synthesis of sitagliptin, the active ingredient in the diabetes drug Januvia™. The award recognized that the "highly innovative and efficient catalytic synthesis for sitagliptin … creates 220 pounds less waste for each pound of sitagliptin manufactured and increases the overall yield by nearly 50 percent."[25] The centerpiece of this synthesis was the discovery "that hydrogenation of unprotected enamines using rhodium salts of a ferrocenyl-based ligand as the catalyst gives β-amino acid derivatives [*e.g.*, sitagliptin] of high optical purity and yield," allowing Merck to avoid the use of costly and wasteful protecting groups (Scheme 12.2). During examination of the U.S. patent application covering this invention, the patent examiner found the use of such catalysts in the reaction to be novel and nonobvious.[26] Accordingly, Merck was awarded U.S. Patent No. 7,468,459.

In 2010, Merck (in collaboration with Codexis) again won the Presidential Green Chemistry Challenge for a further improvement of the sitagliptin synthesis that eliminated the need for a metal catalyst altogether.

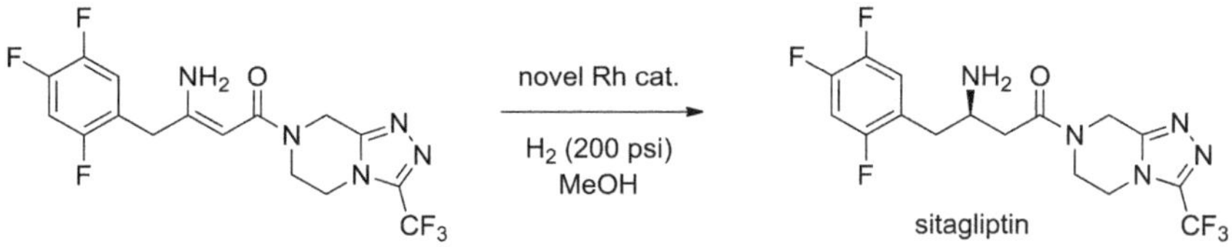


**Scheme 12.2** Merck's asymmetric catalytic hydrogenation of a β-amino acid sitagliptin precursor.

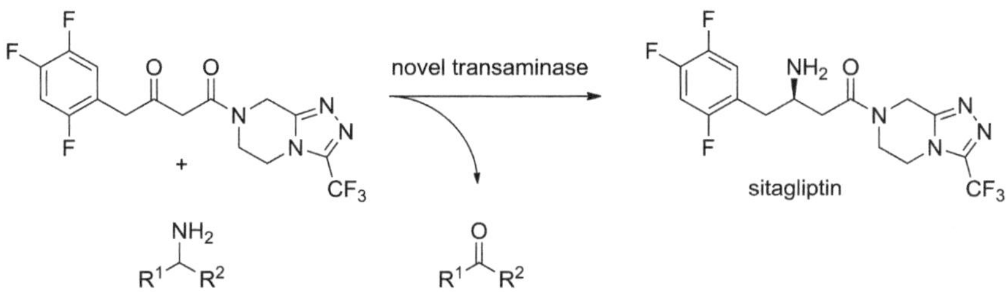


**Scheme 12.3** Merck's improved asymmetric hydrogenation of a sitagliptin precursor using a novel transaminase.

Specifically, Merck and Codexis developed a transaminase biocatalyst that asymmetrically hydrogenated the enamine precursor of sitagliptin (Scheme 12.3). "The streamlined, enzymatic process eliminates the high-pressure hydrogenation, all metals (rhodium and iron), and the wasteful chiral purification step. The benefits of the new process include a 56 percent improvement in productivity with the existing equipment, a 10–13 percent overall increase in yield, and a 19 percent reduction in overall waste generation."[27] Codexis was awarded a patent on the transaminase enzyme itself because the patent examiner found it, with its specific mutations, to be novel and nonobvious.[28]

These patents demonstrate the availability of patent protection for green chemistry innovation involving the preparation of novel reagents leading to an improved process.[29]

### *12.2.1.3 Synthesis of Sertraline*

Sertraline, the active ingredient in Pfizer's antidepressant Zoloft™, is a methylcyclohexylamine. A critical intermediate in the preparation of sertraline is the corresponding imine, which is typically made by the condensation of a tetralone with methylamine in the presence of a dehydration agent such as titanium tetrachloride. The inventors of U.S. Patent No. 6,232,500 ("the '500 patent") discovered a greener method for preparing the key sertraline intermediate without using a dehydration agent such as titanium tetrachloride (Scheme 12.4). This modification both avoided the safety and environmental drawbacks of using such a dehydration agent and

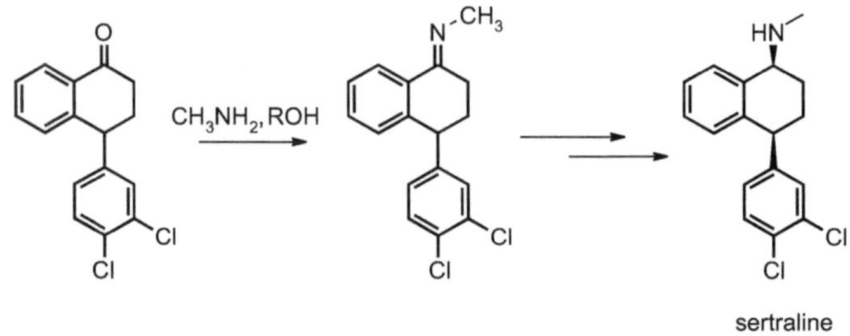


**Scheme 12.4** Pfizer's synthesis of sertraline.

eliminated the need for removal of byproducts (*e.g.*, titanium dioxide or spent molecular sieves) from the reaction mixture. By performing the condensation in particular alcohol solvent media, the inventors were able to shift the condensation equilibrium towards the imine.

As noted during prosecution of the patent application, "[u]se of the alcohol system, as claimed, allows 'one pot' transformation (*i.e.*, naphthaleneone → ketimine → amine), whereas the person of ordinary skill would have expected the complete transformation to involve isolation of each intermediate compound, as well as the use of multiple solvents (aprotic solvents for the ketimine formation; aprotic or protic solvents for the amine transformation)."[30] An inventor explained in a declaration submitted during prosecution that the prior art did not appreciate that such a reaction could be driven to near completion without the use of a dehydrating reagent by relying on the differential solubility characteristics of tetralone and the ketimine in alcohol solvents.[31] The examiner ultimately agreed and allowed the '500 patent to issue.

Based on the advancement claimed in the '500 patent, Pfizer received the Presidential Green Chemistry Award in 2002. "The new process doubles overall product yield, reduces raw material use by 20–60 percent, eliminates the use or generation of approximately 1.8 million pounds of hazardous materials, reduces energy and water use, and increases worker safety."[32]

### *12.2.1.4 New Solid Forms of Pharmaceutical Compounds*

New solid forms of a pharmaceutical compound may possess a unique profile of physical, chemical, or biological properties, and thus may offer advantages with relevance to the field of green chemistry. New solid forms may also offer opportunities from the standpoint of patentability. This section describes these advantages and opportunities with reference to examples of patented pharmaceutical solid forms.

When incorporated into a solid dosage form, a small-molecule pharmaceutical compound generally may exist as a crystalline or amorphous material.[33] Although each case depends on its own facts, new solid forms of a drug compound sometimes merit patent protection.[34] Patents have been obtained on polymorphs, solvates, and amorphous forms of known pharmaceutical compounds. For example, U.S. Patent No. 6,268,489 claims crystalline azithromycin dihydrate, and U.S. Patent No. 7,309,782 claims crystalline azithromycin monohydrate hemi-ethanol solvate. Amorphous forms have also been patented, such as the amorphous form of rifaximin claimed in U.S. Patent No. 7,709,634. Solid forms of active pharmaceutical ingredients may offer potential opportunities for green chemistry improvements. A new solid form, for example, may allow for a more efficient isolation of a final product, improved drying or handling characteristics, or even an ability to use less active ingredient.

Pharmaceutical co-crystals represent another potential opportunity for patentable green chemistry innovation.[35] Co-crystals are solid forms containing an active compound together with a pharmaceutically acceptable co-crystallizing agent.[36] One well-studied class of pharmaceutical co-crystals involves itraconazole, an antifungal drug.[37] Itraconazole was known to be extremely water insoluble in its crystalline free-base form. Itraconazole was therefore marketed as an amorphous form as a way of improving its low water solubility. The amorphous dosage form of itraconazole, however, required administration of up to two capsules three times daily – a burdensome regimen for patients.[38]

Researchers set out to identify improved solid forms with increased bioavailability and absorption, which could decrease the per-dose pill count as well as frequency of dosing. As part of this effort, they screened for new co-crystal forms of itraconazole. The screening revealed a series of co-crystals of itraconazole with pharmaceutically acceptable carboxylic acids, including fumaric acid, succinic acid, L-malic acid, and the different isomers of tartaric acid. A number of these co-crystals demonstrated favorable dissolution rates, ranging from four- to twenty-fold higher than the crystalline itraconazole free base. The dissolution of the L-malic acid co-crystal with itraconazole "rivals the dissolution of the commercial product containing amorphous [itraconazole]."[39] An enhanced dissolution rate can improve absorption and bioavailability, potentially reducing the dosage of the drug. Some of the co-crystals also offered processing advantages; for example, the co-crystal of itraconazole with succinic acid is "particularly easy to filter and dry to a free-flowing powder."[40] Such properties may offer green chemistry advantages, for example by reducing the manufacturing burden or simplifying processing or formulation of the active ingredient.

Co-crystals of itraconazole have been described and claimed in issued patents. For example, U.S. Patent Nos. 7,078,526 and 7,446,107 claim compositions consisting of multicomponent pharmaceutical crystal systems involving itraconazole. These crystal systems include a hydrochloric acid salt of itraconazole that is combined with tartaric acid to form a tartaric acid co-crystal of the itraconazole hydrochloride salt. This co-crystal was claimed using a characteristic differential scanning calorimetry endothermic transition temperature or characteristic X-ray powder diffraction peaks.[41]

Patent protection has been obtained on other pharmaceutical co-crystal materials. For example, U.S. Patent No. 7,927,613 claims co-crystals of carbamazepine with saccharin, celecoxib with nicotinamide, 5-fluorouracil with urea, and aspirin with 4,4′-bipyridine. At least some of these patented co-crystals appear to have properties conferring potential green chemistry advantages; for example, as the patent discloses, Form II of the 1:1 carbamazepine : saccharin co-crystal has improved solubility over previously known forms of carbamazepine, including increased molar solubility and longer solubility in aqueous solutions.[42] As noted above, improved solubility may mean that less drug compound needs to be administered to achieve a comparable therapeutic effect.

## 12.2.2 Court Decisions with Potential Relevance to Green Chemistry Inventions

To date, there are not many court decisions dealing directly with issues of green chemistry in the pharmaceutical industry. This is likely due in part to the fact that green chemistry is still developing in its importance to the industry, and, as noted above, it can take many years from the filing of a patent application to the issuance of a court decision in a patent infringement case. Nevertheless, in deciding patent cases, courts regularly rely on cases involving analogous facts or related fields of science or technology. Accordingly, this section focuses on case law that, if not directly dealing with issues in the field of pharmaceutical green chemistry, is at least pertinent to issues that may arise in that context.

### *12.2.2.1 Patentability of an Improved Chemical Process*

One case with potential relevance to patenting green chemistry technology is *Ecolochem, Inc. v. Southern California Edison Co.*[43] At issue in *Ecolochem* was a patent covering an improved process for purifying water for use in nuclear power plants. The process involved removing dissolved oxygen from water by adding hydrazine and passing the mixture through activated carbon, thereby catalyzing a reaction between the oxygen and hydrazine. The process further entailed passing the water through a mixed-bed ion-exchange resin to remove dissolved contaminants introduced by the activated carbon.

One issue in the case involved whether Ecolochem's patented process was obvious in view of a prior art process for deoxygenation that also used hydrazine but did not involve the use of an ion-exchange resin. The Federal Circuit held that the process was nonobvious and therefore patentable. The court concluded that the prior art taught away from using the patented deoxygenation process, describing alternative methods as energy saving and significantly less expensive. As the court explained, the author of one of the key prior art references testified that a hydrazine-based process was understood at the time as unfavorable, as it led to contaminated deoxygenated water. The court further found no motivation to combine a hydrazine-based process with a mixed-bed ion-exchange resin.

The fact that the patented method overcame great skepticism in the field underpinned the Federal Circuit's decision to uphold the patent covering Ecolochem's deoxygenation process. As the court explained, "the process not only worked, but worked better than expected."[44] Other factors lent additional support to the court's conclusion of nonobviousness. For example, the court concluded that the invention enjoyed substantial commercial success, due in part to the improved filtration properties claimed in the patent.[45] The court also noted that the invention was both acclaimed and copied by others in the field after it was released.[46]

*Ecolochem* provides an important illustration of criteria that may render a new and improved chemical process patentable over a prior art process.[47] In this respect, *Ecolochem* may have relevance to the field of pharmaceutical green chemistry. The factors that the Federal Circuit considered in determining the nonobviousness of the invention at issue in *Ecolochem* are factors that may be considered in connection with the patentability of a new synthetic route or manufacturing process for a new pharmaceutical compound.

### *12.2.2.2 Patentability of Improved Pharmaceutical Solid Forms*

The potential to obtain patent protection for a new solid form with improved green chemistry properties is illustrated by the case of *Glaxo Inc. v. Novopharm Ltd.*[48] This case involves ranitidine hydrochloride, the active ingredient in Glaxo's popular antiulcer medication Zantac™.

In the first few years of its development, ranitidine hydrochloride (Figure 12.1) was produced in only one solid form, later known as Form 1. The patent claiming the compound ranitidine hydrochloride, U.S. Patent No. 4,128,658 (the '658 patent) contained Example 32, in which the hydrochloride salt of ranitidine was generated in denatured alcohol and recrystallized *in situ* upon addition of ethyl acetate as Form 1. The method of making ranitidine hydrochloride was modified over time, and in the midst of clinical trials a new polymorph, Form 2, was discovered. Form 2 had better filtration and drying characteristics than Form 1, making it better suited for commercial applications. Glaxo therefore proceeded with commercialization of Form 2.[49] Glaxo obtained a patent covering Form 2, U.S. Patent No. 4,521,431 (the '431 patent), which had a later expiration date than the '658 patent claiming ranitidine hydrochloride.

Years later, upon expiration of the '658 patent, Novopharm sought to market a generic product containing ranitidine hydrochloride Form 2. Novopharm contended that the '431 patent, which had not yet expired, was

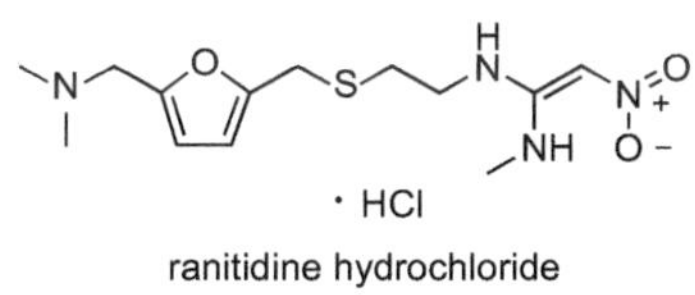


**Figure 12.1** Ranitidine hydrochloride.

invalid under the doctrine of inherent anticipation. A single prior art reference may inherently anticipate a claimed invention even if the reference does not expressly disclose a feature of the invention "if that missing characteristic is necessarily present, or inherent, in the single anticipating reference."[50]

The Federal Circuit disagreed with Novopharm. It concluded that although Example 32 of the earlier '658 patent sometimes produced the Form 2 polymorph, sometimes it did not. During litigation Novopharm's experts performed the process disclosed in Example 32 of the '658 patent thirteen times, and each time they made Form 2.[51] The Federal Circuit noted, however, that when Glaxo first followed Example 32 it produced only Form 1, and during litigation Glaxo's expert also produced Form 1 using the method of Example 32. The court concluded that because Example 32 of the '658 patent does not necessarily produce Form 2, it cannot inherently anticipate the claims of the '431 patent directed to the Form 2 polymorph of ranitidine hydrochloride.

The *Glaxo* case illustrates that new solid forms, such as polymorphs, of an active pharmaceutical ingredient may offer green chemistry advancements as well as commercial benefits and patent exclusivity. Each case, however, depends on its own facts. As *Glaxo* demonstrates, the prior art must be carefully considered in determining the patentability of any particular solid form.

Indeed, a subsequent case demonstrates that even if a new solid form of a pharmaceutical compound is novel, it is not necessarily patentable. In *Pfizer v. Apotex*, Pfizer sought to enforce against Apotex a patent covering amlodipine besylate salt, the active ingredient in Pfizer's Norvasc™ hypertension product.[52] Pfizer had originally developed amlodipine as a maleate salt, but later discovered and switched to the besylate salt form (Figure 12.2). As a Pfizer scientist explained in a declaration submitted to the USPTO, amlodipine besylate possessed a highly desirable combination of physicochemical properties including good solubility, stability, non-hygroscopicity, and processability.[53]

Despite these improved properties, at least some of which may confer green chemistry advantages, the Federal Circuit concluded that the amlodipine besylate salt was obvious in view of the earlier-disclosed amlodipine maleate salt. The court characterized Pfizer's efforts to discover improved amlodipine

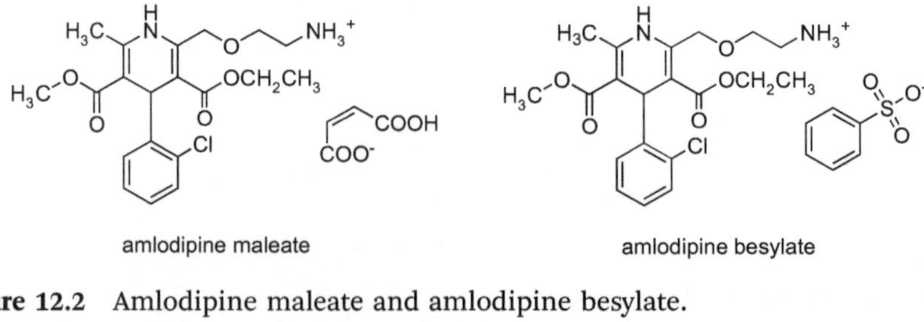


**Figure 12.2** Amlodipine maleate and amlodipine besylate.

salts as "routine, verification testing to optimize selection of one of several known and clearly suggested pharmaceutically-acceptable salts."[54] As a result, the patent covering amlodipine besylate was invalidated for obviousness.

The court in *Pfizer v. Apotex* arguably erected an impediment to patenting green chemistry inventions that improve upon existing pharmaceutical products in ways relating to the processing and manufacture of the pharmaceutical compound. The Federal Circuit's decision characterized as "common-sensical" Pfizer's attempts to create products that are "stronger, cheaper, cleaner, faster, lighter, smaller, more durable, or more efficient."[55] Writing in dissent from the court's denial to rehear the case, one judge wrote that the court improperly elevated therapeutic properties over other properties that may be improved through innovation – including "solubility, stability, hygroscopicity, and processability," which "must also play a role in the analysis of advantages" that may confer patentability.[56] Although the ramifications of *Pfizer v. Apotex* are not fully known, it remains an important decision illustrating potential hurdles in seeking patent protection for a new pharmaceutical salt form possessing potential green chemistry advantages.

## 12.3 Some Practice Tips

The sections above highlight that innovation in the green chemistry space – including developing improved reactions, preparing novel compounds or catalysts for use in such reactions, and preparing new solid forms of known compounds – can in some instances be patentable in the United States. The pharmaceutical chemist should therefore keep in mind some of the following considerations when developing compounds, synthetic routes, and solid forms:

1. A new catalyst, solvent, or reagent useful in the synthesis of a compound is potentially patentable in its own right. To the extent such novel substances lead to an improvement in the synthesis of a pharmaceutically useful compound – whether that be a higher yield, reduced waste, or something else – evidence of such improvements may aid in demonstrating patentability.
2. An improvement in a known synthesis that does not rely on a new catalyst, solvent, or other reagent – such as an improved reaction condition that leads to enhanced yield, or a new synthetic approach with enhanced atom economy – may nonetheless be patentable in its own right. In this situation, however, the measure of the improvement may be critical to the patentability of the innovation. As the *Ecolochem* case demonstrated, evidence such as skepticism in the field that the new process would work, commercial success of the invention, praise by others, and unexpected results may each be used as arguments in favor of the patentability of a new process.
3. A new solid form of a known pharmaceutical compound may, in some instances, merit patent protection. Evidence that the new solid form

possesses unexpected properties that confer specific advantages – including properties with relevance to the field of green chemistry, such as improved processing, dissolution, or solubility – should be considered as part of the patentability determination.

4. While these practice tips may be useful for a pharmaceutical scientist to keep in mind, before one files a patent application a patent attorney ordinarily should be consulted in assessing the likelihood that any particular invention merits patent protection. Communication with one's patent lawyer is also important in deciding whether to publicly disclose an invention, for example through publication or public use, lest one's green chemistry innovation unwittingly becomes prior art.

## 12.4 Conclusion

As the pharmaceutical scientist is well aware, the green chemistry revolution is upon us. In developing new medicinal compounds, it is vital that they be prepared as efficiently and sustainably as reasonably possible. Fortunately, the patent system is capable of rewarding many aspects of innovation, and significant advancements in green chemistry are no exception. In addition to giving a brief background on the patent system, this chapter has shown, both anecdotally and through a look at decisional law in the United States, that green chemistry innovations can be patentable inventions. The pharmaceutical chemist will benefit from keeping these concepts in mind – and using a patent attorney as a resource – when achieving results with potential green chemistry advantages.

## References

1. 35 U.S.C. § 101.
2. Notable exceptions in the field of medicine include inventions involving naturally occurring genes and methods of administering a drug based on measurements of levels of its natural metabolites. *See, e.g., Ass'n Molecular Pathology v. Myriad Genetics, Inc.*, 133 S. Ct. 2107 (2013); *Mayo Collaborative Servs. v. Prometheus Labs., Inc.*, 132 S. Ct. 1289 (2012).
3. *E.g.*, *Diamond v. Diehr*, 450 U.S. 175 (1981) (holding that a process for curing synthetic rubber was patent-eligible); *Knoll Pharm. Co. v. Teva Pharm. USA, Inc.*, 367 F.3d 1381 (Fed. Cir. 2004) (reversing summary judgment of invalidity for a patented combination of hydrocodone and ibuprofen); *Glaxo Inc. v. Novopharm Ltd.*, 52 F.3d 1043 (Fed. Cir. 1995) (holding that a patent covering a new pharmaceutical crystal form of ranitidine was not anticipated by the prior art); *In re Cyclobenzaprine Hydrochloride Extended-Release Capsule Patent Litig.*, 676 F.3d 1063 (Fed. Cir. 2012) (holding that an extended-release pharmaceutical formulation was not obvious over the prior art); *Ortho-McNeil Pharm., Inc. v. Mylan Labs., Inc.*, 520 F.3d 1358 (Fed. Cir. 2008) (holding that claims to a method of using topiramate were nonobvious).

4. As the U.S. Court of Appeals for the Federal Circuit has stated, "there are no per se rules relating the patentability of product and process in either direction. A process yielding a novel and nonobvious product may nonetheless be obvious; conversely, a process yielding a well-known product may yet be nonobvious." *TorPharm, Inc. v. Ranbaxy Pharm., Inc.*, 336 F.3d 1322, 1327 (Fed. Cir. 2003) (citation omitted).
5. 35 U.S.C. § 102.
6. *Schering Corp. v. Geneva Pharm., Inc.*, 339 F.3d 1373, 1380 (Fed. Cir. 2003).
7. 35 U.S.C. § 103.
8. *KSR Int'l Co. v. Teleflex Inc.*, 550 U.S. 398, 406 (2007); *Graham v. John Deere Co.*, 383 U.S. 1, 17-18 (1966).
9. *In re Soni*, 54 F.3d 746, 750 (Fed. Cir. 1995).
10. *See, e.g.*, *Genetics Institute, LLC v. Novartis Vaccines & Diagnostics, Inc.*, 655 F.3d 1291, 1307 (Fed. Cir. 2012).
11. Other objective evidence of nonobviousness includes a long-felt but unmet need for the particular invention claimed, as well as the failure of others to successfully solve the problem which the particular invention solves.
12. 35 U.S.C. § 112.
13. Securing a patent is not always the end of the road at the USPTO. The patent laws include a system for post-grant review of a patent, by which an issued patent may nevertheless be subject to further review—in some cases in an adversarial proceeding—if, for example, a requester demonstrates a "reasonable likelihood that that the [requester] would prevail with respect to at least 1 of the claims challenged in the [request]." 35 U.S.C. § 314(a); *see also* 35 U.S.C. § 324(a) (post-grant review).
14. The patent laws are national laws, and jurisdiction over cases or controversies implicating patent rights usually lies in the federal legal system, as opposed to the court system of any particular state. 28 U.S.C. § 1338(b).
15. In the United States, patent litigation regarding generic biologics, or "biosimilars," is governed by the Biologics Price Competition and Innovation Act (BPCIA), enacted in 2010. Litigation regarding the BPCIA has begun only recently. The FDA approved the first ever biosimilar in the United States on March 6, 2015.
16. United States EPA, Information about the Presidential Green Chemistry Challenge, http://www2.epa.gov/green-chemistry/information-about-presidential-green-chemistry-challenge.
17. Johnson & Johnson, Our Strategic Framework: Our Most Sustainable Products, http://www.jnj.com/caring/citizenship-sustainability/strategic-framework/our-most-sustainable-products.
18. Pfizer, Greener Processes, http://www.pfizer.com/responsibility/protecting_environment/greener_processes.
19. 74 Fed. Reg. 64666 (Dec. 8, 2009).
20. When instituting the pilot program, the United States Secretary of Commerce stated: "By ensuring that many new products will receive

patent protection more quickly, we can encourage our brightest innovators to invest needed resources in developing new technologies and help bring those technologies to market more quickly." *See* USPTO, Press Release (Dec. 7, 2009), *available at* http://www.uspto.gov/news/pr/2009/09_33.jsp.
21. 75 Fed. Reg. 69049 (Nov. 10, 2010).
22. *See* USPTO, Green Petition Report Summary, *available at* http://www.uspto.gov/patents/init_events/green_report_summary20120426.pdf.
23. 76 Fed. Reg. 77979 (Dec. 15, 2011).
24. U.S. Patent No. 6,806,380, at col. 1, *ll.* 39-42 ("A method is desired for the preparation of imidoesters that uses safer hydrogen chloride solutions .... Pre-made hydrogen chloride solutions are also economical because large excesses of solution are not required.").
25. United States EPA, 2006 Greener Synthetic Pathways Award, http://www2.epa.gov/green-chemistry/2006-greener-synthetic-pathways-award.
26. U.S. Patent Application No. 10/549,425, Statement of Reasons for Allowance ("A search of the prior art failed to uncover a reference that teaches the recited reaction with the catalyst of formula III. The search also failed to uncover a reference that provides a basis to modify a known reaction of a compound of formula II to include the catalyst of formula III.").
27. United States EPA, 2010 Greener Reaction Conditions Award, http://www2.epa.gov/green-chemistry/2010-greener-reaction-conditions-award.
28. U.S. Patent No. 8,293,507 (Oct. 23, 2012).
29. Professor Yi Tang of the University of California, Los Angeles and coworkers were awarded U.S. Patent No. 8,211,664 (July 3, 2012) for a biocatalytic method of synthesizing simvastatin, the active ingredient in Merck's Zocor™. In granting the 2012 Presidential Green Chemistry Award for a greener synthetic pathway, the EPA noted that this technology is "practical and cost-effective" and "avoids the use of several hazardous chemicals including *tert*-butyl dimethyl silane chloride, methyl iodide, and *n*-butyl lithium." United States EPA, 2012 Greener Reaction Conditions Award, http://www2.epa.gov/green-chemistry/2012-greener-reaction-conditions-award. In the USPTO, the patentability of the invention was premised on the use of a novel acyltransferase enzyme to make the drug without the need to purify its precursor.
30. U.S. Patent Application No. 09/380,562, Amendment dated May 30, 2000, at 9-10.
31. *See* U.S. Patent Application No. 09/380,562, Declaration of Juan C. Colberg, Ph.D. dated December 4, 2000, at ¶¶ 6-17.
32. United States EPA, 2002 Greener Synthetic Pathways Award, http://www2.epa.gov/green-chemistry/2002-greener-synthetic-pathways-award.
33. Some compounds may crystallize into multiple forms. A crystal form may consist solely of a drug compound, either as an unionized species or as a salt with one of many pharmaceutically acceptable salt-forming acids or bases. A crystal form may contain the drug compound alone or may further contain a solvent such as water.

34. W. Cabri *et al.*, *Org. Process Res. Dev.*, 2007, **11**, 64.
35. (a) A. V. Trask, *Mol. Pharmaceutics*, 2007, **4**, 301; (b) A. V. Trask *et al.*, *Chem. Commun.*, 2004, 7, 890.
36. P. Vishweshwar *et al.*, *J. Pharm. Sci.*, 2006, **95**, 499.
37. Remenar *et al.*, *J. Am. Chem. Soc.*, 2003, **125**, 8456.
38. U.S. Patent No. 7,446,107 (Nov. 4, 2008), at cols. 5–6.
39. Remenar *et al.*, *J. Am. Chem. Soc.*, 2003, **125**, 8456.
40. Remenar, *supra* note 39, at 8456.
41. U.S. Patent No. 7,078,526 (July 18, 2006), at claims 1 & 2.
42. U.S. Patent No. 7,927,613 (Apr. 19, 2011), at Ex. 32.
43. 227 F.3d 1361, 1365 (Fed. Cir. 2000).
44. *Ecolochem*, 227 F.3d at 1374.
45. *Ecolochem*, 227 F.3d at 1378.
46. *Ecolochem*, 227 F.3d at 1380 (noting, however, that this factor may not be entitled to great weight in the obviousness determination).
47. *Ecolochem* was decided before the Supreme Court's most recent obviousness decision, *KSR Int'l Co. v. Teleflex Inc.*, 550 U.S. 398 (2007). *KSR* endorsed a less rigid test for obviousness, and case law continues to refine the ways in which obviousness applies to inventions in the pharmaceutical field. *Otsuka Pharm. Co. v. Sandoz, Inc.*, 678 F.3d 1280, 1298 (Fed. Cir. 2012); *see also* Andrew V. Trask, *"Obvious To Try": A Proper Patentability Standard in the Pharmaceutical Arts?*, 76 Fordham L. Rev. 2625 (2008), *available at* http://ir.lawnet.fordham.edu/flr/vol76/iss5/9.
48. *Glaxo Inc. v. Novopharm Ltd.*, 52 F.3d 1043 (Fed. Cir. 1995).
49. *Glaxo*, 52 F.3d at 1046.
50. *See, e.g.*, *Schering Corp. v. Geneva Pharm., Inc.*, 339 F.3d 1373, 1377 (Fed. Cir. 2003).
51. *Glaxo*, 52 F.3d at 1047.
52. *Pfizer, Inc. v. Apotex, Inc.*, 480 F.3d 1348 (Fed. Cir. 2007).
53. *Pfizer*, 480 F.3d at 1356.
54. *Pfizer*, 480 F.3d at 1371.
55. *Pfizer*, 480 F.3d at 1371 (quoting *DyStar Textilfarben GmbH v. C.H. Patrick Co.*, 464 F.3d 1356, 1368 (Fed. Cir. 2006)).
56. *Pfizer*, 480 F.3d at 1384 (Rader, J., dissenting from denial of rehearing *en banc*).

CHAPTER 13

# *The Business Case for Green Chemistry in Drug Discovery*

JOSEPH M. D. FORTUNAK,*[a] MARTINS O. EMEJE,[b] HARRIET KAMMENDI,[c] ERMIAS L. TILAHUN[d] AND XIANG S. WANG[d]

[a] Department of Chemistry, Howard University, 525 College Street NW, Washington, DC 20059, USA; [b] Center for Nanomedicine and Biophysical Drug Delivery, National Institute for Pharmaceutical Research and Development, Abuja, Nigeria; [c] Kandih Group, 3920 Blackburn Lane, Burtonsville, MD 20866, USA; [d] College of Pharmacy, Howard University, 4th and College Streets NW, Washington, DC 20059, USA
*Email: jfortunak@comcast.net

## 13.1 Introduction and Background

Drug discovery has been defined in previous chapters as all of the activities leading up to the identification of a new chemical entity (NCE) for advancement into drug development.[1] Drug development is defined as all of the activities required to demonstrate that a specific NCE is safe, effective, and can be reproducibly manufactured with appropriate quality for human use in the treatment, mitigation, prevention, or diagnosis of a disease.[2] Ideas and discussion of why and how one would use green chemistry (GC) for drug discovery have been presented throughout previous chapters of this book. This chapter has a focus on the business justification of GC – the advantages in terms of corporate profitability that accrue to the use of GC in the drug discovery process.

RSC Drug Discovery Series No. 46
Green Chemistry Strategies for Drug Discovery
Edited by Emily A. Peterson and Julie B. Manley

Published by the Royal Society of Chemistry, www.rsc.org

Most drugs on the market today were discovered and developed by large companies that specialize in medicines and healthcare – *i.e.,* the global pharmaceutical industry. Drug discovery in the global pharmaceutical industry is an expensive process that requires sizeable teams of world-class scientists from many different disciplines working together over an extended period of time. The Tufts Center for the Study of Drug Development ("Tufts") is widely recognized within the industry as "the center" for high-quality data on the actual costs of bringing drugs to the market. A very recent (2014), extensive study from the center estimates that the cost of bringing an NCE to the point of marketing approval is about \$2.6 billion.[3] Of this amount, out-of-pocket drug discovery costs were estimated to average about \$430 million for each NCE approval.[3] The cost of drug failures is accounted for in this figure – *i.e.,* not every discovery program identifies a drug candidate, and not every NCE is approved for marketing. Historical figures from 1995–2007 showed that the probability of technical success (PTS) for any NCE entering the first phase (Phase I) of human clinical trials was about 11.8%. The out-of-pocket expense for drug discovery can be reduced using GC, as has been abundantly discussed in preceding chapters. The low PTS for NCEs entering drug development, however, suggests that if GC can improve the success rate of drug candidates entering development, this is an area of prime importance.

The Tufts figures have been hotly debated. Because the Tufts numbers are taken from "self-reporting" figures gathered from interviews with industry personnel, their veracity and how they are derived has been questioned. As a counterpoint, Light and Warburton have provided a ridiculously low estimate of industry costs for bringing a new drug to the market of \$56 million.[4] A simple, alternative way of estimating the cost of new drugs is to divide industry R&D expense by the number of NCEs that are approved for marketing. Using this approach and working from publicly-available figures on corporate R&D expense, average company R&D spend per new drug approval ranged from \$4 to 11 billion from 1997 to 2011, with the average overall cost of bringing an NCE to the market being over \$5 billion.[5]

These huge numbers raise the temptation to view the "big pharma" approach to drug discovery and development as being on the verge of failure. This would, from one perspective, argue that the pharmaceutical industry is a victim of its own success, all of the inexpensive drug targets have been worked through and only the very challenging (and very expensive) ones remain. One much different model of drug discovery, often practiced in less-developed economies, is the use of traditional knowledge. The use of natural products extracts or phytomedicines for disease treatment represents a simple form of a "reverse pharmacology" or RP approach to drug discovery.[6] Phytomedicines and RP can be very efficient, but this is a very opportunistic approach, in that the scope of phytomedicines is limited by the materials available. One cannot, for instance, expect that natural products screening will provide the rapid, inexpensive discovery of effective treatments for viral diseases such as HIV/AIDS and hepatitis B or C. RP is demonstrably very efficient – when it works. Rational drug discovery, through an understanding of the etiology of the disease state and identifying a molecular target for selective intervention, possesses a wider applicability for identifying treatments for major unmet medical needs. Within the "global pharmaceutical" model for drug discovery, however, substantial evidence is available for utilizing GC to improve the efficiency of the discovery process.

In this chapter on a few occasions statistics on new drug approvals are cited from the US Food and Drug Administration (FDA). This is not meant to slight the European Medicines Authority (EMA) or any other strict regulatory authority. The advent of global electronic filings in a common format and the Prescription Drug User Fee Act (PDUFA) in the United States has led to a global regulatory environment in which new drug approvals in the USA and Europe most often coincide very closely.[7,8] Given that US FDA approvals allow access to the world's largest single pharmaceutical market without additional individual country approval requirements, we have elected to cite US FDA statistics and the US FDA CDER (Center for Drug Evaluation and Research) web site on occasion throughout. Whenever possible, drugs are referred to by their common generic (USAN) rather than proprietary (trademark) names. When trademark/copyrighted names are used to refer to drugs we have duly noted this by the ™ designation.

## 13.2 Drug Discovery

Academics, private industry, and governments (*e.g.*, national institutes of health) are major sources of world-class innovation in medicinal sciences and human health. Academic or "basic" research largely serves the purpose of generating fundamental new scientific knowledge for its own sake. Global pharmaceutical companies are corporate entities whose purpose is to make a profit for their stakeholders through improving human health. Pharmaceutical companies are nearly exclusively the drivers of major innovation in new drug discovery. Only large pharmaceutical companies are able to focus the huge amount of resources across multiple scientific disciplines required to consistently bring major new medicines to the market.[9] A major challenge which GC can in part address is the need to accomplish more with a reduced expenditure of resources, so as to enhance the sustainability of the pharmaceutical industry.

Table 13.1 illustrates how we have elected to delineate the major steps in the process of discovering an NCE as part of our analysis. An initial stage of target identification (*i.e.*, selecting the site of therapeutic intervention or the mechanism of action of a new drug candidate) relies heavily on genomics, proteomics, and bioinformatics, but is not heavily chemistry-intensive. This is followed by a second discovery stage of target validation, high-throughput screening of drug libraries for biological activity, and the generation of chemical "hits" or structural classes of compounds that possess sufficient biological activity to be potentially interesting. A third stage of drug discovery involves the optimization of leads through structure-based design,

**Table 13.1** Major steps in the drug discovery process.

| Stage | Steps | | |
|---|---|---|---|
| 1. Modest chemistry involvement | *Genomics*: bioinformatics, gene arrays, gene identification, DNA sequence maps | *Functional genomics*: Gene transcriptions & translation, mutations and polymorphism, protein–protein interactions, metabo- and toxico-genomics | *Target identification and selection*: protein structure, target evaluation, target druggability and selectivity |
| 2. Heavy chemistry involvement | *Target validation*: Assay development, binding assays, cellular assays, animal models | *Screening*: chemoinformatics, compound libraries, high-throughput screening, QSAR | *Hit structures*: Multiple compounds with diverse molecular structures that meet initial selection criteria |
| 3. Chemistry centered | *Hit-to-lead*: SAR and QSAR, chemoinformatics | *Lead optimization*: toxicology, pharmacokinetics, physiochemical properties | *Drug candidate selection*: drug metabolism, total ADME profile, sub-chronic toxicology |

computational techniques, and chemoinformatics. This final major stage of drug discovery also involves in-depth evaluation and final structural optimization to arrive at NCEs with acceptable ADME (absorption, distribution, metabolism, and elimination) properties and evidence of safety in animal studies (toxicology). This end stage set of activities for drug discovery results (in successful programs) in the selection of a single molecule or a very small number of molecules that go forward into development. The second and third stages of this process for drug discovery are where chemistry-intensive activities take place.

GC that reduces the cost and resources needed to discover new drugs is important. In assessing the business case for GC, improving the success rate of NCEs in development is of huge importance. Since the success rate of NCEs (*i.e.*, about 11.8%) entering development is so low, GC techniques that improve the success rate of drugs in development is a substantial business driver. The early identification of NCEs likely to fail due to unacceptable toxicity and the design of drug molecules to target the active site for drug delivery are two ways that readily come to mind.

The Boston Consulting Group published a study (2013) of 842 drugs brought into development by 27 different pharmaceutical innovator companies between 2004 and 2011.[10] Three factors were shown to have a high positive correlation with success in drug development. These were: (1) patents obtained per R&D spend, (2) citations per publication, and (3) early termination of projects. Of these factors, the early termination of projects was by far the most highly correlated with success. This is a study outcome that would be predicted by many in the industry. Patents and citations correlate with the generation of valuable intellectual property and the quality of published R&D as recognized by peers. The adage of "fail fast, fail often," has long been a mantra in pharmaceutical innovator companies. The early termination of failing NCEs is a hallmark of an organization that efficiently gathers information needed to make high-quality decisions. The contributions that GC can make to improving success rates and early termination of failures are a most telling argument in favor of GC.

The initial development phase for an NCE includes longer-term (most often one-month) animal toxicology studies, selection of a route of synthesis, the first, often kilogram-scale synthesis under Good Manufacturing Practice (GMP), and a battery of other studies that support the filing of an Investigational New Drug Application (INDs or INDA) to initiate human clinical trials. Toxicology studies support the identification of signal toxicities and target organs likely to be important in human exposure. Effective *in silico* analysis for potential toxicity and selectivity of drug action is becoming increasingly important for the early identification of potential drug failures. Figures from the Tufts study suggest that 92% of NCEs fail to advance to a marketing approval, while 11.83% of drugs in development are approved; this yields a simple calculation of the success rate for NCEs in this transition phase of about 68% (*i.e.*, $8\% \div 11.83\% = 68\%$).

### 13.2.1 What Does Drug Discovery Really Cost?

Drug R&D costs are debated hotly. The 2014 Tufts study pegged the average cost of bringing a new drug to the point of FDA approval at \$2.558 billion.[3] This study by DiMasi *et al.* arrived at a figure of \$430 million as the average out-of-pocket spend on drug discovery for each NCE approved. This figure includes allowances for the failure rate of NCEs entering development. The direct, out-of-pocket costs for bringing a drug through discovery and development, to the point of FDA approval, were estimated as \$1.395 billion. A substantial adjustment (\$1.163 billion) for the cost of capitalization of money brings the total R&D figure to \$2.558 billion. When estimating the true costs of business investment, it is standard practice to apply a correction factor for money invested in advance of new product launches. For the pharmaceutical industry, this factor essentially estimates both out-of-pocket costs and the additional, "hidden" cost of spending money several years in advance of any possible sales from a new drug launch. The rate of inflation as well as standard estimates for the cost of "lost opportunity" apply to the costs of capitalization. Lost opportunity is a correction factor (typically about 3–5%) to adjust for investment gains that a company forgoes while investing in new products with long development timelines. Although this estimation process may seem to contain an element of "sleight-of-hand" this is a standard business practice, and is not unique to the pharmaceutical industry. A sobering figure from this analysis is that, while the out-of-pocket costs for bringing a new drug through discovery average \$430 million, the fully capitalized cost of this \$430 million is actually \$1.1 billion, because this money is spent so far in advance of product launch. The \$430 million figure was estimated from information covering the time period 1995–2007. More recent costs were not included because an assumption was made that a minimum of five years will elapse between the time of identifying an NCE to enter development and an FDA approval.

### 13.2.2 Green Chemistry as a Business Driver

This chapter addresses major ways in which GC can drive the business of drug discovery. One of these is perfectly obvious – by reducing direct (*i.e.*, out-of-pocket) expense. Reducing laboratory expenses and reducing the number of people needed in discovery teams to identify an optimal drug candidate are clear ways of reducing direct expense. Speeding the time to drug approvals, however, will also improve the business of drug discovery. Reducing the time that NCEs spend in R&D affects the capitalization cost of an NCE, thereby improving business performance. Increasing the success rate of drugs in R&D is another major way of improving the business of pharmaceuticals. Examples are discussed of improving the probability of technical success (PTS) in drug discovery with the drugs tenofovir alafenamide fumarate (TAF) and sofosbuvir (SOF). A case study on the development of the herbal medicine Niprisin™ is also discussed. Niprisin™ and other phytomedicines have both reduced development timelines and substantially higher than normal PTS when entering development because they are identified from a body of traditional knowledge that includes a history of prior human use, greatly reducing the odds of failure due to unacceptable toxicity.

The reduction of out-of-pocket expense is an area that we are not addressing in great detail, as several previous chapters address strategies to incorporate green chemistry that also reduce out-of-pocket costs. Berkeley Cue has pointed out that the cost of disposing of solvent waste in discovery is about \$15 $kg^{-1}$.[11] This greatly exceeds the cost of purchasing most solvents for bulk use, where, for example, methanol presently costs approximately \$650 $ton^{-1}$ and dichloromethane is less than \$5 $kg^{-1}$.[12] Improved atom economy, reduced solvent use, reduction of hazards, reduction in the number of synthetic steps, and recycling of solvents all can reduce out-of-pocket costs. An excellent review of this topic, "Sustainable practices in medicinal chemistry: current state and future directions," was recently published in the *Journal of Medicinal Chemistry*.[13] A web site devoted to green and sustainable techniques in drug discovery has been launched, providing an online venue for discussions about GC applied to the discovery process.[14]

The major cost component of drug discovery is strongly correlated with the number of people in a drug discovery program. Research scientists represent a very substantial investment in compensation. An even larger proportion of personnel cost is the expense of keeping a scientist working in a laboratory. Laboratory facilities are notoriously expensive due to their design and functional requirements. Consumable supplies, information technology services, waste disposal, and space allocations add to these expenses. The fully-burdened cost of holding a research scientist in a drug discovery organization is generally about $400 000 per year.[15] No matter where R&D costs are allocated, however, it is certainly true that drug discovery requires many person years of effort. In the various companies one of us worked in (GSK, Dupont Merck, and Abbott Labs) from 1983 to 2004, the average number of person years required to discover an NCE for a new therapeutic indication was typically about 100–120. Discussions with colleagues working in the drug discovery seem to indicate that about 120–150 person years are a very typical "spend" to arrive at a drug candidate for challenging molecular targets. Given that people are for the most part more expensive than technology, the better that GC is able to reduce the number of person-years required to discover an NCE, the more important it is as a business advantage.[16]

## 13.3 Green Practices in Drug Discovery

### 13.3.1 Advanced Technologies

Considering GC as good business practice, some common discovery practices such as combinatorial chemistry and solid-phase synthesis are techniques aligned with the principles of green chemistry. Combinatorial chemistry meets several of the twelve principles of GC. Combinatorial chemistry utilizes polymer supports to perform reactions with minimal solvents and enables the preparation of chemical libraries – the simultaneous synthesis of hundreds or thousands of related compounds by relatively straightforward techniques. Several compounds prepared by combinatorial chemistry are often tested in a single assay, reducing the amount of effort and resources devoted to isolation and purification of compounds during initial biological screening. However, combinatorial chemistry does have drawbacks – including some related to GC. Many reactions are not compatible with the polymers used as solid support. This correspondingly limits the variety of molecular families that can be synthesized in this manner. Combinatorial chemistry often uses excesses of expensive reagents to assure the complete conversion of polymer-bound substrates. Cleavage of compounds from the polymer support also requires solvent and reagent use, and compounds must usually be purified to assure reliable assay results. Individual compounds must also be re-synthesized and retested to confirm their structure and assay for compounds of interest. Thus, while combinatorial chemistry is useful, it is not universally applicable and is not clearly green in every case. Overall, combinatorial chemistry must be carefully evaluated as to the overall GC advantages of preparing any specific library of compounds.

Flow reactors are micro-sized reactors that can be operated continuously, and are increasingly being used as a green, flexible alternative to

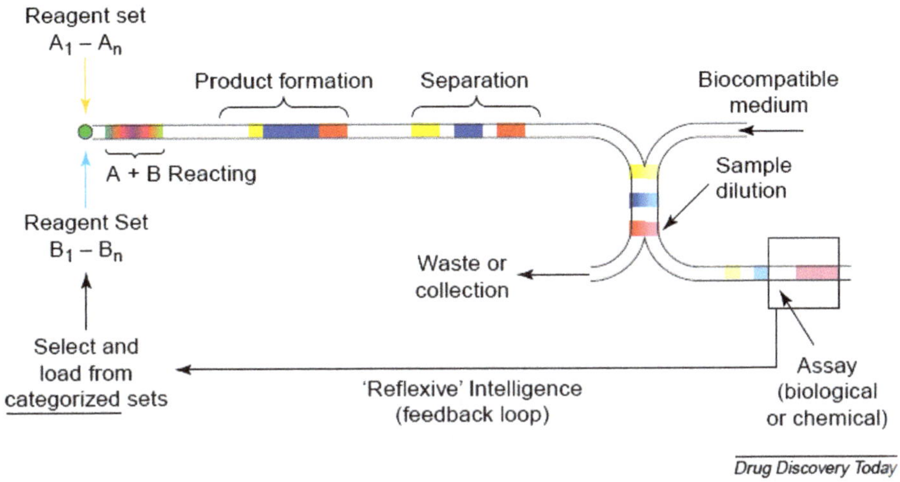


**Figure 13.1** Flow chemistry approach for combinatorial chemistry in drug discovery. Taken from ref. 22.

combinatorial chemistry.[17] Many reactions can be carried out in flow that are not possible using solid-phase synthesis. Flow reactors are adaptable for parallel synthesis, and allow for the rapid preparation of compound libraries in a time frame that rivals combinatorial chemistry. Flow synthesis does not have the combinatorial chemistry disadvantages of requiring separation of mixtures of compounds to confirm biological activity. The basic design of these reactors allows for extremely efficient mixing and heat transfer, precise timing and control of stoichiometry for reagent additions, minimization of solvent use, short reaction times, and the efficient telescoping of reaction steps. Flow techniques have been used for the rapid synthesis of chemical libraries for medicinal chemistry. The basic design of these reaction platforms has even been adapted to simplify the process of carrying out combinatorial chemistry.[18–22] The combination of flow techniques with a bioassay system can even be used (Figure 13.1) to assay compounds immediately as they are synthesized.[22]

## 13.3.2 Green Synthetic Methodologies

Drug discovery creates new molecules with selective, potent, biological activity. This challenge is not readily compatible with the principles of GC. Dan Levy has ably laid out the dichotomy between drug discovery and inherently sustainable chemistry.[23] The twelve principles of GC were developed with chemical manufacturing in mind, rather than being primarily directed towards R&D. The first principle of GC, for example, relates to the minimization of hazardous substances. If hazardous substances are not used in drug R&D, this limitation on the reagents and scaffolds available for drug design would substantially restrict the range of novel molecules available for testing. In discovery research, however, large volumes of waste solvent are generated. Techniques to reduce solvent use and minimize the environmental impact of discovery chemistry are discussed elsewhere in this book. Solvent selection and green reagents guides have been published by different groups (*e.g.*, GSK, Sanofi) to inform this process.[24–26]

The fuller impact of GC during drug discovery only becomes evident later – during drug development. As syntheses are scaled up for Phase I human studies, efficiency becomes increasingly important. If one takes into account the people and time required for process development, the cost of raw materials, and pilot plant expenses, synthesizing from several hundred grams up to perhaps 20 kg of active pharmaceutical ingredient (API) for Phase I development can have a fully-allocated cost of $500 000 per kg or more.[27] Scheme 13.1 illustrates the difference between inefficient and efficient (*i.e.*, Green) chemistry for Phase I development. Efavirenz (EFV) is a drug for the treatment of HIV/AIDS, possessing a single chiral center. The first development synthesis of EFV utilized the medicinal chemistry synthesis, employing the resolution of racemic drug through a camphanic acid amide. This synthesis (approximately 4 kg) generated about 350 kg of waste per kg of API. A series of close analogues of EFV entered Phase I development approximately two years later. The synthesis route used by the discovery chemists to make these analogues utilized an asymmetric addition step to generate a chiral center with a high degree of enantioselectivity. This chemistry allowed a large number of analogues to be rapidly produced by the discovery program, reducing the person-years of effort required for identifying a new NCE. After a second-generation drug candidate was identified, the route of synthesis for scale-up generated only about 40 kg of waste per kg of API, using

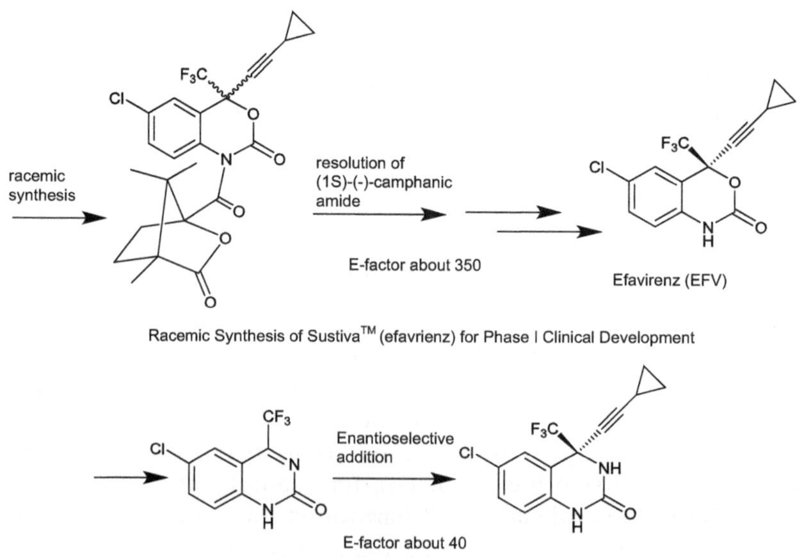


**Scheme 13.1** Impact of greener chemistry on early stage development of EFV (efavirenz) and fast-follower analogues.

very nearly identical starting materials. The difference in time required for producing kilogram amounts of drug was also strikingly different; the camphanic acid route used for EFV drug discovery required about three months for scale-up while the asymmetric addition route used in discovering analogues required about four weeks for pilot plant scale-up.[27] The current commercial synthesis of EFV utilizes chemistry very similar to the asymmetric addition step used in the "analogues" discovery as shown in Scheme 13.1, with an *E*-factor of about 30–35 for the end-to-end manufacturing route.[28,29]

The true impact of green chemistry in synthesis can be difficult to see. A detailed examination of the differences between synthetic processes used in drug discovery *versus* development will often reveal profound differences in efficiency – even for chemistries that look very similar on paper. The Pfizer discovery *versus* development routes for synthesizing the drug candidate CP-809,101-42 illustrates this very nicely.[30] The discovery synthesis required only the three steps in which bonds were broken/formed (Scheme 13.2).

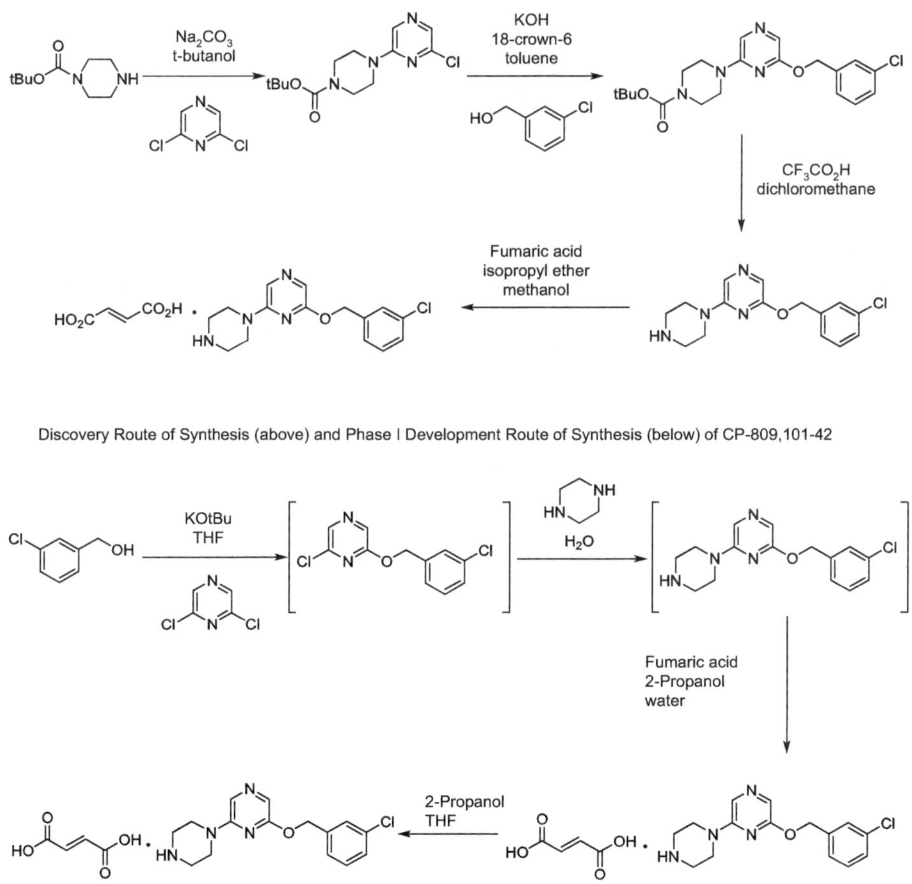


**Scheme 13.2** Pfizer synthesis routes for CP-809,101-42.

The modified route of synthesis for development changed the order of Steps 1 and 2 in the process, thereby eliminating the need for using a Boc-protected form of piperazine (rather than piperazine itself) as a raw material for synthesis. By telescoping together individual steps to eliminate the workup of intermediates, eliminating chromatography, and reducing solvent usage in individual reactions, the development group dramatically reduced the *E*-factor of this synthesis from 1180 to 43! This profound difference in the amount of waste generated during synthesis speaks eloquently for the introduction of GC into synthesis as early as possible in the drug discovery process.

### *13.3.2.1 Organocatalysis*

Green synthetic methodologies of many types (using water as solvent, C–H bond activation) have been discussed so far in this book. One particular synthetic methodology (organocatalysis, OC) is highlighted here to show how GC can impact not only the efficiency but also the speed with which drug candidates can be identified. Asymmetric centers with defined chirality are a hallmark of natural products and many new drug candidates. OC utilizes low-molecular weight organic compounds (usually simple amino acids or derivatives) for which a metal is not part of the critical catalytic step. Current OC is focused on asymmetric reactions with chiral catalysts.[31] Organocatalysis generally replaces multiple steps in a synthesis, and is therefore greener than alternative approaches. Scheme 13.3 compares the "discovery route" for generating the core structure of the HIV protease inhibitor darunavir (DRV) *versus* a much shorter, more efficient organocatalytic process that introduces two chiral centers in a single step with complete (<0.2% of any other stereoisomer) control of stereoselectivity.[32,33]

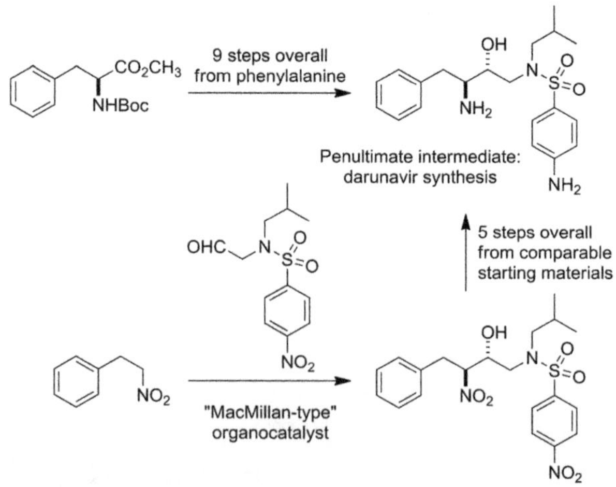


**Scheme 13.3** Commercial and organocatalytic syntheses of the final intermediate of darunavir.

Another example of OC illustrates the potential for synthetic methodology to drive the rate of drug discovery. A collaboration between the Merck discovery organization and Professor David MacMillan's research group at Princeton University combines cascade OC with an affinity screening process. This enables the rapid synthesis of many molecules with multiple chiral centers, combined with an almost instantaneous protein affinity assay that identifies the most active molecules present. This combination of techniques is a powerful tool for the efficient drug discovery of biologically active compounds, potentially reducing very substantially the person-years of effort required to arrive at chiral drug candidates.[34]

#### 13.3.2.2 Diversity-Oriented Synthesis

The demonstration of organocatalysis being harnessed to speed the process of drug synthesis and screening is actually an example of diversity-oriented synthesis (DOS). Combinatorial chemistry faces a challenge to its business utility in that, despite its undeniable efficiency for synthesis, no demonstrable increase in drug candidates is broadly attributable to its use.[35] DOS is a recent, promising attempt to improve combinatorial chemistry. The objective of DOS is to produce high diversity in synthetic molecules, beginning from building blocks that inherently possess biologic activity. A library of molecules containing a common molecular scaffold is first synthesized, just as in traditional combinatorial chemistry. These compounds are then used to generate new libraries. Utilizing chemical biology, the building blocks for DOS contain a series of "privileged motifs", or structural elements that are predisposed to be biologically active.[36]

Drug discovery most often involves screening small molecules for binding to a protein target. Phenotypic drug discovery involves screening small molecules against a biological pathway in cells or organisms, regardless of a protein target. One goal of DOS is to synthesize collections of small molecules capable of perturbing any disease-related biological pathway, leading eventually to the identification of new, therapeutic protein targets. Koh's research group has illustrated the DOS approach to drug discovery beginning from the privileged motif of benzopyran (Scheme 13.4). By subjecting a small number of functionalized benzopyran rings to multiple iterations of well-defined transformations (*e.g.*, Diels–Alder, Stille, or click chemistry), they generated a library of 22 molecules with high skeletal diversity. This library was then further elaborated to rapidly synthesize 284 molecules with a wide set of structural diversity.[37]

## 13.4 *In Silico* Methods for Drug Testing

### 13.4.1 *In Silico* Identification of Target Structures

Small-molecule drug candidates are evaluated for their overall quality by various measures including binding potency, selectivity, and ADME

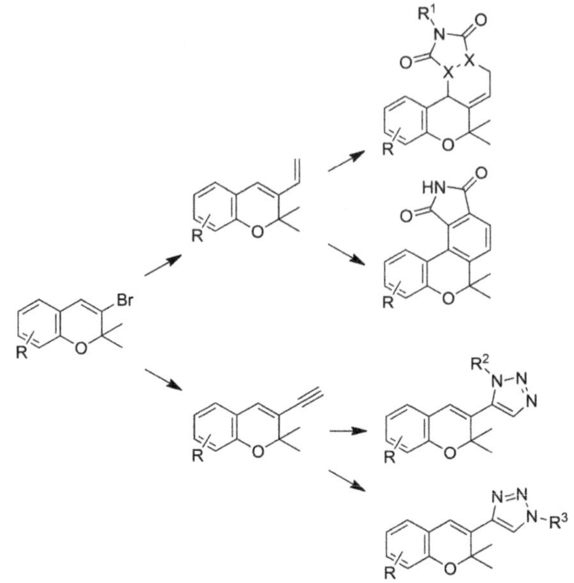

**Scheme 13.4** Koh's benzopyran diversity-oriented synthesis (DOS).

properties. This predictive assessment of the overall "druggability" of a molecule is a key measure of probable success in clinical trials. These characteristics are under the control of chemists during optimization of lead compounds and at compound selection. The term *in silico* (in silicon) is taken to mean "performed by a computer". *In silico* methods are employed extensively to discover drug candidates, identify new therapeutic uses for known drugs and to predict off-target side effects (*i.e.*, toxicities). The green chemistry application of *in silico* techniques is illustrated in multiple ways by (1) identifying likely toxic liabilities of molecules before they are even tested, (2) optimizing molecular design and binding fit to protein targets, and (3) predicting the likelihood of overall physicochemical properties of drug candidates. The traditional view that drugs only bind to a single target has dramatically changed in recent years with the discovery that promiscuity of drug binding *in vivo* may be very common.[38] *In silico* technologies consisting mainly of structure-based drug discovery (SBDD) and chemoinformatics are heavily used to improve development success rates in advance of human and animal testing (Figure 13.2).[39–42]

SBDD provides powerful tools for identifying starting points (hits) for synthetic medicinal chemistry. Molecular docking techniques based on three-dimensional structures and compound libraries play a major role in developing small-molecule therapeutics.[39] It is well recognized that SBDD can reduce time and cost for drug discovery.[42] Structural information and the nature of target/ligand interactions can be obtained from sources such as crystallography, spectroscopy, and bioinformatics. SBDD can also predict hit compounds that are not identifiable by other means, and assist in

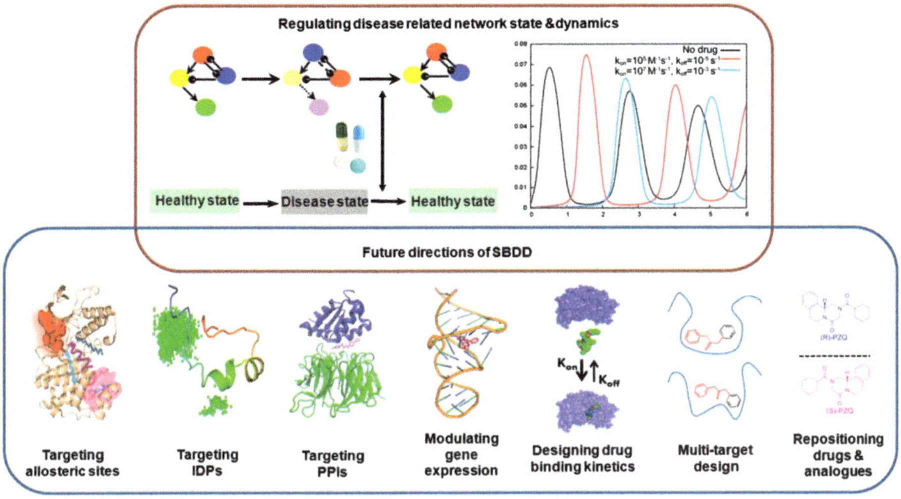


**Figure 13.2** The systems-based drug intervention and new directions for SBDD. *Top*: illustration of disease intervention at the systems level to regulate disease-related network state and dynamics. *Bottom*: list of important new directions for SBDD. (Abbreviations used: IDPs, intrinsically disordered proteins; PPIs, protein–protein interactions.)[39–42]

hit-to-lead and lead optimization. There are a large number of compounds whose discovery has been assisted by or based upon SBDD. Drugs for the treatment of HIV/AIDS are particularly well-known for this, including HIV protease inhibitors, non-nucleoside inhibitors of HIV-1 reverse transcriptase (NNRTIs), integrase inhibitors, and fusion inhibitors.[43] Table 13.2 illustrates SBDD applied to the discovery of NNRTIs. All of the compounds in Table 13.2 underwent similar *in silico* evaluations to optimize their multipoint binding as an element of structural design.

The emergence of drug resistance from mutated HIV viruses has also been anticipated by *in silico* methods.[44] A powerful structure-based technique called molecular hybridization can be applied to crystallographic structures to assist in the design of optimized drug candidates.[45] This approach has generated novel chemical scaffolds for hit NNRTI drug candidates (Table 13.2). The NNRTI drugs etravirine and rilpivirine were designed through *in silico* analysis to possess structural components providing (1) optimal binding in the NNRTI binding pocket, (2) conformational flexibility to bind tightly to both the "wild type" RT as well as to many different mutations, and (3) long half-life. These novel molecules were undeniably designed primarily as "good medicinal chemistry". Designing NCEs that are much more potent and have a longer half-life than the drugs they replace, however, reduces the dosage of drugs and is therefore inherently a green element of drug discovery. Using *in silico* techniques to arrive at molecular designs and then demonstrating the real predictive value of these molecules is arguably an inherently green approach to drug discovery. These drugs were shown to have

**Table 13.2** Investigational NNRTIs developed by SBDD approaches.

| Compound | Structure | $IC_{50}$ (μm) | Date | Origin | Status | Ref. |
|---|---|---|---|---|---|---|
| F3385-2590 | 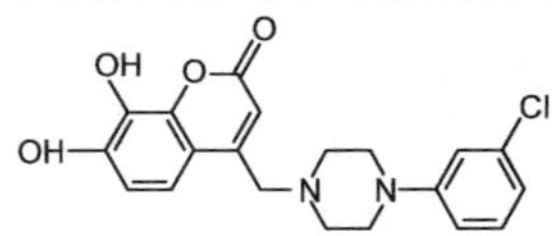 | 0.8 | 2014 | Rutgers University | Pre-clinical | 43 |
| Etravirine (TMC125) | 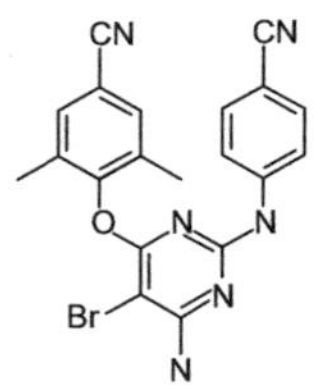 | 0.0009 | 2008 | Tibotec Pharma | On market | 44 |
| Rilpivirine (TMC278) | 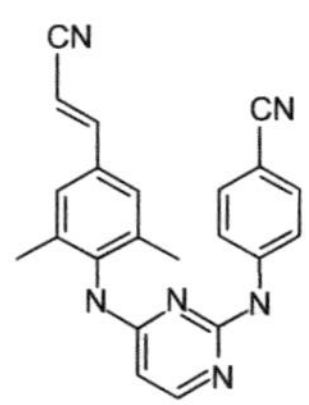 | 0.00007 | 2011 | Tibotec Pharma | On market | 45 |

| | | | | | | |
|---|---|---|---|---|---|---|
| Unnamed | 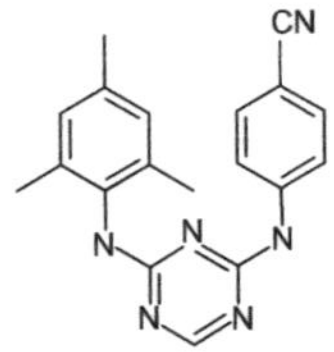 | 0.009 | 2014 | Yale University | Pre-clinical | 46 |
| 26-trans | 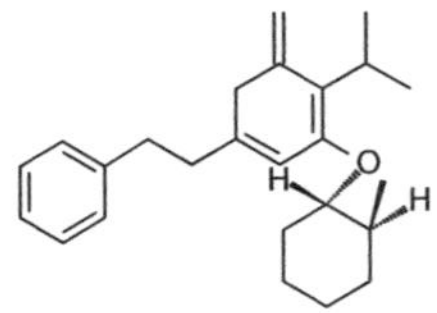 | 0.003 | 2013 | Peking University | Pre-clinical | 47 |
| Unnamed | 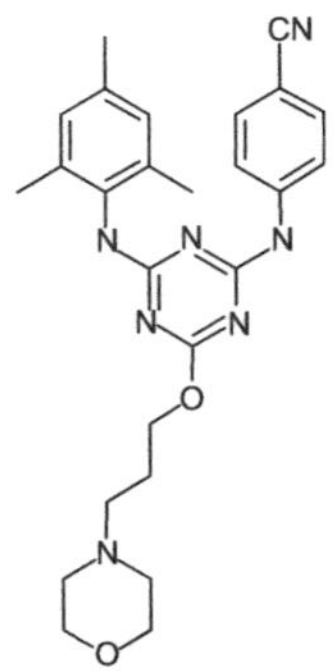 | 0.014 | 2014 | Yale University | Pre-clinical | 46 |

superior binding and improved pharmacokinetics *versus* the approved drugs nevirapine, efavirenz, and delavirdine.[44,45]

### 13.4.2 Chemoinformatic Identification of Off-target Side Effects and Toxicity

Chemoinformatics is defined as "...the mixing of those information resources to transform data into information and information into knowledge for the intended purpose of making better decisions faster in the area of drug lead identification and optimization."[46,47] The similarity ensemble approach (SEA) is a chemoinformatics tool for predicting protein similarities as well as the mode of target identification and off-target binding.[48] SEA compares protein targets by the similarity of molecules that bind to them, with prediction confidence expressed as expectation values adapted from basic local alignment search tool (BLAST) algorithms (Figure 13.3). This approach predicts side-effects and new indications for many drugs.[48] In one study, the SEA testing of 656 drugs *versus* 73 side-effect targets identified 1042 predictions for potential adverse drug interactions which were tested for true biological relevance. This resulted in the identification of 499 previously-unknown drug-to-side-effect targets. Remarkably, 151 new

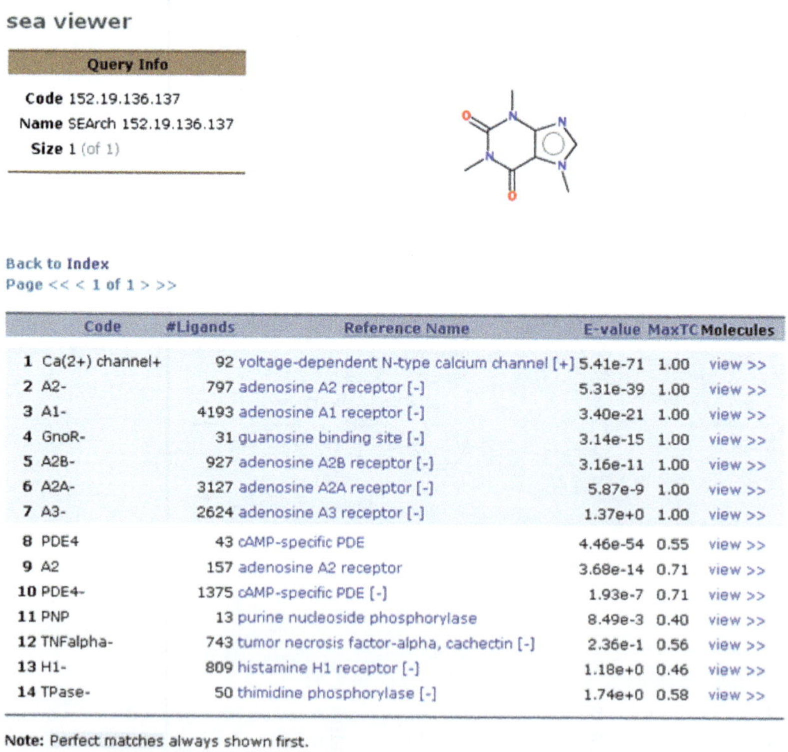
sea viewer

Query Info

**Code** 152.19.136.137
**Name** SEArch 152.19.136.137
**Size** 1 (of 1)

Back to Index
Page << < 1 of 1 > >>

| | Code | #Ligands | Reference Name | E-value | MaxTC | Molecules |
|---|---|---|---|---|---|---|
| 1 | Ca(2+) channel+ | 92 | voltage-dependent N-type calcium channel [+] | 5.41e-71 | 1.00 | view >> |
| 2 | A2- | 797 | adenosine A2 receptor [-] | 5.31e-39 | 1.00 | view >> |
| 3 | A1- | 4193 | adenosine A1 receptor [-] | 3.40e-21 | 1.00 | view >> |
| 4 | GnoR- | 31 | guanosine binding site [-] | 3.14e-15 | 1.00 | view >> |
| 5 | A2B- | 927 | adenosine A2B receptor [-] | 3.16e-11 | 1.00 | view >> |
| 6 | A2A- | 3127 | adenosine A2A receptor [-] | 5.87e-9 | 1.00 | view >> |
| 7 | A3- | 2624 | adenosine A3 receptor [-] | 1.37e+0 | 1.00 | view >> |
| 8 | PDE4 | 43 | cAMP-specific PDE | 4.46e-54 | 0.55 | view >> |
| 9 | A2 | 157 | adenosine A2 receptor | 3.68e-14 | 0.71 | view >> |
| 10 | PDE4- | 1375 | cAMP-specific PDE [-] | 1.93e-7 | 0.71 | view >> |
| 11 | PNP | 13 | purine nucleoside phosphorylase | 8.49e-3 | 0.40 | view >> |
| 12 | TNFalpha- | 743 | tumor necrosis factor-alpha, cachectin [-] | 2.36e-1 | 0.56 | view >> |
| 13 | H1- | 809 | histamine H1 receptor [-] | 1.18e+0 | 0.46 | view >> |
| 14 | TPase- | 50 | thimidine phosphorylase [-] | 1.74e+0 | 0.58 | view >> |

**Note:** Perfect matches always shown first.

**Figure 13.3** Snapshot of target predictions by the similarity ensemble approach (SEA).

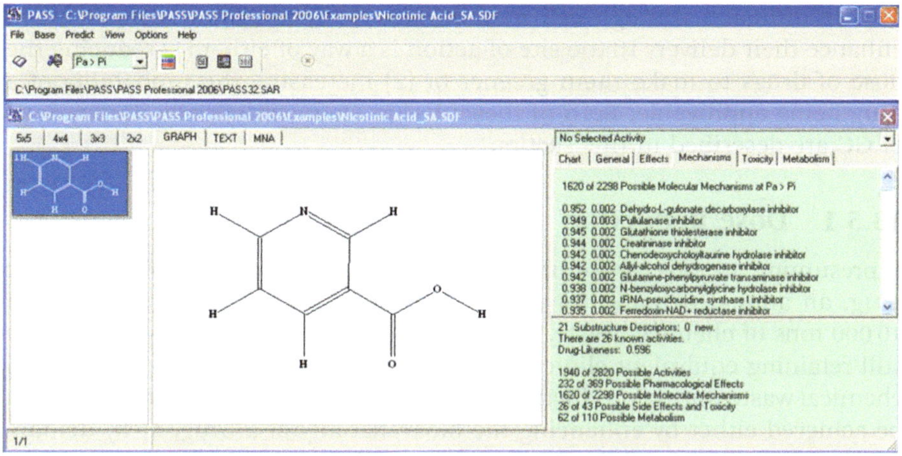


**Figure 13.4** Snapshot of pharmaceutical and toxicological profile predicted by PASS software.

drug-target associations were ultimately confirmed by further testing. This resulted in 247 significant links to adverse drug reactions in patients. This computational evaluation results in the design of less hazardous drugs (and chemicals), addressing GC principle number 4. This approach also inherently incorporates the GC applications to drug discovery discussed earlier – *i.e.*, increasing the success rate of NCEs in development and reducing the effort required to discover new, approvable drug candidates. The potential of this computational method for early prediction of toxic liabilities and the reduction of failures in drug development due to drug–drug interactions or toxicities is enormous.[49]

The concept of a biological activity spectrum is the basis for the PASS (prediction of activity spectra for substances) software.[50] Other approaches predict the chemical interactions of a drug with a specific biological target, PASS predicts a broad profile of pharmaceutical and toxicological properties, or a "biological activity spectrum" of a molecule. In PASS, predictions are based on an SAR analysis compared with a "training set" of over 60 000 compounds. PASS utilizes multilevel neighborhoods of atoms (MNAs) as the chemical structure descriptors and the algorithm of the activity spectra estimation as the training procedure. The output of predictions is presented by the activity spectrum, which is a ranked list of the probabilities of specific type of activity being noted (Figure 13.4).

## 13.5 Targeting the Active Site

Drugs must reach their site of biological activity in order to be useful. Many drugs fail in development because their pharmacokinetic properties (particularly absorption) are inadequate to provide sufficient concentrations at the site of action. Most drugs arrive at the active site by dissolution and passive diffusion through cell membranes. Designing molecules so as to enhance their delivery to the site of action is a way of either (1) reducing the dose of drugs to make them greener or (2) increasing the probability of a drug being approvable. Both of these design characteristics as illustrations of GC are described in this section.

### 13.5.1 Dose Reduction

If presuming an *E*-factor for pharmaceutical production of 100 for a specific drug, an annual demand for API of 100 tons results in the generation of 10 000 tons of chemical waste. Reducing the *dose* of this drug by 25% (while still retaining equivalent clinical efficacy) will then reduce the generation of chemical waste during production by 2500 tons per year. Dose reductions can be achieved either by enhancing the bioavailability of a drug, or by demonstrating that a drug is clinically effective at a lowered dose. One example of GC that was identified during drug development but was not implemented for many years is that of the dose of efavirenz (EFV; proprietary name Sustiva™). EFV is approved as a 600 mg daily adult dose. In Phase II clinical trials EFV demonstrated equivalent antiviral activity at 200, 400, or 600 mg daily doses.[51] When EFV was launched in 1998, HIV drugs were dosed at the maximum tolerated dose to minimize the potential for viral resistance. Hence, EFV was approved at a daily dose of 600 mg. EFV is now typically dosed in combination with one NRTI drug such as emtricitabine (FTC) or lamivudine (3TC), and with the NNRTI drug tenofovir disoproxil fumarate (TDF). Triple-combination cocktails containing EFV are now the most-widely used drug regimens for the treatment of HIV/AIDS; demand for EFV in low- and middle-income countries is presently about 1500 metric tons per year.[52a] Recent clinical trials in several hundred people showed that EFV is equivalent in its antiviral activity at 400 and 600 mg daily doses in triple-combination dosing.[52b] At the 400 mg daily dose, EFV demonstrated reduced side effects, so that a higher percentage of patients continued on therapy to achieve effective viral suppression. If the 400 mg daily dose is uniformly adopted for use, this would reduce the demand for EFV by about 500 tons per year. The *E*-factor for producing EFV is about 35. The reduction in the generation of chemical waste from this dosing switch, therefore, would amount to about 17 500 tons per year.

### 13.5.2 Conjugation of Drugs with Target Recognition Elements

Most drugs fail in development because of inadequate bioavailability, unacceptable pharmacokinetics, lack of efficacy, or unacceptable toxicity.[53] These effects are often related. Unacceptable toxicities, for example, arise when drugs are not sufficiently selective in targeting the site of biological activity. If scientists were capable of selectively delivering drugs to the active site, the success rate of drugs in development would increase dramatically. Advanced drug delivery techniques to improve bioavailability or to target drugs for delivery to an active site are a broad area of pharmaceutical research, and examples of successful applications are plentiful. The use of drug–polymer conjugates as in the case of Interferon-alpha bound to poly(ethylene glycol) (pegylated interferon or PEG-INF) is one advance in drug conjugates. PEG-INF was until recently part of the standard of care for the treatment of hepatitis C.[54] Abraxane™ is a protein-bound (albumin) formulation of the anticancer drug taxol that demonstrates significantly improved rates of response *versus* the use of unbound taxol in the combination treatment of non-small cell lung cancers (NSCLCs) and pancreatic cancer.[55]

Monoclonal antibody (mAb)–drug conjugates consist of a small-molecule drug, a low-molecular weight linker, and a monoclonal antibody. The mAb is incorporated as a means of actively targeting the site of disease in the human body. These drugs are represented by the anticancer drugs mylotarg™ (approved in 2000, although later withdrawn from the market),[56] adcetris™ (approved 2011),[57] and kadcyla™ (ado-trastuzumab emtansin; approved 2013).[58] Over 40 INDs have been filed for the use of drug–antibody conjugates for treating human diseases, but most of these have been filed since 2008 so the demonstration of widespread advantage in using such treatments is still underway.[59] In addition to providing a business case for GC by improving drug success rates in development, however, this approach also incorporates principle four of the twelve GC principles; by targeting the active site, drugs are inherently safer for use.

### 13.5.3 Pharmacokinetic Enhancers

Another technique for "greening" drug discovery is the use of drugs as pharmacokinetic enhancers (PKEs). The ability of a drug to be available at the site of action is governed by its properties of absorption, distribution, metabolism, and elimination (ADME). The concentration of a drug in blood plasma is most often the best indirect measure of its ability to reach the site of therapeutic activity. The absorption properties of a drug can sometimes be enhanced by the choice of formulation and solid-state (*e.g.*, crystalline *versus* amorphous) form. A drug's distribution properties are largely inherent to its specific molecular structure. Drugs are largely metabolized by specific enzymes. Drug metabolism has long been viewed as inherently unchangeable for a given drug. Drug–drug interactions that change the metabolic pattern (concentration *versus* time profile) of specific drugs in blood plasma are well known. These have long been viewed as undesirable, because significant, unexpected changes of drug concentrations can be dangerous or even fatal. Drugs that inhibit metabolizing enzymes in the body as "pharmacokinetic enhancers" (PKEs), however, have been used as a successful element of making drug discovery greener by allowing the successful dose reduction of multiple drugs.

Cobicistat (Tybostat™; or COBI)[60] was approved by the US FDA in combination use (with elvitegravir, tenofovir disoproxil fumarate, and emtricitabine) for the treatment of HIV/AIDS in 2012 as Stribild™, or "the Quad Pill". Cobicistat is an inhibitor of the cytochrome P450 (CYP) 3A enzyme, most particularly of the sub-enzyme CYP 3A4. CYP 3A4 is a primary metabolizing enzyme for many drugs. By virtue of down-regulating CYP 3A4, COBI inhibits the metabolism, and enhances the concentration and half-life of many drugs in blood plasma. As a component of Stribild™, COBI presumably allows for the daily dose of elvitegravir (ELV; 150 mg per day) to be much lower than would otherwise be the case; it is known that ELV is primarily metabolized by CYP 3A4.[61]

The use of PKEs has precedent. The HIV-1 protease inhibitor drug ritonavir (RTV) has been used at sub-therapeutic doses to "boost" the blood levels of other protease inhibitors since 1996 (Figure 13.5).[62] As it turns out, RTV not only inhibits the metabolism of drugs by CYP 3A4, but it also inhibits P-glycoprotein (P-gp) efflux transporter proteins. These proteins recognize xenobiotics (foreign molecules) with broad specificity. P-gp proteins actively transport drugs and other xenobiotics out of cells.[63] This technique for improving the ADME properties of protease inhibitors has been an important success in the treatment of HIV/AIDS. COBI has no significant antiviral activity, and is being tested as a PKE for other drugs with different mechanisms of action and other therapeutic uses. Typical adult doses for RTV as a PKE are 100 or 200 mg per day. RTV is also being used as a PKE to boost blood levels of Abbott Labs direct-acting antiviral (DAA) drugs for the treatment of hepatitis C.

A natural product responsible for the pungency of black pepper is also known to be a potent PKE. Piperine (Figure 13.6) is present as 5–10% of the dried weights of commercial black, and white, pepper. Piperine is an inhibitor of CYP 3A4 and an inhibitor of P-gp proteins. Piperine also inhibits the production of glucuronic acid in the stomach.[64] Glucuronic acid commonly conjugates to free acid and hydroxyl groups on molecules, thereby enhancing their water solubility and enabling them to be readily eliminated through renal clearance.

Piperine is co-dosed with rifampicin (rifamycin) for the treatment of tuberculosis (TB). Rifampicin is a potent inducer of CYP 3A4 metabolism. Chronic dosing of rifampicin results in the slow reduction of rifampicin concentrations in blood plasma over time by the up-regulation of CYP 3A4 (self-induction). After about six weeks, patients typically have about 70% of

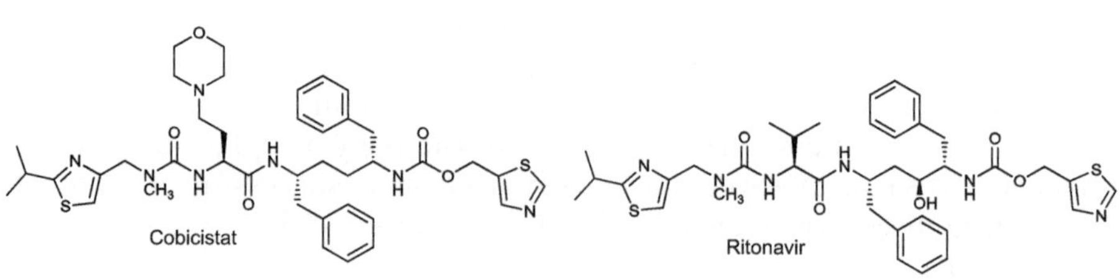


**Figure 13.5** Structures of the pharmacokinetic-enhancers (PKEs) cobicistat and ritonavir.

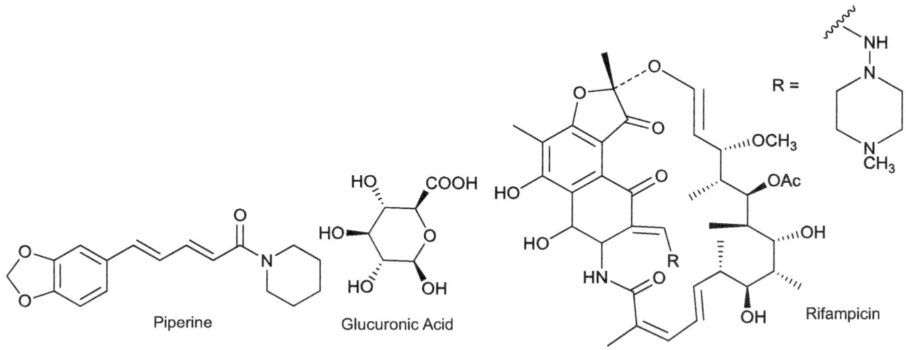


**Figure 13.6** Structures of piperine, glucuronic acid, and rifampicin.

the maximum blood levels of rifampicin that are achieved upon dosing for 10–14 days. Rifampicin without a PKE is dosed at 450 mg per day to effectively treat TB. Daily doses of 200 mg of rifampicin given with 10 mg of piperine actually provide significantly higher blood levels of rifampicin than 450 mg daily doses without piperine.[65] This dramatic effect also appears to reduce the resistance of TB mycobacterium to rifampicin, enabling its effective use for drug-resistant TB. Piperine is being investigated as a pharmacokinetic enhancer for several other purposes.

## 13.5.4 Prodrugs

Principle two of GC speaks to atom economy in designing chemicals, while principle eight is to avoid the use of derivatives. Prodrugs are derivatives of active drug molecules that are metabolized to the active drug. Prodrugs are almost always of higher molecular weight than the active drug, and are, therefore, contrary to these GC principles. The use of prodrugs under appropriate circumstances is, however, an interesting "backwards" twist on this principle. Creating derivatives of drugs can make them much greener than they might be otherwise, and they can increase the PTS of drug candidates as recently demonstrated by Gilead Sciences. Prodrug moieties can enhance the delivery of bioactive molecules to the site where they are active. The use of a specific ProTide (prodrug–nucleotide) moiety attached to the drug tenofovir (*i.e.*, tenofovir alafenamide fumarate, TAF; Figure 13.7) has enabled the greening of a drug by a huge dose reduction. The use of the same ProTide moiety in the case of the drug sofosbuvir (Sovaldi™; Figure 13.8) has meant the difference between a molecule that is inactive in cellular antiviral assays *versus* a drug that has been approved as a breakthrough new therapy for the treatment of hepatitis C.[66,67]

Acyclic nucleoside phosphonates are a class of drugs effective against a wide variety of DNA viruses and retroviruses. Acyclic nucleoside phosphonates are structurally represented by the compound tenofovir and its

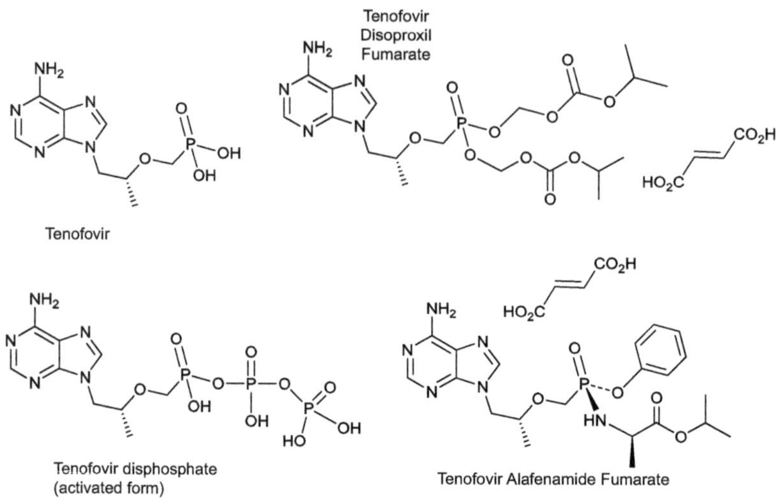


**Figure 13.7** Structures of tenofovir, tenofovir diphosphate, tenofovir disoproxil fumarate (TDF), and tenofovir alafenamide fumarate (TAF).

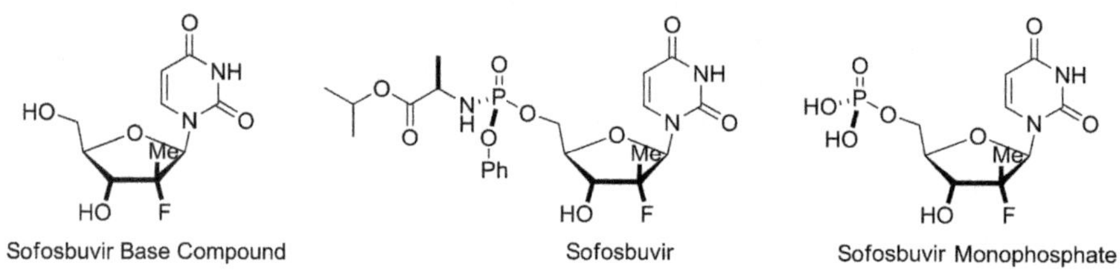


**Figure 13.8** Sofosbuvir base compound, sofosbuvir, and sofosbuvir monophosphate.

prodrug derivative tenofovir disoproxil fumarate (TDF). Tenofovir is an effective inhibitor of hepadnavirus and retroviruses *in vitro*. Tenofovir has excellent aqueous solubility, but is poorly lipophilic ($\log P = 1.3$ at pH 6.5) and is poorly absorbed.[68] The conversion of tenofovir into its corresponding disoproxil (bis-isopropoxycarbonyloxymethyl) ester increased the lipophilicity of tenofovir, enhancing its oral dosing utility. TDF is now widely used as a component of first-line therapy for HIV/AIDS. The global demand for TDF for HIV/AIDS and the treatment of hepatitis B is currently about 900–1200 tons per year.

The synthetic process for TDF generates approximately 50 kg of chemical waste for every kilogram of product formed.[69] The mechanism by which TDF becomes an active drug involves more than a single step. TDF must first be hydrolyzed to tenofovir. Tenofovir is activated inside the cell as its diphosphate in a process that successively involves two different phosphorylating enzymes. TDF is not an ideal prodrug. The disoproxil esters of TDF are cleaved indiscriminately by various hydrolyzing enzymes in the body, after which tenofovir in circulating blood plasma is readily eliminated. This is why the daily dose of TDF is 300 mg. Tenofovir as its alafenamide (alanine–phenoxide) prodrug (TAF) is highly resistant to hydrolysis in blood plasma. Inside a cell TAF is cleaved by two specific enzymes and phosphorylated to provide the active drug. In two Phase III trials, TAF was shown to be clinically equivalent to 300 mg of TDF at a dose of 10 mg per day.[70] Using different HIV drug combinations (*i.e.*, with efavirenz + lamivudine or efavirenz + emtricitabine) the daily dose of TAF is expected to be 30 mg. The synthesis of TAF is similar in *E*-factor to the synthesis of TDF.[71] Supposing a 90% reduction in daily dose (from 300 mg to 30 mg) the global demand of TAF as a replacement for TDF would be about 90–120 tons per year. At an *E*-factor of approximately 50, this would save about 40 500–54 000 tons of treatable chemical waste per year.

ProTide (pronucleotide) derivatives are designed to deliver nucleotide monophosphate derivatives directly into a virally-infected cell. In the case of sofosbuvir (Figure 13.8, Sovaldi™; SOF) – a newly-approved therapy for the treatment of hepatitis C – this makes the difference between an inactive molecule and a "breakthrough" new therapy. The base compound without the phosphoramidate prodrug was tested and found to be essentially inactive in cellular (phenotypic) assays against hepatitis C.[66,67] Incorporation of the ProTide prodrug moiety, however, enhances the absorption of SOF *versus* the base compound. The ProTide moiety was deliberately designed by McGuigan to provide nucleotide drugs in their mono-phosphorylated form upon hydrolysis.[72,73] This overcomes the problem that the first-step phosphorylation of the base compound by the enzyme thymidylate kinase can be very slow. Since SOF is hydrolyzed intracellularly to its mono-phosphate, thymidylate kinase is not required for intracellular activation and SOF is highly active.[66,67]

## 13.6 Reverse Pharmacology

"Reverse pharmacology" (RP) integrates experiential and clinical identification of natural medicines (traditional knowledge) into drug discovery, utilizing a process that is quite different from rational drug design.[72,73] This approach identifies natural products or mixtures of natural products that are "hits" based on traditional knowledge or, less often, by screening in biological assays. This constitutes a green approach to drug discovery since it takes advantage of the existing, cultural, and ethnobotanical knowledge of a region to short-cut the very lengthy, expensive, resource-intense process of genomics, proteomics, target identification, target validation, and high-throughput screening that is used in rational drug design.

The "hits" identified by RP are further developed into drug candidates by interdisciplinary exploratory studies including safety and activity testing, combined with experimental and clinical research. During clinical development, mixtures of compounds obtained as crude extracts are often used for human dosing, but indiscriminate dosing of patients is not allowed. Although the stage of drug discovery/development at which compound isolation takes place is moved later into the development phase, the classical activities of isolating compounds and elucidating their structure are still a part of RP. The correlation of API content with biological activity, the development and validation of analytical methods, process validation, and the setting of specifications for the (multiple) active ingredients present are all steps in the development process for phytomedicines.

### 13.6.1 Drugs from Traditional Knowledge

Phytomedicines, or herbal medicines, are plants, herbs, plant extracts, and purified natural products used for the prevention or treatment of disease. Phytomedicines are most often identified and derived from traditional knowledge. The regulatory requirements for approving phytomedicines are substantially reduced *versus* NCEs, although the label claims for efficacy are also much more cautious. More than 80% of people in underdeveloped countries depend on herbal medicines.[74,75] In recent years, plant sources have received much attention in the treatment of various diseases.[76–78] The World Health Assembly adopted a resolution (1989) supporting national traditional medicines programs as important to the health of individuals and communities.[79] The World Health Organization has also issued guidance on (1) quality assessment, (2) pharmacovigilance, and (3) Good Manufacturing Practices for herbal medicines. A major issue with appropriate use and assuring the quality of herbal medicines still exists as countries are only beginning to regulate the registration, approval, use, and standardization of such products.[80–82]

Herbal medicines must be safe, efficacious, and affordable to be useful. An important component of affordability relates to how herbal medicines are prepared. Extracts that contain a single, highly-purified component present in very minor amounts in plants harvested by destructive methods are counter to the tradition of herbal medicines. Gathering of herbal medicines is most often carried out by harvesting of fruits, seeds, immature shoots, or leaves in a manner that is not destructive of the entire plant. The processing techniques used for herbal medicines are most often similar to those used for processing spices. The active components of herbal medicines represent rather large amounts (several percent) by weight of their respective plant sources and their structures are often rather similar to those present in spices such as curcumin, resveratrol, and carotenoids. Phytomedicines are typically purified by short, simple processes such as aqueous or aqueous alcoholic decoction, followed by solvent removal. This is a large difference from pharmaceutical natural products derived from plants, such as the vinca alkaloids, taxol, or camptothecins. These are typically present as much less than 1% of the dried weight of their respective plant sources and are purified by methods that are quite resource intensive and highly inefficient. These techniques for herbal medicines derive from the resources available to traditional healers, and are inherently much greener for this reason.

Reverse pharmacology and herbal medicines are of major interest in Sub-Saharan Africa. A major center for herbal medicines exists at the National Institute for Pharmaceutical R&D (NIPRD) in Nigeria. For many years NIPRD has been engaged in a systematic approach to gathering, classifying, extracting, testing, and identifying the therapeutic elements present in plant sources. A quality Management System is in place at NIPRD, and Standard Operating Procedures (SOPs) for phytomedicines are followed for botanical identification, cultivation and gathering, extraction, testing, specifications development, and pharmaceutical formulation of phytomedicines.[75,83–85] A multidisciplinary team of scientists is employed to carry out this mission. One successful project that resulted in the herbal medicine Niprisan™ for the treatment of sickle-cell anemia is described in some detail below.

### 13.6.2 Selection of Drug Candidates

Niprisan™ is a combined extract of four plants that was traditionally used locally in Nigeria to prevent sickle cell disease (SCD).[75,77] SCD is caused by a point mutation in the gene coding for the beta globin chain of the hemoglobin tetramer. The mutated hemoglobin (HbS) that arises from this genetic defect has altered surface charge properties, with profound effects on the solubility of the protein, the physiology of oxygen transport, and on the well-being of the individual. In un-mutated hemoglobin (HbA) ionized glutamate residues govern the shape and charge distribution of the protein. In the HbS form, valine residues replace glutamate, creating hydrophobic protrusions. Mutated HbS aggregates in chainlike, polymeric structures. These insoluble fibers distort red blood cells into an elongated, "sickled" shape. These cells do not pass freely through arterial and venous capillaries. This causes impaired circulation, tissue and organ damage, severe anemia, and increased mortality. *In vitro* studies showed that Niprisan™ possesses potent anti-sickling effects. One or more of the active ingredients contained in Niprisan™ enhances the solubility of deoxy-HbS, retarding polymerization and increasing oxygen affinity.[75] Controlled clinical trials conducted in Nigeria showed that Niprisan™ significantly reduces vaso-occlusive events in SCD patients.[75,78,85]

During early development, NIPRD evaluated various extracts from the medicinal plants used to prepare this herbal product. Niprisan™ was eventually standardized by extracting various seeds, fruits, and stem biomass from *Pterocarpus osun, Sorghum bicolor, Piper guineensis*, and *Eugenia caryophyllum*. Data from *in vivo* tests in laboratory animals and using blood from sickle cell patients showed that the product is effective at preventing cell sickling. $LD_{50}$ values in mice and rats of 900 mg $kg^{-1}$ (i.p.) and 5 g $kg^{-1}$ P.O showed substantial safety margins.[75,77,85] Sub-chronic studies indicated the absence of major tissue and organ toxicities. Since thousands of people were already taking crude extracts, an initial human safety study was carried out in only 20 patients.[77] The standardized extract was tested in laboratory animals and blood from sickle cell patients before human use.

### 13.6.3 Clinical Development

Clinical Phase II/III trials began in March 1994, using a capsule dosage form. This represented about an 18 month timeframe from the initiation of a discovery program at NIPRD. In that time, the product was identified, extracts were standardized, assays were developed, safety assessment was carried out, an initial isolation/manufacturing process was identified, and clinical supplies were prepared by filling of capsules with lyophilized extracts. The extraction process developed for obtaining Niprisan™ is quite efficient. The gathering of materials for extraction is not destructive of the cultivated plants. The extracts are obtained by steeping for several hours in large-scale fixed equipment using about four liters per kilogram of 1 : 1 aqueous ethanol. After filtration and distillation/lyophilization to remove solvent, the dried Niprisan™ extract represents about 20% of the collected biomass. One kilogram of Niprisan™ is, therefore, obtained utilizing about 5 kg of biomass, and 10 L each of water and ethanol for a total *E*-factor of about 25.

A randomized, double-blind, placebo-controlled, crossover trial was conducted as part of the Niprisan™ approval process.[77,85] Patients with homozygous sickle cell disease and three or more painful or vaso-occlusive crises per year were randomized into two study groups. One group took the test drug (12 mg per kg body weight) daily for six months before crossing over to placebo for another six months. Another group took placebo for six months before crossing over to active drug for six months. There was a one month washout period between patient crossovers. The daily use of Niprisan™ reduced the mean number of crises per person-month by 55% from placebo – a highly significant improvement ($\rho < 0.05$). Patients taking Niprisan™ generally rated their health as better and reported less sickness and absenteeism *versus* placebo as well. Apart from headache (12%), patients reported no significant adverse effects.[75,77,85] It was concluded that Niprisan™ was efficacious for the prophylactic management of SCD. Notably, SCD patients can, to a substantial degree, detect and predict symptoms of impending SCD attacks. Subsequent prophylactic use of Niprisan™ with chronic, intermittent dosing by approximately 90 000 people in Nigeria has resulted in the reduction of SCD attacks by 98%.

### 13.6.4 Isolation and Identification of Active Constituents

Herbal extracts almost always contain several compounds with varying biological properties. It is sometimes difficult to attribute biological activity to a single particular compound. It is still useful to isolate active ingredients, however, to (1) serve as a marker for quality control and standardization and (2) to evaluate the potential to develop a single-agent treatment. Experience at NIPRD has shown that some phytomedicines are more active as crude extracts than as purified entities. Additional experience as shown – as is the case with Niprisan™ – that some extracts contain multiple, closely-related molecules with similar biological activity. Niprisan™ was approved for marketing by the Nigerian National Agency for Food and Drug Administration and Control (NAFDAC) in 2006.[86] The original sponsor of the NAFDAC marketing application, XeChem International, estimated that the total discovery and development costs of Niprisan™ were approximately $28 million (out-of-pocket) even considering the $8 million cost of building a manufacturing facility to produce Niprisan capsules.[87] This figure is much lower than the figures discussed earlier in this chapter for rational drug discovery and development.

### 13.6.5 Conclusions on Reverse Pharmacology for Drug Discovery

Several broad questions are often posed when comparing reverse pharmacology to rational drug discovery. These questions are:

1. Are medicines from traditional knowledge/reverse pharmacology safe?
2. Do these medicines really qualify as being green?
3. Are these medicines truly effective?
4. Why would anyone practice reverse pharmacology when rational drug design is available?

Rational drug discovery costs huge amounts of money. Emerging economies, especially those in Sub-Saharan Africa, cannot afford to extensively adopt this approach to discovering and developing new medicines for their unique health needs. The example of using RP for drug discovery in SCD is particularly appropriate because it must be appreciated that – with the huge associated costs of bringing a new drug to the market – global pharmaceutical originator companies cannot be investing heavily in SCD and other diseases that qualify for the designation of neglected tropical diseases (NTDs). Reverse pharmacology is a much less expensive approach to drug discovery. Although RP does not possess the same scope for discovering new medicines as rational drug design, it is traditionally practiced – as we have seen – in an efficient, green manner. Failure rates in development are much lower for RP drug candidates than for rational drug design, because evidence of human safety from traditional use is an important element of identifying hits for development. Clinical trials using herbal medicines and an RP approach are admittedly not as rigorous as those for rational drug discovery. Large, multi-center controlled trials with several thousand patients are incredibly expensive and difficult to manage. This level of resources is just not available to develop herbal medicines for NTDs. Yet, it is possible to design and successfully execute placebo-controlled, crossover studies to demonstrate clinical efficacy for herbal medicines. Ultimately, drugs derived from RP are more accessible and affordable than those from pharmaceutical originator companies. As an ultimate goal, one would wish for RP to be carried out with just as much rigor and oversight as for rational drug design. Until equivalent resources are available to devote to the development of new treatments for NTDs, however, some compromises must be made to provide new medicines for these significant, unmet medical needs. In truth, this only emphasizes the need for GC in drug discovery, as making any and all valid forms of drug discovery much more efficient is the only means by which effective medicines will be made available to all of the more than seven billion people alive today.

## 13.7 Conclusions

This chapter has presented a series of varying perspectives on the business case for Green Chemistry in drug discovery. Each of these is meant to be a short survey of the grounding principles of operation, and an assessment of where these approaches are in terms of their present state of being green. Reducing out-of-pocket expense and reducing the human effort required for drug discovery are both considered to be green as well as being good business. Other processes that improve the PTS of drugs in development, shorten development time, or that predict drug failures in advance of expensive animal and human testing also represent GC. Reverse pharmacology can be stunningly effective for drug discovery when traditional knowledge is available to guide candidate selection. Rational drug discovery is very expensive for most challenging drug targets. The proliferation of approaches to generating new protein targets and of synthesizing and selecting new drug candidates holds great promise for improving mankind's ability to provide new medicines for unmet medical needs, in a manner that is consistent with alleviating human suffering and decreasing mortality on a global scale. Although "big pharma" and rational drug design are often viewed with pessimism, the contents of this book show that the industry is making consistent and important progress towards a new generation in drug discovery. This will be a generation in which sustainability is achieved not only for the current paradigm of medicines for the rich, but also of medicines for all in need.

## References


1. D. C. Swinney and J. Anthony, *Nat. Rev. Drug Discovery*, 2011, **10**, 507.
2. For a short discussion and definition of the term "drug development," see: http://www.definitions.net/definition/Drug%20development (Accessed December 12, 2014).
3. The Tufts University study has not been published as of December, 2014. For a summary of the findings and results of this study, see: http://csdd.tufts.edu/news/complete_story/pr_tufts_csdd_2014_cost_study (Accessed December 12, 2014).
4. D. W. Light and R. Warburton, *Biosocieties*, 2011, **6**, 1.
5. Published online, 10 February, 2012. See: http://www.forbes.com/sites/matthewherper/2012/02/10/the-truly-staggering-cost-of-inventing-new-drugs/ (Accessed November 2, 2014).

6. See: http://www.answers.com/Q/What_is_reverse_pharmacology (Accessed December 12, 2014).
7. The requirements and contents for the Common Technical Document (CTD) format for regulatory filings across multiple countries are published online. See: http://www.fda.gov/Drugs/DevelopmentApprovalProcess/FormsSubmissionRequirements/ElectronicSubmissions/ucm153574.htm (Accessed December 1, 2014).
8. Much of the information publicly available about the US Prescription Drug User Fee Act (PDUFA) is published online. PDUFA fees fund hiring of US FDA field inspectors and other experts to speed the review of marketing applications. See: http://www.fda.gov/ForIndustry/UserFees/PrescriptionDrugUserFee/default.htm (Accessed December 1, 2014).
9. For a reasonably detailed illustration of the major scientific disciplines involved in the various stages of drug discovery, see http://www.dementiatoday.com/alzheimers-disease-hope-through-drugs/ (Accessed November 11, 2014).
10. M. Ringel, P. Tollman, G. Hersch and U. Schulze, *Nat. Rev. Drug Discovery*, 2013, **12**, 901.
11. B. W. Cue, in *Green Techniques for Organic Synthesis and Medicinal Chemistry*, ed. W. Zhang and B. W. Cue, Jr., John Wiley & Sons, Hoboken, New Jersey, USA, 1st edn, 2012, ch. 20, p. 563.
12. Figures obtained from Infodrive India; see: www.infodriveindia.com and are quoted for exports from India to the USA as of July, 2014. Pricing does not include the cost of freight.
13. M. C. Bryan, B. Dillon, L. G. Hamman, G. J. Hughes, M. E. Kopach, E. A. Peterson, M. Pourashaf, I. Raheem, P. Richardson, D. Richter and H. F. Sneddon, *J. Med. Chem.*, 2013, **56**, 6007.
14. See: https://www.linkedin.com/groups/Green-Chemistry-in-Drug-Discovery-4008985?split_page=2.
15. This figure is taken from personal history from one of us (JF) as well as from interviews with several drug discovery managers currently working on the pharmaceutical industry.
16. As a justification for this statement we would offer: although capital equipment, represented by the current prices of an HPLC (roughly $90 000) or a modest synthesis flow reactor ($85 000+) are expensive, these items are useful for many years. Even if only one person uses an HPLC, with a useful life-span of about ten years, this amounts to an out-of-pocket expense of $9000 per year. At an FTE cost of $400 000 per year, the purchase of an HPLC would be justified if is improved an individual's productivity by about 2.25%! We would argue that the difference in productivity of a bench chemist with *versus* without an HPLC is much greater than 2.25%.
17. See: http://syrris.com/applications/flow-chemistry/flow-basics (Accessed December 12, 2014).
18. C. Wiles and P. Watts, *Future Med. Chem.*, 2009, **1**(9), 1593.
19. C. Wiles, *Spec. Chem. Mag.*, 2009, **29**(7), 40–41.

20. C. Wiles and P. Watts, *Org. Process Res. Dev.*, 2008, **12**, 1001.
21. See: http://www.labbulletin.com/articles/FlowSyn-AUTO-LF-The-automatic-choice-for-Combinatorial-Flow-Chemistry#.VHUGH8lTyIE (Accessed December 12, 2014).
22. P. Watts and S. J. Haswell, *Drug Discovery Technol.*, 2003, **8**(13), 586.
23. See: http://organicchemistry-educationandindustry.blogspot.com/2010/04/green-chemistry-and-drug-discovery.html (Accessed December 12, 2014).
24. R. K. Henderson, C. Jimeénez-González, D. J. C. Constable, S. R. Alston, G. G. A. Inglis, G. Fisher, J. Sherwood, S. P. Binks and A. D. Curzons, *Green Chem.*, 2011, **13**, 854.
25. J. P. Adams, C. M. Alder, I. Andrews, A. M. Bullion, M. Campbell-Crawford, M. G. Darcy, J. D. Hayler, R. K. Henderson, C. A. Oare, I. Pendrak, A. M. Redman, L. E. Shuster, H. F. Sneddon and M. D. Walker, *Green Chem.*, 2013, **15**, 1542.
26. D. Prat, O. Pardigon, H.-W. Flemming, S. Letestu, V. Ducandas, P. Isnard, E. Guntrum, T. Senar, S. Ruisseau, P. Cruciana and P. Hosek, *Org. Process Res. Dev.*, 2013, **17**(12), 1517.
27. Personal recollections from JF with assistance from James Sherbine and Rodney L. Parsons, Jr.
28. B. Jiang and Y. Si, *Amino Alcohol Ligand and it use in Preparation of Chiral Propargylic Tertiary Alcohols and Tertiary Amines via Enantioselective Addition Reaction.* US Pat. 7,439,400 issued 21 October, 2008.
29. R. B. Bollu, N. R. Ketavarapu, V. S. K. Indukuri, S. R. Gorantla and S. Chava, *Efficient Process to Induce Enantioselectivity in Procarbonyl Compounds*, US Patent Application 2012/0264933, filing date 22 June, 2012.
30. D. B. Damon, B. P. Jones, R. W. McLaughlin and H. A. Watons, Jr., Abstract # xxxx, 2003, 226th National Meeting of the American Chemical Society, NY.
31. B. List, *Chem. Rev.*, 2007, **107**(12), 5413.
32. A. K. Ghosh, D. L. Zawsona and H. Mitsuya, *Bioorg. Med. Chem.*, 2007, **15**, 7576.
33. J. M. D. Fortunak and C. King, unpublished results.
34. See: http://www.princeton.edu/chemistry/macmillan/research/ (Accessed December 1, 2014).
35. D. J. Newman and G. M. Cragg, *J. Nat. Prod.*, 2007, **70**(3), 461.
36. L. Zhang, M. Zheng, F. Zhao, Y. Zhai and H. Liu, *ACS Comb. Sci.*, 2014, **16**(4), 184.
37. M. Zhu, B. J. Lim, M. Koh and S. B. Park, *ACS Comb. Sci.*, 2014, **14**(2), 124.
38. B. L. Roth, D. J. Sheffler and W. K. Kroeze, *Nat. Rev. Drug Discovery*, 2004, **3**(4), 353.
39. S. Kalyaanamoorthy and Y. P. Chen, *Drug Discovery Today*, 2011, **16**(17–18), 831.
40. L. Chen, J. K. Morrow, H. T. Tran, S. S. Phatak, L. Du-Cuny and S. Zhang, *Curr. Pharm. Res.*, 2012, **18**(9), 1217.

41. D. B. Kitchen, H. Decornez, J. R. Furr and J. Bajorath, *Nat. Rev. Drug Discovery*, 2004, **3**(11), 935.
42. J. Pei, N. Yin, X. Ma and L. Lai, *J. Am. Chem. Soc.*, 2014, **136**, 11556.
43. D. Himmel, N. S. Myshakina, T. Ilina, A. Van Ry, W. C. Ho, M. A. Parniak and E. Arnold, *J. Mol. Biol.*, 2014, **426**, 2617.
44. E. B. Lansdon, K. M. Brendza, M. Hung, R. Wang, S. Mukund, D. Jin, G. Birkus, N. Kutty and X. Liu, *J. Med. Chem.*, 2010, **53**, 4295.
45. X. Chen, P. Zhan, D. Li, E. De Clerq and X. Liu, *Curr. Med. Chem.*, 2011, **18**(3), 359.
46. A. C. Mislak, K. M. Frey, M. Bollini, W. L. Jorgensen and K. S. Anderson, *Biochmica et. Biophysica Acta*, 2014, **1840**(7), 2203.
47. A. Li, y. Ouyang, Z. Wang, Y. Cao, X. Liu, L. Ran, C. Li, L. Li, L. Zhang, K. Qiao, W. Xu, Y. Wang, Z. Zhang, C. Tian, Z. Liu, S. Jiang, Y. Shao, Y. Du, L. Ma, X. Wang and J. Liu, *J. Med. Chem.*, 2013, **56**, 3593.
48. M. J. Keiser, B. L. Roth, B. N. Armbruster, P. Ernsberger, J. J. Irwin and B. K. Shoichet, *Nat. Biotechnol.*, 2007, **25**(2), 197.
49. E. Lounkine, M. J. Keiser, S. Whitebread, D. Mikhailov, J. Hamon, J. L. Jenkins, P. Lavan, E. Weber, A. K. Doak, S. Cote, B. K. Shoichet and L. Urban, *Nature*, 2012, **486**(7043), 361.
50. V. Poroikov, D. Akimov, E. Shabelnikova and D. Filimonov, *SAR QSAR Environ. Res.*, 2001, **12**(4), 327.
51. USFDA, Center for Drug Evaluation and Research: Medical Review, Application No. 20-972, pg. 24. Available online at: www.accessdata.fda.gov/.../nda/98/20972clinical_review.pdf (Accessed December 13, 2014).
52. (a) Personal communication, December 17, 2014; Henk den Besten, Partners for Supply Chain Management Systems, Arlington, VA; (b) R. Puls *et al.*, *Lancet*, 2014, **383**(9927), 1474.
53. R. Frank and R. Hargreaves, *Nat. Rev. Drug Discovery*, 2003, **2**(3), 566.
54. R. P. Ward and M. Kugelmas, *Am. Fam. Physician*, 2005, **72**(4), 656.
55. W. J. Gradishar, *Expert Opin. Pharmacother.*, 2006, 7(8), 1041.
56. P. F. Bross, J. Beitz, G. Chewn, X. H. Chen, E. Duffy, L. Kieffer, S. Roy, R. Sridhara, A. Rahman, G. Williams and R. Pazdur, *Clin. Cancer Res.*, 2001, 7(6), 1490.
57. C. Vaklavas and A. Forero-Torres, *Ther. Adv. Hematol.*, 2012, **3**(4), 209.
58. Niculescu-Duvaz, *Curr Opin. Mol. Ther.*, 2010, **12**(3), 350.
59. B. A. Teicher and J. H. Doroshow, *N. Engl. J. Med.*, 2012, **367**(19), 1847.
60. E.-I. Lepist, T. K. Phan, A. Roy, L. Tong, K. MacLennan, B. Murray and A. S. Ray, *Antimicrob. Agents Chemother.*, 2012, **56**(10), 5409.
61. For the USFDA labelling information on the Quad Pill see: http://google2.fda.gov/search?q=stribild&client=FDAgov&site=FDAgov&lr=&proxystylesheet=FDAgov&requiredfields=-archive%3AYes&output=xml_no_dtd&getfields=* (Accessed December 5, 2014).
62. B. Turk, *Nat. Rev. Drug Discovery*, 2006, **5**, 785.
63. S. G. Aller, J. Yu, A. Ward, Y. Weng, S. Chittaboina, R. Zhuo, P. M. Harrell, Y. T. Trinh, Q. Zhang, I. L. Urbatsch and G. Chang, *Science*, 2009, **323**(5922), 1718.

64. R. K. Bhardwaj, H. Glaeser, L. Becquemont, U. Klotz, S. K. Gupta and M. F. Fromm, *J. Pharmacol. Exp. Ther.*, 2002, **302**, 645.
65. D. D. Bhatt, M. R. Trivedi, I. A. Modi and B. M. Khamar, *Pharmaceutical Composition of Rifampicin*, WO Patent Application 2011/010214, International filing date 22 July, 2010.
66. E. Murakami, H. Bao, M. Ramesh, T. R. McGrayer, T. Whitaker, H. M. Micolochick Steuer, R. F. Schinazi, L. J. Stuyver, A. Obikhod, M. J. Otto and P. A. Furman, *Antimicrob. Agents Chemother.*, 2007, **51**, 503.
67. M. J. Sofia, D. Bao, W. Chang, J. Du, D. Nagarathnam, S. Rachakonda, P. G. Reddy, B. S. Ross, P. Wang, H.-R. Zhang, S. Bansal, C. Espiritu, M. Keilman, A. M. Lam, H. M. Micholochick Steuer, C. Niu, M. J. Otto and P. A. Furman, *J. Med. Chem*, 2010, **53**, 7202.
68. E. De Clercq, in *Antiviral Drugs: From Basic Discovery through Clinical Trials*, ed. W. M. Kazmierski, John Wiley & Sons, Inc., Hoboken, New Jersey, USA, 1st edn, 2011, ch. 7, pp. 211–243.
69. D. H. B. Ripin, D. S. Teager, J. Fortunak, S. M. Basha, N. Bivins, C. N. Boddy, S. Byrn, K. K. Catlin, S. R. Houghton, S. T. Jagadeesh, K. A. Kumar, J. Melton, S. Muneer, L. N. Rao, R. V. Rao, P. C. Ray, N. G. Reddy, R. M. Reddy, K. C. Shekar, T. Silverton, D. T. Smith, R. W. Stringham, G. V. Subbaraju, F. Talley and A. Williams, *Org. Process Res. Dev.*, 2010, **14**, 1194.
70. See: http://www.drugs.com/clinical_trials/gilead-s-investigational-tenofovir-alafenamide-taf-based-single-tablet-hiv-regimen-meets-48-week-16650.html (Accessed December 13, 2014).
71. W. A. Lee, G. X. He, E. Eisenberg, T. Cihlar, S. Swaminathan, A. Mulato *et al.*, *Antimicrob. Agents Chemother.*, 2005, **49**, 1898.
72. M. Serpi, K. Madela, F. Pertusati and M. Slusarczyk, *Curr. Protoc. Nucleic Acid Chem.*, 2013, 15.5.1–15.5.15, (Wiley Online Library).
73. C. McGuigan, R. M. Pathirana, J. Balzarini and E. De Clerq, *J. Med. Chem.*, 1993, **36**, 1048.
74. B. Patwardhan and R. A. Mashelkar, *Drug Discovery Today*, 2009, **14**, 804.
75. C. Wambebe, H. Khamofu, J. A. F. Momoh, M. Ekpeyong, B. S. Audu, O. S. Njoku, E. A. Bamgboye, R. N. Nasipuri, O. O. Kunle, J. I. Okogun *et al.*, *Phytomedicine*, 2001, **8**(4), 252.
76. M. L. Willcox, B. Graz, J. Falquet, C. Diakite, S. Giani and D. Diallo, *Malaria J.*, 2011, **10**(Suppl 1), S8.
77. K. Perampaladas, H. Masum, A. Kapoor, R. Shah, A. S. Daar and P. A. Singer, *BMC Int. Health Hum. Rights*, 2010, **10**(Suppl. 1), S11.
78. E. W. Iyamu, E. A. Turner and T. Asakura, *Br. J. Hematol.*, 2003, **122**, 1001.
79. See: http://www.who.int/mediacentre/factsheets/fs134/en/ (Accessed October 6, 2014).
80. World Health Organization, *WHO Guidelines for Assessing Quality of Herbal Medicines with Reference to Contaminants and Residues*, WHO Press, 20 Avenue Appia, 1211 Geneva 27, Switzerland, 2007, p. 118.

81. World Health Organization, *WHO Guidelines on Safety Monitoring of Herbal Medicines in Pharmacovigilance Systems*, WHO Press, 20 Avenue Appia, 1211 Geneva 27, Switzerland, 2004, p. 82.
82. World Health Organization, *WHO Guidelines on Good Manufacturing Practices (GMP) for Herbal Medicines*, WHO Press, 20 Avenue Appia, 1211 Geneva 27, Switzerland, 2007, p. 92.
83. M. O. Emeje, A. Izuka, C. Y. Isimi, S. I. Ofoefule and O. O. Kunle, *Pharm. Dev. Technol.*, 2011, **16**(2), 170.
84. M. O. Emeje and O. O. Kunle, *J. Nutraceut. Funct. Med. Foods*, 2005, **4**, 147.
85. C. O. Wambebe, E. A. Bamgboye, B. O. Bardu, H. Khamofu, J. A. Momoh, M. Ekpeyong, B. S. Audu, S. O. Njoku, N. R. Nasipuri, O. O. Kunle *et al.*, *Curr. Ther. Res.*, 2001, **62**, 26.
86. National Agency for Food and Drugs, Nigeria (NAFDAC); see: http://nafdac.org.ng/ (Accessed December 13, 2014).
87. Personal communication, Ramesh Pandey, Chairman and CEO of XeChem International, December 16, 2006.

# Subject Index